Collaborative Design and Planning for Digital Manufacturing

Lihui Wang • Andrew Y.C. Nee
Editors

Collaborative Design and Planning for Digital Manufacturing

Springer

Lihui Wang, PhD, PEng
University of Skövde
PO Box 408
541 28 Skövde
Sweden

Andrew Y.C. Nee, PhD, DEng
National University of Singapore
9 Engineering Drive 1
117576 Singapore
Singapore

ISBN 978-1-84882-286-3 e-ISBN 978-1-84882-287-0

DOI 10.1007/978-1-84882-287-0

A catalogue record for this book is available from the British Library

Library of Congress Control Number: 2008939925

Cover design: eStudio Calamar S.L., Girona, Spain

Printed on acid-free paper

9 8 7 6 5 4 3 2 1

springer.com

Preface

Manufacturing has been one of the key areas that support and influence a nation's economy since the 18th century. Being the primary driving force in economic growth, manufacturing constantly serves as the foundation of and contributes to other industries with products ranging from heavy-duty machinery to hi-tech home electronics. In past centuries, manufacturing contributed significantly to modern civilisation and created momentum that is used to drive today's economy. Despite various revolutionary changes and innovations in the 20th century that contributed to manufacturing advancements, we are continuously facing new challenges when striving to achieve greater success in beating the global competition.

Today, manufacturers are competing in a dynamic marketplace that demands short time-to-market and agility in production. In the 21st century, manufacturing is gradually shifting to a distributed environment with increasing dynamism. In order to win a competition, locally or globally, customer satisfaction is treated with the highest priority. This has led to mass customisation and even more complex manufacturing processes, from shop floor to every level along the manufacturing supply chain. At the same time, outsourcing has forged a multi-tier supplier structure involving numerous small-to-medium-sized enterprises, where highly-mixed products in small batch sizes are handled simultaneously in manufacturing operations. Moreover, unpredictable issues like job delay, urgent-order insertion, fixture shortage, missing tools, and even machine breakdown are challenging manufacturing companies, and adding high uncertainty to the fluctuating environment. Engineers often find themselves in a situation that demands adaptive planning and scheduling capability in dealing with daily operations in such a dynamic manufacturing environment.

Targeting the uncertainty issue in manufacturing, research efforts have shifted to improving the adaptability and responsiveness of manufacturing operations in the so-called digital manufacturing environment. The digital manufacturing approach offers manufacturers the ability to digitally represent entire operations on their computers, aiming at producing products safely, ergonomically, efficiently, and right the first time. Digital manufacturing supports a variety of day-to-day activities from collaborative product design to manufacturing execution control. It is further facilitated by advanced information technology (IT) and artificial intelligence (AI) tools in dealing with complex and dynamic issues in the distributed environment, targeting manufacturing uncertainty.

Thanks to recent advancements in AI and IT, manufacturing research has progressed to a new level in adaptive decision making and trouble shooting, in order to address those problems encountered in today's manufacturing with increasing globalisation and outsourcing. While research and development efforts have been translated into a large volume of publications and impacted the present and future practices in manufacturing, there exists a gap in the literature for a focused collection of works that are dedicated to the collaborative design and planning for digital manufacturing. To bridge such a gap and present the state of the art to a broad readership, from academic researchers to practising engineers, is the primary motivation for this book.

Targeting digital manufacturing, Chapter 1 presents a systematic approach in designing and deploying various computing tools with a scalable hardware/software platform. A toolbox dubbed Watchdog Agent® consisting of modularised embedded algorithms for signal processing and feature extraction, performance assessment, diagnostics and prognostics for machinery prognostic applications, is introduced. Focusing on collaborative design, Chapter 2 reports on a design framework that allows efficient flow of design and manufacturing information across mechanical and electrical domains for the development of mechatronic systems. Constraints between mechanical and electrical design domains are classified, modelled, and bi-directionally propagated to provide automated feedback to designers of both engineering domains. The cross-discipline information sharing is further extended in Chapter 3, where a unified feature scheme is proposed to support entity associations and propagation of modifications across a product's lifecycle. For collaborative product design and development in today's decentralised environment, integration of suppliers into the process chain of an OEM (original equipment manufacturer) is investigated in Chapter 4. A web-based tool called CyberStamping is developed to realise the collaborative supplier integration for automotive product development. Extending the design scope to system level, Chapter 5 introduces a method for designing the structure of reconfigurable manufacturing systems for a contract manufacturer based on the use of co-operative co-evolutionary agents. The aim is to determine the structure of a reconfigurable manufacturing system that can be converted from one configuration to another to manufacture the different products of the customers of the contract manufacturer. With customer satisfaction in mind, Chapter 6 describes a conceptual framework of a web-based platform for supporting collaborative product review and customisation within a virtual/augmented reality environment. It can also be used to demonstrate the product to end users.

Linking to collaborative planning, Chapter 7 introduces a reference model of manufacturing process planning for extended enterprises. It represents a workflow modelling strategy and a reference architecture that enable collaborative process management. Furthermore, Chapter 8 proposes an adaptive setup planning approach for solving uncertainty issues in job-shop operations. It loosely integrates scheduling functions during setup planning, and utilises a two-step decision-making strategy for generating machine-neutral and machine-specific setup plans at each stage. In order to allocate operations on machines, Chapter 9 uses an auction-based heuristic with dual objectives of minimising the make span and maximising the system throughput. It is supported by an agent-oriented architecture.

Extending the scope to collaborative product development, Chapter 10 presents a web-based rapid prototyping system. The workflow and overall system architecture are described in detail. Adopting a multi-agent approach, Chapter 11 describes the methodology towards a desktop assembly factory, including 3D representation of individual physical agents, assembly features, and clustering algorithms. Within the context, the physical agents are empowered by intelligent software agents.

Information sharing is crucial for collaborations in digital manufacturing, and can offer added value in achieving global optimisation. Chapter 12 addresses this issue using STEP and XML for information sharing between different systems with standard data accessing interfaces. Moreover, in Chapter 13, a web-based Kanban system in various environments, from manufacturing cells to virtual enterprises, is developed. It allows decision makers to plan and manage production flows of a virtual enterprise more effectively.

In order to achieve real-time traceability, visibility and interoperability in shop floor planning and control, Chapter 14 proposes to use workflow management as a mechanism to facilitate an RFID-enabled real-time reconfigurable manufacturing system, whereas the workflow of production processes is modelled as a network and agents are used to wrap web services. Similarly, Chapter 15 reports on a web-based approach for manufacturing management and control. The proposed methodology integrates engineering and manufacturing management through an ERP software tool. Finally, in Chapter 16, performance measures of distributed manufacturing systems are investigated. These measures would help enterprises evaluate alternative configurations/architectures of a particular distributed manufacturing system and choose the one to meet their goal.

Altogether, the sixteen chapters provide an overview of some recent research efforts towards collaborative design, planning, execution control and manufacturing management for digital manufacturing in the 21st century. They are believed to make significant contributions to the literature. With the rapid advancement of information and communication technologies, we believe that the subject area of this book will continue to be a very active research field for many years to come.

Taking this opportunity, the editors would like to express their deep appreciation to all the authors for their significant contributions to this book. Their commitment, enthusiasm, and technical expertise are what made this book possible. We are also grateful to the publisher for supporting this project, and would especially like to thank Mr Anthony Doyle, Senior Editor for Engineering, for his assistance and earnest co-operation, both with the publishing venture in general and the editorial details. We hope that readers will find this book informative and useful.

London, Canada
Singapore
October 2008

Lihui Wang
Andrew Y.C. Nee

Contents

List of Contributors

J. V. Abellán
Department of Industrial Systems
Engineering and Design
Universitat Jaume I
Av. Vicent Sos Baynat s/n. 12071
Castellón
Spain

Alberto J. Álvares
Dep. Engenharia Mecânica e Mecatrônica
Universidade de Brasília
CEP 70910-900, Brasília, DF
Brazil

Jonathan Bankston
The G. W. Woodruff School of
Mechanical Engineering
Georgia Institute of Technology
Savannah, GA 31407
USA

Ningxu Cai
Department of Mechanical and Materials
Engineering
The University of Western Ontario
London, ON N6A 5B9
Canada

F. Frank Chen
Department of Mechanical Engineering
University of Texas at San Antonio
San Antonio, TX 78249
USA

G. Chen
School of Mechanical and Aerospace
Engineering
Nanyang Technological University
Singapore

Kenway Chen
The G. W. Woodruff School of
Mechanical Engineering
Georgia Institute of Technology
Savannah, GA 31407
USA

Kwai-Sang Chin
Department of Manufacturing Engineering
and Engineering Management
City University of Hong Kong
Hong Kong
China

George Chryssolouris
Laboratory of Manufacturing System
and Automation
Department of Mechanical Engineering
and Aeronautics
University of Patras
Greece

Armando Colombo
Schneider Electric
Seligenstadt, P&T HUB,
63500 Steinheimer Street 117
Germany

Q. Y. Dai
Faculty of Information Engineering
Guangdong University of Technology
Goangzhou, Guangdong
China

José L. N. de Souza Jr
Dep. Engenharia Mecânica e Mecatrônica
Universidade de Brasília
CEP 70910-900, Brasília, DF
Brazil

A. Estruch
Department of Industrial Systems
Engineering and Design
Universitat Jaume I
Av. Vicent Sos Baynat s/n. 12071
Castellón
Spain

Mervyn Fathianathan
The G. W. Woodruff School of
Mechanical Engineering
Georgia Institute of Technology
Savannah, GA 31407
USA

Hsi-Yung Feng
Department of Mechanical Engineering
The University of British Columbia
Vancouver, BC V6T 1Z4
Canada

Joao C. E. Ferreira
Dep. Engenharia Mecânica
Universidade Federal de Santa Catarina
CEP 88040-900, Florianopolis, SC
Brazil

Oscar Ho
Department of Industry and Manufacturing
Systems Engineering
The University of Hong Kong
Hong Kong
China

George Q. Huang
Department of Industry and Manufacturing
Systems Engineering
The University of Hong Kong
Hong Kong
China

Carlos Insaurralde
Tampere University of Technology
FIN-33101 Tampere, P.O. Box 589
Finland

Hongbo Lan
School of Mechanical Engineering
Shandong University
Jinan, Shandong 250061
China

Edzel Lapira
NSF I/UCR Centre for Intelligent
Maintenance Systems
University of Cincinnati
Cincinnati, OH 45221
USA

José L. Martinez Lastra
Tampere University of Technology
FIN-33101 Tampere, P.O. Box 589
Finland

Jay Lee
NSF I/UCR Centre for Intelligent
Maintenance Systems
University of Cincinnati
Cincinnati, OH 45221
USA

Lin Li
S. M. Wu Manufacturing Research Centre
University of Michigan
Ann Arbor, MI 48109-2125
USA

Linxia Liao
NSF I/UCR Centre for Intelligent
Maintenance Systems
University of Cincinnati
Cincinnati, OH 45221
USA

Ji Ma
Department of Mechanical and Materials
Engineering
The University of Western Ontario
London, ON N6A 5B9
Canada

Y.-S. Ma
Department of Mechanical Engineering
University of Alberta
Edmonton, Alberta T6G 2E1
Canada

Dimitris Mavrikios
Laboratory of Manufacturing System
and Automation
Department of Mechanical Engineering
and Aeronautics
University of Patras
Greece

Jun Ni
S. M. Wu Manufacturing Research Centre
University of Michigan
Ann Arbor, MI 48109-2125
USA

Jitesh H. Panchal
The G. W. Woodruff School of
Mechanical Engineering
Georgia Institute of Technology
Savannah, GA 31407
USA

M. K. Pandey
Department of Industrial Engineering
and Management
Indian Institute of Technology
Kharagpur, 721302
India

Menelaos Pappas
Department of Mechanical Engineering
and Aeronautics
University of Patras
Greece

Xiaoli Qiu
Department of Mechanical Engineering
Southeast University
Nanjing, Jiangsu 210096
China

F. Romero
Department of Industrial Systems
Engineering and Design
Universitat Jaume I
Av. Vicent Sos Baynat s/n. 12071
Castellón
Spain

Dirk Schaefer
The G. W. Woodruff School of
Mechanical Engineering
Georgia Institute of Technology
Savannah, GA 31407
USA

Sanjay Kumar Shukla
Department of Mechanical Engineering
University of Texas at San Antonio
San Antonio, TX 78249
USA

H. R. Siller
Department of Industrial Systems
Engineering and Design
Universitat Jaume I
Av. Vicent Sos Baynat s/n. 12071
Castellón
Spain

Konstantinos Smparounis
Laboratory of Manufacturing System
and Automation
Department of Mechanical Engineering
and Aeronautics
University of Patras
Greece

Dunbing Tang
College of Mechanical and Electrical
Engineering
Nanjing University of Aeronautics and
Astronautics
Nanjing
China

Evandro L. S. Teixeira
Department of Hardware Development
Autotrac Commerce and
Telecommunication
CEP 70910-901, Brasília, DF
Brazil

G. Thimm
School of Mechanical and Aerospace
Engineering
Nanyang Technological University
Singapore

M. K. Tiwari
Department of Industrial Engineering
and Management
Indian Institute of Technology
Kharagpur, 721302
India

C. Vila
Department of Industrial Systems
Engineering and Design
Universitat Jaume I
Av. Vicent Sos Baynat s/n. 12071
Castellón
Spain

M. I. M. Wahab
Department of Mechanical and Industrial Engineering
Ryerson University
Toronto, ON M5B 2K3
Canada

Hung-da Wan
Department of Mechanical Engineering
University of Texas at San Antonio
San Antonio, TX 78249
USA

Lihui Wang
Centre for Intelligent Automation
University of Skövde
PO Box 408
541 28 Skövde
Sweden

Evangelos Xanthakis
Laboratory of Manufacturing System and Automation
Department of Mechanical Engineering and Aeronautics
University of Patras
Greece

Frank J. Xu
E-Business Technology Institute
The University of Hong Kong
Hong Kong
China

Xun Xu
Department of Mechanical Engineering
School of Engineering
University of Auckland
New Zealand

Nathan Young
Georgia Institute of Technology
Savannah, GA 31407
USA

YingFeng Zhang
School of Mechanical Engineering
Xi'an Jiaotong University
Xi'an, Shaanxi
China

Saeed Zolfaghari
Ryerson University
Toronto, ON M5B 2K3
Canada

1

Informatics Platform for Designing and Deploying e-Manufacturing Systems

Jay Lee[1], Linxia Liao[1], Edzel Lapira[1], Jun Ni[2] and Lin Li[2]

[1]NSF I/UCR Centre for Intelligent Maintenance Systems
University of Cincinnati, Cincinnati, OH 45221, USA
Emails: jay.lee@uc.edu, liaol@email.uc.edu, Lapiraer@email.uc.edu

[2] S. M. Wu Manufacturing Research Centre
University of Michigan, Ann Arbor, MI 48109-2125, USA
Emails: junni@umich.edu, lilz@umich.edu

Abstract
e-Manufacturing is a transformation system that enables manufacturing operations to achieve near-zero-downtime performance, as well as to synchronise with the business systems through the use of informatics technologies. To successfully implement an e-manufacturing system, a systematic approach in designing and deploying various computing tools (algorithms, software and agents) with a scalable hardware and software platform is a necessity. In this chapter, we will first give an introduction to an e-manufacturing system including its fundamental elements and requirements to meet the changing needs of the manufacturing industry in today's globally networked business environment. Second, we will introduce a methodology for the design and development of advanced computing tools to convert data to information in manufacturing applications. A toolbox that consists of modularised embedded algorithms for signal processing and feature extraction, performance assessment, diagnostics and prognostics for diverse machinery prognostic applications, will be examined. Further, decision support tools for reduced response time and prioritised maintenance scheduling will be discussed. Third, we will introduce a reconfigurable, easy to use, platform for various applications. Finally, case studies for smart machines and other applications will be used to demonstrate the selected methods and tools.

1.1 Introduction

The manufacturing industry has recently been facing unprecedented challenges of ever changing, global and competitive market conditions. Besides sales growth, manufacturing companies are also currently looking for solutions to increase the efficiency of their manufacturing processes. In order to attain, or retain, a favourable market position, companies must allocate resources reasonably and provide products and services of the highest possible quality. Maintenance practitioners, plant managers, and even shareholders are becoming more interested in production asset performance in manufacturing plants. To meet customers' needs brought by e-commerce, many manufacturers are trying to optimise their supply chains and introduce maintenance, repair and operations (MRO) systems, while they are still facing the problem of costly production or service downtime due to unforeseen equipment failure. No matter if the challenge is to attain a shortened lead time, improved productivity, more efficient supply chain, or near-zero-downtime performance, the greatest asset of today's enterprises is the transparency of information between manufacturing operations, maintenance practitioners, suppliers and customers.

Currently, product design and manufacturing operations within a company seem to be completely separate entities because of the lack of real-time lifecycle information that is fed from the latter to the former. They should, however, share a cohesive working relationship to ensure that products are manufactured according to the design, and that designers consider manufacturing operation capabilities and limitations so they can generate a product design that is 'fit for production'. Another example is the inability of assembly plants to investigate their suppliers' operations to determine the processing time and quality of a product before it is actually ordered and shipped. When machine reliability in the suppliers' factory is made available, product quality information can be inferred, and necessary lead times can be projected. This capability is extremely useful in terms of quality assurance as well as lessening inventory.

Increased customer demands on product quality, delivery and service are forcing companies to transform their manufacturing paradigms into a highly collaborative design that seeks to use engineering-based tools to convert and fuse data from virtually any part of the manufacturing environment into information that management can use to make efficient, timely decisions. This philosophy is the guiding principle on which e-manufacturing is founded. e-Manufacturing aims to address the shortcomings present in traditional factory operations to achieve predictive, near-zero-downtime performance that will integrate the various levels of the company. This integration is to be supported by a reliable communication system (both web-enabled and tether-free supported technologies) that will allow seamless data and information flow within a factory [1.1].

As manufacturing systems become more complex and sophisticated, the reliability of individual machines and pieces of equipment becomes increasingly crucial as the breakdown of one machine may result in halting the whole production line in a manufacturing facility. Thus, near-zero-downtime functionality without production or service breakdowns is becoming a necessity for today's enterprises. e-Manufacturing includes the ability to monitor the plant floor assets, predict the variation of product quality and determine performance loss of any machine or component for dynamic rescheduling of production and maintenance operations, and synchronisation with related business services to achieve a seamless integration between manufacturing and higher level enterprise systems [1.1].

e-Manufacturing should integrate seamlessly with existing information systems, such as enterprise resource planning (ERP) [1.2, 1.3], supply chain management (SCM) [1.4], customer relation management (CRM) [1.5] and manufacturing execution system (MES) [1.6], to provide information transparency in order to achieve the maximum benefit. The challenge is that most enterprise information systems are not well integrated or maintained. As data and information can be transmitted anywhere at any time in an e-manufacturing environment, the value of e-manufacturing enables decision making among manufacturers, product designers, suppliers, partners and customers. The role of e-manufacturing, as well as its relationship with existing information systems, is illustrated in Figure 1.1.

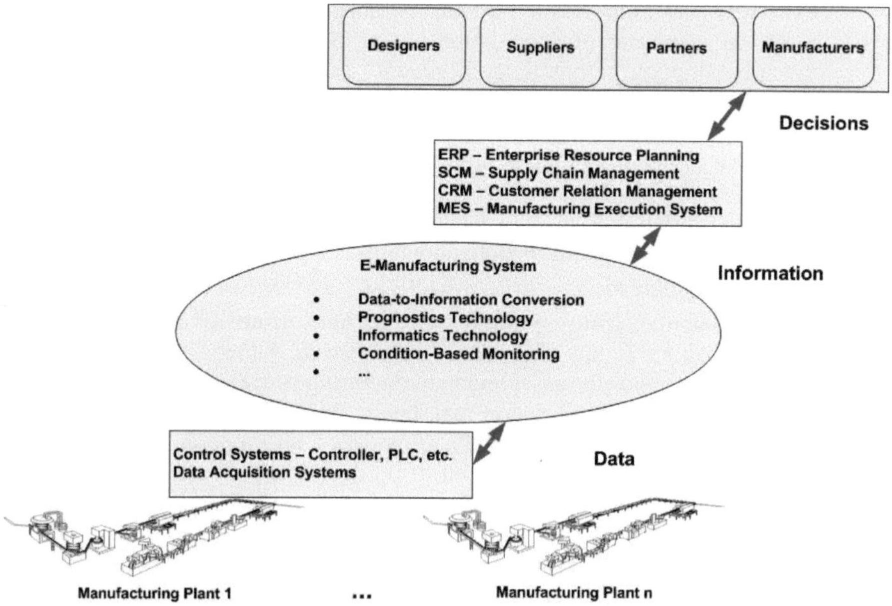

Figure 1.1. E-manufacturing system and its relationship with other information systems

To introduce the e-manufacturing concept into manufacturing enterprises, several fundamental tools need to be developed [1.1, 1.7, 1.8].

- *Data-to-information conversion tools*

  Currently, most state-of-the-art manufacturing, mining, farming, and service machines (*e.g.* elevators) are actually quite 'smart' in themselves; many sophisticated sensors and computerised components are capable of delivering data concerning their machine's status and performance. The problem that this sophistication creates is that a large amount of machine condition-related data is collected. The data is so abundant that the field engineers and management staff are not able to make effective use of it to accurately detect the degradation status of equipment, not to mention being able to track the degradation trend that will eventually lead to a catastrophic failure. A set of data-to-information conversion tools is necessary to convert machine data into performance-related information to provide real-time health indicators/ indices for decision makers to effectively understand the performance of the machines and make maintenance decisions before potential failures occur, which prevents waste in terms of time, spare parts and personnel, and ensures the maximum uptime of equipment.

- *Prediction tools*
  With regard to the reliability of assets, several maintenance approaches exist, such as reactive maintenance and preventive maintenance. Most equipment maintenance strategies today are either purely reactive (reactive maintenance) or blindly proactive (preventive maintenance), both of which can be extremely wasteful. Usually, a machine does not fail suddenly without some measurable process of degradation occurring. Some companies are moving towards adopting a "predict-and-prevent" maintenance methodology, which aims to provide warning of an impending failure on a particular piece of equipment, allowing maintenance to be conducted only when there is definitive evidence of such a failure. Advanced prediction tools are necessary to predict the degradation trend and performance loss, which can provide valuable information for decision makers to make the right decisions before failure, and therefore unscheduled downtime, can occur.
- *Decision support tools*
  In an e-manufacturing environment, data and information can be accessed from anywhere at any time due to web-based, tether-free technology. To effectively monitor the asset and manufacturing process performance, a set of optimisation tools for decision making, as well as easy-to-use and effective visualisation tools to present the prognostics information to achieve near-zero-downtime performance, need to be developed. These decision support systems should be computer-based and integrated with control systems and maintenance scheduling.
- *Synchronisation tools*
  In recent years, the concepts of e-diagnostics and e-maintenance have been gaining attention in various industries. Several case studies [1.9–1.11] and maintenance system architectures [1.12–1.15] have been proposed and studied. Although the necessary devices exist, a continuous and seamless flow of information throughout entire processes has not been effectively implemented, even though the potential cost-benefit for companies is great. Sometimes, it is because the available data is not rendered in a useable or instantly understandable form; a problem that can be solved by using data-to-information conversion and prediction tools. More often, no infrastructure exists for delivering the data over a network, or for managing and storing the data, even if the devices were networked. Synchronisation tools need to be developed to provide seamless information flow and provide online and on-time access to prognostics information, decision support tools and other information systems such as ERP systems.

e-Manufacturing also utilises informatics technologies, *e.g.* tether-free communication, B2B (business to business), B2C (business to customer), industrial Ethernet, XML (extensible markup language), TCP/IP (transmission control protocol/Internet protocol), UDP (user datagram protocol), and SOAP (simple object access protocol), to integrate information and decision making among data flow (of machine/process level), information flow (of factory and supply system level) and cash flow (of business system level) [1.8] systems. For large scale and distributed applications, e-manufacturing should also function as a scalable information platform to combine data acquisition, local data evaluation and wireless communication in one module that is able to provide the right information to the right person at the right time.

There are many issues for designing and deploying e-manufacturing systems. This chapter will focus on describing one of the most important issues in prognostics design for data-to-information conversion through an informatics platform in the e-manufacturing environment. The remainder of the chapter is organised as follows: Section 1.2 introduces a systematic methodology (5S) in designing a prognostics system to convert data to information for manufacturing applications. Section 1.3 describes in detail the informatics platform, which contains a modularised prognostics toolbox and decision support for data-to-information conversion, combined with automatic tool selection capabilities and reconfigurable software architecture, which are implemented on a scalable hardware platform. Section 1.4 gives three industrial case studies to demonstrate the effectiveness of reconfiguring the prognostics toolbox and hardware platform for various manufacturing applications. Section 1.5 concludes the chapter and provides some interesting directions of future work.

1.2 Systematic Methodology in Prognostics Design for e-Manufacturing

1.2.1 Overview of 5S Methodology

To introduce prognostics technologies for e-manufacturing in manufacturing applications, the concepts of "technology-push" and "need-pull" [1.16] are borrowed from the engineering R&D management literature.

1. In the "technology-push" (TP) approach, the design of a prognostics system is driven by technology. Technology drives a sequence of design and implementation events and exploration of the feasibility of adopting this design, and eventually leads to applications and diffusion of the technology developed.
2. In the "need-pull" (NP) approach, in which the design of the prognostics is driven by customer/market needs. Prognostics technology is introduced due to the low satisfaction level of the existing system or the needs to serve the new market needs. Technologies are then incorporated and developed to meet the aforementioned gaps.

For example, if a company's purposes for introducing prognostics focus on increasing competitiveness in the market or improving the asset availability, but they have no clue where to start; the TP approach can be applied. Usually, a large amount of data is available but it is not known which component or which machine is the most critical, and on which prognostics technologies should be applied. The most appropriate way to determine these critical components is to perform data streamlining. Once cleaned, the data then needs to be converted to information, which can be used for many different purposes (health condition assessment, performance prediction or diagnosis), by various prognostics tools (*e.g.* statistical models and machine learning methods). As part of this procedure, different technologies are explored to examine the feasibility and benefit of introducing prognostics and ultimately seek to bring useful information to decision makers. In a situation in which a company knows exactly what prognostics functions (*e.g.* a chart that can show the risks of different systems to prioritise the decision making or a curve that can show the trend of the degradation) are most important, and what machines or components are of the greatest interest. The NP approach can then be applied. Based on different needs for prognostics functions and target monitoring objectives, different prognostics technologies are selected to fit the applications and present the appropriate information that is of great value to the decision makers.

In manufacturing systems, decisions need to be made at different levels; the component level, machine level and system level, as shown in Figure 1.2. Visualisation tools for decision making at different levels can be designed to present prognostics information. The functionalities of the four types of visualisation tools are described as follows:

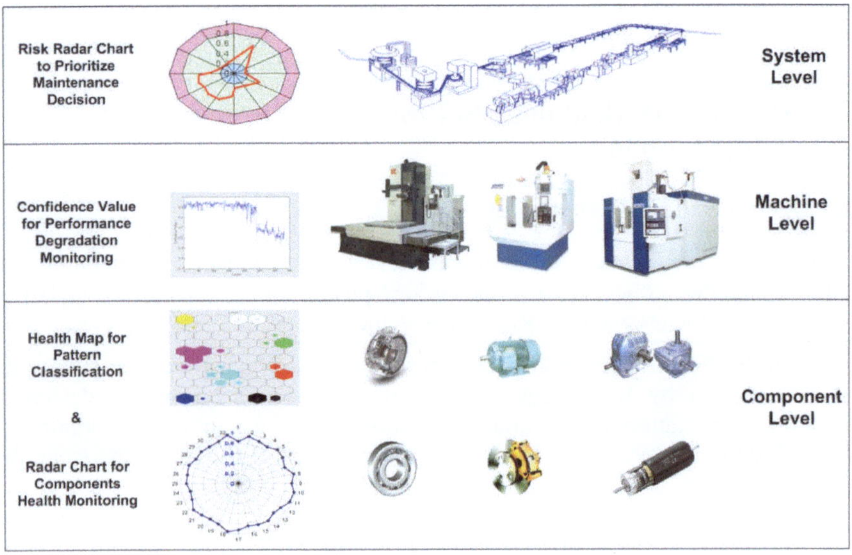

Figure 1.2. Decision making at different levels

- *Radar chart for components health monitoring* – A maintenance practitioner can look at this chart to get an overview of the health of different components. Each axis on the chart shows the confidence value of a specific component.
- *Health map for pattern classification* – A health map is used to determine the root causes of degradation or failure. This map displays different failure modes of the monitored components by presenting different failure modes in clusters, each indicated by a different colour.
- *Confidence value for performance degradation monitoring* – If the confidence value (0: unacceptable, 1: normal, between 0~1: degradation) of a component drops to a low level, a maintenance practitioner can track the

historical confidence value curve to find the degradation trend. The confidence value curve shows the historical/current/predicted confidence value of the equipment. An alarm will be triggered when the confidence value drops under a preset unacceptable threshold.

- *Risk radar chart to prioritise maintenance decision* – A risk radar chart is a visualisation tool for plant-level maintenance information management that displays risk values, indicating equipment maintenance priorities. The risk value of a machine (determined by the product of the degradation rate and the value of the corresponding cost function) indicates how important the machine is to the maintenance process. The higher the risk value, the higher the priority given to that piece of equipment for requiring maintenance.

For the TP approach, data can be converted to useful information through the exploration of the feasibility of different computing tools at different levels, and then the appropriate visualisation tools will be selected to present the information. For the NP approach, visualisation tools can be selected first in the case when the goals for prognostics are clearly defined. Then, different computing tools can be selected according to different visualisation tools for decision making that are required at different levels.

At the lowest level, namely the *component level*, a radar chart and health map can be used to present the degradation information of components (*e.g.* gearboxes, bearings, and motors) and diagnosis results, respectively. To generate the radar chart, data collected from each component need to be converted to confidence value (CV) ranging from 0 to 1. The health condition of each monitored component can be easily examined from the CV at each axis on the radar chart. For example, Fast Fourier transform and wavelet packet energy can be used to deal with the stationary and non-stationary vibration data, respectively, and extract features from the raw data. After the raw data is transformed into a feature space, logistic regression and statistical pattern recognition can be used to convert data into information (CV) based on the data availability. The health map can be generated by using the self-organising map to provide diagnostics information. At the *machine level*, all the health assessment information for the component of a machine will be fused by assigning weight to each component according to its importance to the performance of that machine. The result is an overall evaluation of the health condition of the machine over time, which is presented in a CV curve. At the *system level*, all the prognostics information is gathered from the machine level. A risk radar chart is used to prioritise the maintenance decision making based on the risk value of each machine. The risk value for each machine is calculated by multiplying the degradation rate of the machine with the cost/loss function of the machine, which not only shows the performance degradation but also shows how much it will cost when downtime is incurred. Therefore, maintenance scheduling can be prioritised by examining the risk values on the risk radar chart.

In all, no matter which approach applies and at which level the decisions must be made, the key issue is how to convert data to prognostics information to assess and predict asset performance in order to achieve near-zero-downtime performance. This chapter will present a 5S systematic step-by-step methodology for prognostics design utilising different computing tools for different applications in an e-manufacturing environment.

As shown in Figure 1.3, 5S stands for *Streamline*, *Smart Processing*, *Synchronise*, *Standardise*, and *Sustain*. 5S is a systematic step-by-step methodology that sorts out useful data from the raw datasets and converts data to performance-related information, which will eventually be fed back to closed-loop product lifecycle design, via a scalable embedded informatics platform following industrial standards. Each *S* is elaborated in the following sections.

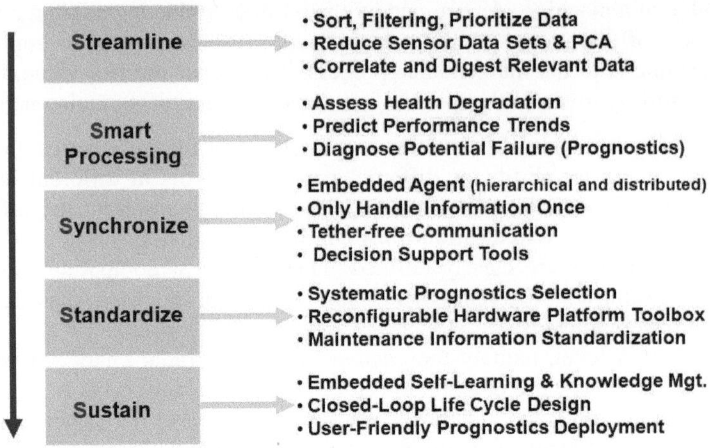

Figure 1.3. 5S methodology

1.2.2 The 1st S – Streamline

The purpose of *Streamline* is to identify critical components and prioritise the data to ensure the accuracy of the next S; *Smart Processing*. Identifying the critical components on which the prognostics should be performed is the first key step by deciding which components' degradation has significant impact to the system's performance or costs a lot when the downtime happens. Also in the real world, data collected from multiple sensors are not necessarily ready to use due to the missing data, redundant data, or noise or even sensor degradation problems. Therefore, instead of directly getting into prognostics, it is necessary to streamline the raw data before processing it. There are three fundamental elements of Streamline:

- *Sort, Filter and Prioritise Data*, which focus on identifying the critical components from maintenance records, historical data and human experience. A powerful method for identifying critical components is to create a four quadrant chart (shown in Figure 1.4) that displays the frequency of failure *vs*. the average downtime of failure. Basically, when the data is graphed in this way, the effectiveness of the current maintenance strategy can be seen. There is one horizontal and one vertical line drawn on the graph, to make four quadrants. They are numbered 1–4 starting with the upper right and moving counter-clockwise. Quadrant 1 contains the component failures that occur most frequently and result in the longest downtime. Typically, there should not be any failures occurring in this quadrant because they should have been noticed and fixed during the design stage. These failures could be a manufacturing defect or improper use generating significant downtime. In Quadrant 2 are components with a high frequency of failure, but short length of downtime for each failure, so the recommendation for these failures is to have more spare parts on hand. Quadrant 3 contains components with a low frequency of failure and low average downtime per failure, which means that the current maintenance practices are working for these failures or parts and requires no changes. In Quadrant 4, lie the most critical failures because they cause the most downtime per occurrence, even if they do not occur very often. This is where the prognostics should be focused. An example is shown in Figure 1.4, which shows that cable, encoder, motor and gearbox are critical components on which prognostics should be focused, in this example case.

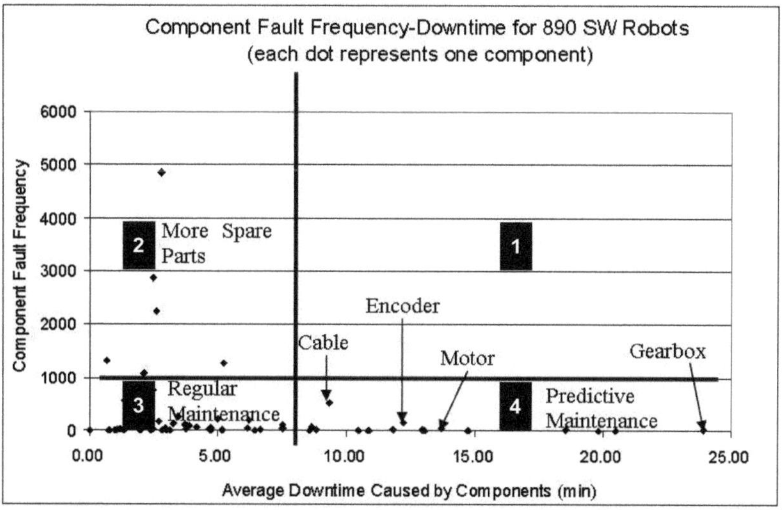

Figure 1.4. Four quadrant chart for identifying critical components

- *Reduce Sensor Data & PCA (Principal Component Analysis)*, which aims at variable/feature selection (which selects variable/feature subset that are relevant to its focus while ignoring the rest), instance selection (which selects appropriate instance subsets to train the mathematical models to achieve acceptable testing accuracy, while ignoring all others) and statistical methods (*e.g.* PCA, *etc.*) to reduce the number of necessary input sensors and reduce the required calculation time for real-time applications.
- *Correlate and Digest Relevant Data*, which focuses on utilising different plots, data processing methods (*e.g.* denoising, filtering, missing data compensation) to find the correlation between datasets and avoid the influence of irrelevant data. In real applications, some data might be trivial for health assessment and diagnosis, the existence of which can tend to increase the computational burden and impair the performance of the models

(*e.g.* classifiers). Efforts are necessary, therefore, to be made before the further analysis to ensure the accuracy of the mathematic models. Several basic quality control (QC) tools, such as check sheet, Pareto chart, flow chart, fish bone diagram, histogram, scatter diagram and control chart, can contribute to this end.

1.2.3 The 2nd S – Smart Processing

The second S, which is *Smart Processing*, focuses on computing tools to convert data to information for different purposes such as health assessment, performance prediction and diagnosis in manufacturing applications. The data-to-information conversion process and a modularised computing toolbox will be further described in Section 1.3.1.

There are three fundamental elements of Smart Processing:

- *Evaluate Health Degradation*, which includes methods to evaluate the overlap between the most recently obtained feature space and those observed during normal operation. This overlap is expressed through the CV, ranging between zero and one, with higher CVs signifying a high overlap, and hence a performance closer to normal [1.17, 1.18].
- *Predict Performance Trends*, which is aimed at extrapolating the behaviour of process signatures over time and predicting their behaviour in the future, in order to provide valuable information for decision making before failures occur.
- *Diagnose Potential Failure*, which aims at analysing the patterns embedded in the data to find out what previously observed fault has occurred and the potential failure mode that provides a reference for taking maintenance action.

There are two important issues that need to be addressed when applying the second S to various applications, namely tool selection and model selection. The purpose of tool selection is to prioritise different computing algorithms and select the most appropriate ones based on application properties and input data attributes. After suitable tools are selected, the next problem is to determine the appropriate parameters for each tool/model, in order to balance model complexity and testing errors, ensuring the accuracy for usage in manufacturing applications.

The smart processing procedure is illustrated in Figure 1.5. Data is obtained from several resources (*e.g.* from the embedded sensors on the machines, from maintenance database and from manually input working conditions) and further transformed into multiple-regime features by selecting the appropriate computational tools for signal processing and feature extraction. In the feature space, health indices are calculated by statistically detecting the deviation of the feature space from the baseline by choosing the appropriate computational tools for health assessment/evaluation. Future machine degradation trends are predicted based on the health indices by selecting appropriate performance prediction tools; a dynamic health feature radar chart, which shows the health condition of the critical components, is then presented for the users' reference using representations, such as the CV curve for performance degradation assessment, a health map for failure mode pattern classification and a radar chart for component degradation monitoring and so on.

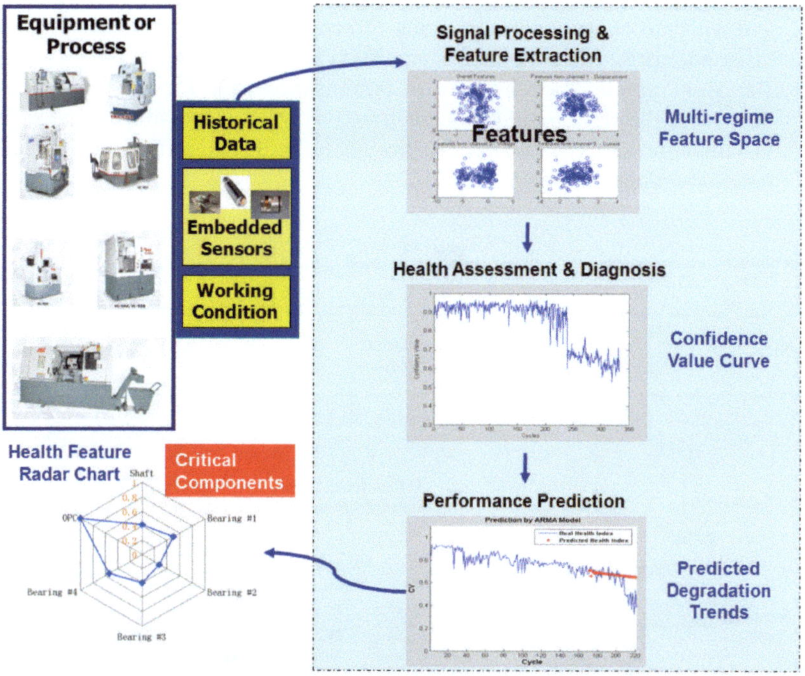

Figure 1.5. Data-to-information conversion process

1.2.4 The 3rd S – Synchronise

Synchronise is the third S of the 5S methodology. It integrates the results of the first two S's (streamline and smart processing) and utilises advanced technologies, such as embedded agents and tether-free communication, to realise prognostics information transparency between manufacturing operations, maintenance practitioners, suppliers and customers. Decision makers can then make use of decision support tools based on the delivered information to assess and predict the performance of machines in order to make the right maintenance decisions, before failures can occur. The prognostics information can be further integrated in the enterprise asset management system, which will greatly improve productivity and asset utilisation.

There are four fundamental elements for Synchronise:

- *Embedded Agents (Hierarchical and Distributed)* include architecture with both a hardware and software platform to facilitate data-to-information conversion and information transmission.
- *Only Handle Information Once (OHIO)* is a principle for dealing with the prognostics information, specifically converting from the equipment data to

information only once. This involves sorting and filtering the information in order to decide whether to discard it or if the maintenance practitioners need to make maintenance decisions right away.

- *Tether-free Communication* is a communication channel that provides online and on-time access to prognostics information, decision support tools and other enterprise information systems such as ERP system.
- *Decision Support Tools* are a set of optimisation tools for decision making (to determine the right maintenance actions) as well as easy-to-use and effective visualisation tools to present the prognostic information to achieve near-zero breakdown performance.

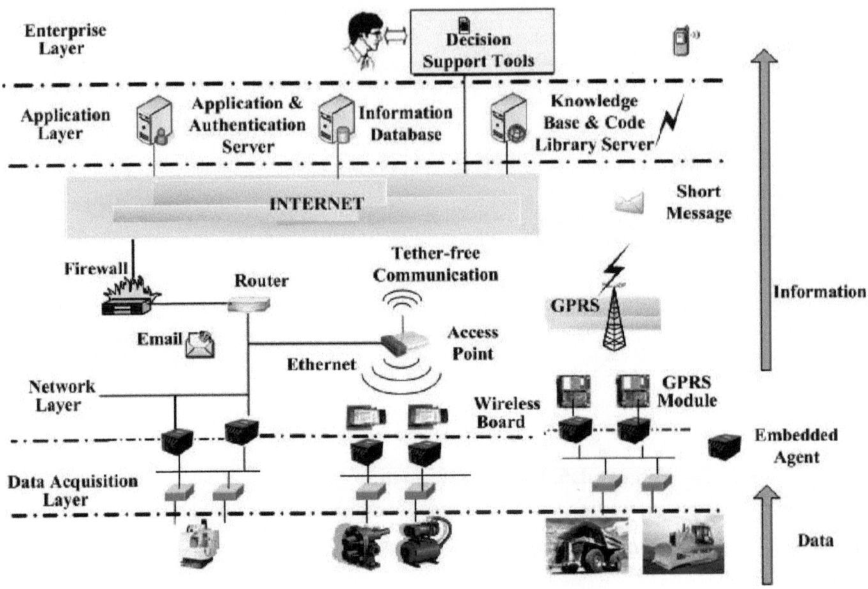

Figure 1.6. Example for the infrastructure of *Synchronise*

A four-layer infrastructure for larger scale industry application as an example of Synchronise is illustrated in Figure 1.6. The *data acquisition layer* consists of multiple sensors that obtain raw data from the components of a machine or machines in different locations as well as other connections (*e.g.* Ethernet and industrial bus) to obtain data from the plant. The embedded agents locate between the data acquisition layer and the network layer. The machine data will be processed locally and converted into performance-related information before it is sent to the Internet. The *network layer* will utilise either traditional Ethernet connections, or wireless connections for communication between the embedded agents, or for sending short messages (SM) to an engineer's mobile phone via general packet radio service (GPRS). Each embedded agent can communicate and collaborate to finish a certain task, which also provides redundancy to the whole system. All the embedded agents communicating through the Ethernet cable or wireless access point are connected to a router to exchange information with the Internet because of the security issues.

Also, a firewall is put between the Internet and the router to provide secure communication and protect the embedded agents from the outside vicious attack. The *application layer* contains application and authentication server, information database and knowledge base and code library server. The application and authentication server provides services between the enterprise users' requests and the database. It also verifies the identity and rights when an end user tries to get the access to the information stored in the database. The information database contains all the asset health information including the performance degradation information, historical performance and basic information (*e.g.* location and serial number) of the assets and so on. The knowledge base and code library server contains rules such as how to select algorithms for data processing, health assessment and prognostics and so on. It also has a depository for all the algorithm components that can be downloaded to the embedded agents when the monitoring task or environmental changes. The *enterprise layer* offers a user-friendly interface for decision makers to access the asset information via the web-based human machine interface (HMI).

1.2.5 The 4th S – Standardise

Standardise has great impacts for enterprises, especially in terms of deploying large scale information technology applications. The implementation of those applications can benefit from a standardised open architecture, information sharing interface and plant operation flow, which brings cost-effective information integration between different systems that can aid in realising the implementation of e-manufacturing. Basically, Standardise includes the following three fundamental elements:

- *Systematic Prognostics Selection Standardisation* defines the unified procedures and architecture in a prognostics system, for instance, the six-tier prognosis architecture defined by MIMOSA OSA-CBM [1.19] (data acquisition, data manipulation, state detection, health assessment, prognostic assessment and advisory generation).
- *Platform Integration and Computing Toolbox Standardisation* focuses on the integration and modularisation of different hardware platforms and computing tools within the information system of a company. Incorporates standards for system integration, so that modules can be developed independently, but also easily adopted in a current information system, due to their interchangeability.
- *Maintenance Information Standardisation* includes enforced work standards in a factory for recording and reporting machine/system failures (or abnormalities), maintenance actions, *etc.*, as complete and timely as possible. This information will be valuable for developing a robust prognostics system, as well as to improve the existing prognostics system through online learning approaches.

1.2.6 The 5th S – Sustain

For the sake of sustainability, information should be embedded in the product serving as an informatics agent to store product usage profiles, historical data, middle-of-life (MOL) and end-of-life (EOL) service data, and to provide feedback to designers and lifecycle management systems. *Sustain* includes the following fundamental items:

- *Closed-loop Lifecycle Design* means to provide feedback to designers based on the integrated prognostics information converted from data in the presence of real-time or periodical health degradation assessment information, performance prediction information and remaining useful life (RUL) information. A system integrated with embedded prognostics for service and closed-loop design is illustrated in Figure 1.7.
- *Embedded Self-learning and Knowledge Management* means to perform the prognostics in manufacturing plants at different levels (component level, machine level, and system level) with least human intervention, which will provide self-learning capabilities to continuously improve the design quality of the products or processes and function as a knowledge base to enhance the six-sigma design, reliability design and serviceability design as well.
- *User-friendly Prognostics Deployment* focuses on the perspective of the end-users or customers who need rapid easy-to-use and cost-effective deployment for prognostics. The deployment involves dynamic procedures such as simulation, evaluation, validation, system reconfiguration and user-friendly web interface development.

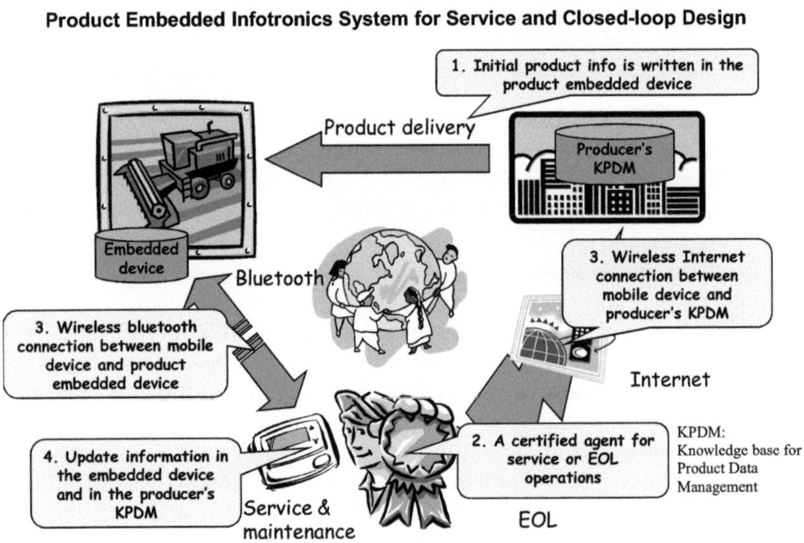

Figure 1.7. Embedded lifecycle information and closed-loop design [1.20]

1.3 Informatics Platform for Implementing e-Manufacturing Applications

The proposed informatics platform for e-manufacturing (namely the Watchdog Agent) is an integration of both software and hardware platforms for assessing and predicting the performance of equipment with decision support functions based on the input from the multiple sensors, historical data and operation conditions. The software platform is an agent-based reconfigurable platform that converts machinery data into performance-related information using a modularised toolbox of prognostics algorithms, combined with automatic tool selection capabilities. The hardware platform is an industrial PC that combines the data acquisition, computation and Internet connectivity capabilities to provide support for information conversion and integration. Figure 1.8 shows the structure of the Watchdog Agent platform.

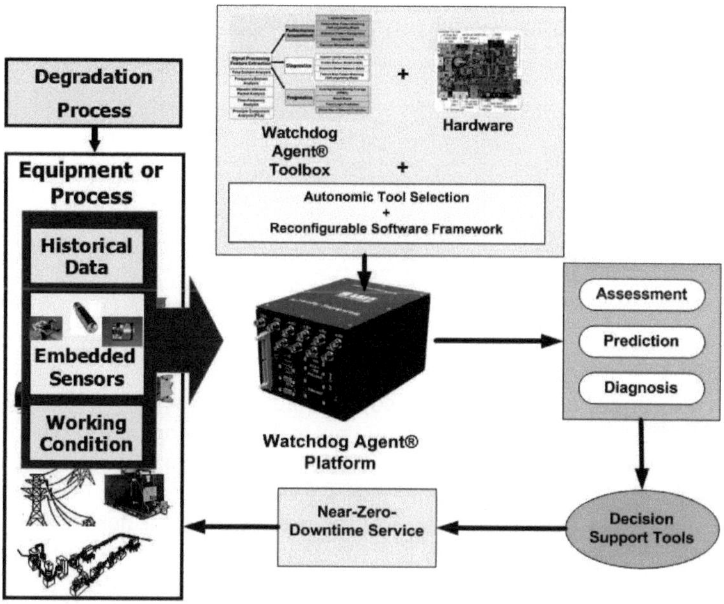

Figure 1.8. Watchdog Agent informatics platform for implementing e-manufacturing

1.3.1 Modularised Prognostics Toolbox – Watchdog Agent Toolbox

The modularised computing toolbox, dubbed the Watchdog Agent, developed by NSF I/UCRC for Intelligent Maintenance Systems (IMS) is shown in Figure 1.9. It consists of computational tools for the four areas of signal processing and feature extraction, health assessment, health diagnosis, and performance prediction.

- *Signal Processing and Feature Extraction Tools*

  Signal processing and feature extraction tools are used to decompose the multi-sensory data into performance-related feature space. The time domain analysis directly uses the waveform for analysis and often involves the comparison of two different signals, *e.g.* time synchronous average (TSA) [1.21]. The fast Fourier transform (FFT) algorithm, which is a typical tool in frequency domain analysis, decomposes or separates the waveform into a sum of sinusoids of different frequencies. Wavelet packet transform (WPT) using a rich library of redundant bases with arbitrary time-frequency resolution enables the extraction of features from signals that combine non-stationary and stationary characteristics [1.22]. PCA is a commonly used statistical method for reducing the dimensionality by transforming the original features into a new set of uncorrelated features.

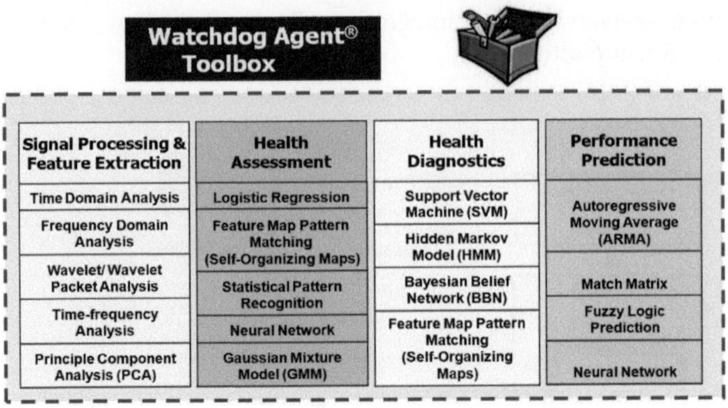

Figure 1.9. Watchdog Agent toolbox

- *Health Assessment Tools*

  Health assessment tools are used to evaluate the overlap between the most recent feature space and that during normal product operation. This overlap is continuously transformed into a CV ranging from 0 to 1 (that indicates abnormal and normal machine performance, respectively) over time, which evaluates the deviation of the recent behaviour from normal behaviour or baseline. Logistic regression is a function that can easily represent the daily maintenance records as a dichotomous problem. The goal of logistic regression is to find the best fitting model to describe the relationship between the categorical characteristic of dependent variable and a set of independent variables [1.23]. Statistical pattern recognition is a method to calculate the system's confidence value or probability of failure by calculating the overlap between the current feature distribution and the baseline feature distribution. The self-organising map is an unsupervised learning neural network that provides a way of representing multi-dimensional feature space in a one- or two-dimensional space while preserving the topological properties of the input space. Neural network is an ideal tool to model complex systems that involve non-linear behaviour and unstable processes. The Gaussian mixture model is a type of density model, which comprises a number of Gaussian functions that are combined to provide a multi-modal density, to be able to approximate an arbitrary distribution to within an arbitrary accuracy [1.24].

- *Health Diagnosis Tools*

  Health diagnosis tools are used to analyse the patterns embedded in the data to find out what previous observed fault has happened. SVM (support vector machines) is usually employed to optimise a boundary curve in the sense that the distance of the closest point to the boundary curve is maximised [1.21], which projects the original feature space to a higher dimensional space by kernel functions and is able to separate the original feature space by a linear hyper-plane in the projected space. The hidden Markov model is an extension of the Markov model that includes cases where the observations are probabilistic functions of the states rather than the states themselves [1.25], which can be used for fault and degradation diagnosis on non-stationary signals and dynamic systems. Bayesian belief network (BBN) is a directed graphic model for probabilistic modelling and Bayesian methods, which can be used to explicitly represent the independencies among the random variables of a domain in which the variables are either discrete or continuous. BBN is a powerful diagnostic tool that can handle large probability distributions, especially for a complex system with a large number of variables.

- *Performance Prediction Tools*

  Performance prediction tools are used to extrapolate the behaviour of equipment signals over time and predict their behaviour in the future. An autoregressive moving average model (ARMA) is used for modelling and predicting future values in a time series of data, which is applicable to linear time-invariant systems whose performance features display stationary behaviour and short-term prediction. A fuzzy logic-based system has a structured knowledge representation in the form of fuzzy IF-THEN rules that are described using linguistic terms and hence are more compatible with human reasoning process than the traditional symbolic approach [1.26]. Match matrix is an enhanced ARMA model that utilises the historic data from different operations, which is fully described in [1.27]. Match matrix excels at dealing with high dimensional feature space and can provide better long-term prediction than ARMA. Neural network is an ideal tool to model non-linear behaviour and unstable processes, which can better capture the dynamic characteristics of the data and could provide more accurate long-term prediction results.

1.3.2 Automatic Tool Selection

The Watchdog Agent toolbox contains a comprehensive set of computational tools to convert data to information and predict the degradation and performance loss. Nevertheless, a common problem is how to choose the most appropriate tools for a predetermined application. Traditionally, tool selection for a specific application is purely heuristic, which is usually not applicable if expert knowledge is lacking, and could be time-consuming for complex problems. In order to automatically benchmark and be able to recommend different tools for various applications, a quality function deployment (QFD) based algorithm selection method is utilised for automatic algorithm selection. QFD provides a structured framework for concurrent engineering, where the ‘voice of the customer’ is incorporated into all phases of product development [1.28]. The purpose is to construct the affinity between the user’s requirements or application conditions and the most appropriate tools.

Table 1.1. HOQ example for automatic tool selection

Process properties \ Watchdog Agent algorithms	Time-Frequency Analysis	Wavelet Packet Energy	Fast Fourier Transform	Principal Component Analysis	Logistic Regression	Self-organising Maps	Statistical Pattern Recognition	Neural Network
Non-stationary	H	H	L	H	×	×	×	×
Stationary	L	L	H	H	×	×	×	×
High frequency	H	H	H	H	×	×	×	×
Low frequency	L	L	H	H	×	×	×	×
Sufficient expertise	×	×	×	H	H	H	M	L
Insufficient expertise	×	×	×	H	L	H	L	H
Low cost implementation	L	M	H	H	H	H	L	L

Table 1.2. A QFD tool selection example

Criteria	User input	Algorithms	Rank
Stationality	Very stable	FFT	1
Impact	Smooth	Time-Frequency Analysis	5
Computation	Low power	Wavelet Packet Energy	4
Amount of data	Limited	AR Filter	3
Data dimension	Vector > 1D	Expert Extraction	2
Expertise	Unavailable	Logistic Regression	3
Prediction span	Short term	Statistical Pattern Recognition	1
		Self-organising Maps	4
		CMAC Pattern Match	2
		Match Matrix Prediction	2
		ARMA Modelling	1
		Recurrent Neural Networks	3
		Fuzzy Logic Prediction	4
		Support Vector Machines	1
		Hidden Markov Model	2
		Bayesian Belief Network	3

Each tool in the Watchdog Agent toolbox is assigned a house of quality (HOQ) [1.29] representing the correlation of the tool with the specific application conditions, such as data dimension (*e.g.* scalar or multi-dimensional), characteristics of the signal (*e.g.* stationary or non-stationary), and system knowledge (*e.g.* enough or limited). Table 1.1 shows a HOQ example for feature extraction and performance assessment tools for automatic tool selection.

The QFD method is used to calculate a final weight for each tool under the constraints of user-defined conditions for ranking the appropriateness of the tools. The tool that is chosen as the most applicable tool has the highest final weight. An example is illustrated in Table 1.2.

In this example, the process is stationary and without impact; therefore, FFT was considered the best choice for signal processing/feature extraction in comparison to time-frequency analysis and wavelet packet. For the same reason, ARMA model and match matrix prediction are more appropriate than neural networks. Due to the lack of expert knowledge, statistical pattern recognition is ranked higher than self-organising maps. Because of the limited historical data, support vector machine is a better candidate for diagnosis than Bayesian belief network, which requires a large amount of data to provide the prior and conditional probabilities.

1.3.3 Decision Support Tools for the System Level

Traditionally, decision support for maintenance is defined as a systematic way to select a set of diagnostic and/or prognostic tools to monitor the condition of a component or machine [1.30]. This type of decision support is necessary because different diagnostic and prognostic tools provide different ways to estimate and display health information, which was described in Section 1.3.1. Therefore, users need a method for selecting the appropriate tool(s) for their monitoring purposes. To address this problem, the automatic tool selection component of the Watchdog Agent has been developed, as described in Section 1.3.2.

However, decision support is also required on the plant floor or 'system' level. Even though the proper monitoring tools can be selected for each machine, users still require a systematic way to decide how to schedule maintenance while considering the effect of an individual machine on system performance. In a manufacturing system, the high degree of interdependency among machines, material handling devices and other process resources requires fast, accurate maintenance decisions at various levels of operation. Because of the dynamic nature of manufacturing systems, maintenance decision problems are often unstructured and must be continuously reviewed due to the changing status of the system. For example, if the predictive monitoring algorithms from five different machines predict that each machine will break down within the following week, users need to know how to quickly and properly assign priority to each machine as well as how to schedule maintenance time to minimally affect production.

For plant-level operations, the main objective of design, control and management of manufacturing systems is to meet the production goal. However, the actual production performance often differs from the designated productivity target because of low operational efficiency, mainly due to significant downtime and frequent machine failures. In order to improve the system performance, two key factors need to be considered: (1) the mitigation of production uncertainties to reduce unscheduled downtime and increase operational efficiency, and (2) the efficient utilisation of the finite factory resources on the throughput-critical sections of the production system by detecting bottlenecks. By considering these two factors, manufacturers can improve productivity, minimise the total cost of operation, and enhance their corporation competitiveness. The plant-level decision-making process considers not only the static system performance in the long term but also the dynamics in the short term. For example, in the system illustrated in Figure 1.10, machines A and B that perform the same task in parallel and have the same capacity will have the same importance to production. However, when the buffer after machine A is filling up for any reason, machine A will become less critical with respect to machine B, because a breakdown in machine A will not affect the system production as much as a breakdown in machine B, due to the buffer effect. Therefore, the dynamic production system status, which is not used in the long term, needs to be considered in the priority assignment in the short term.

Combined with the technologies developed in the short term and the long term, the framework for the plant-level joint maintenance and production decision support tool is developed to efficiently improve system performance as illustrated in Figure 1.11.

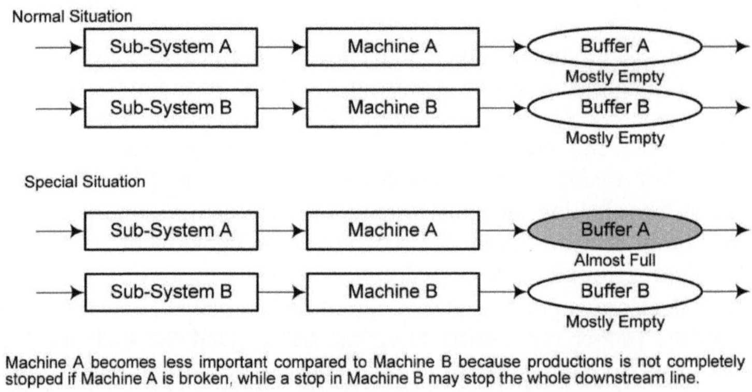

Figure 1.10. Buffer effects on machine importance

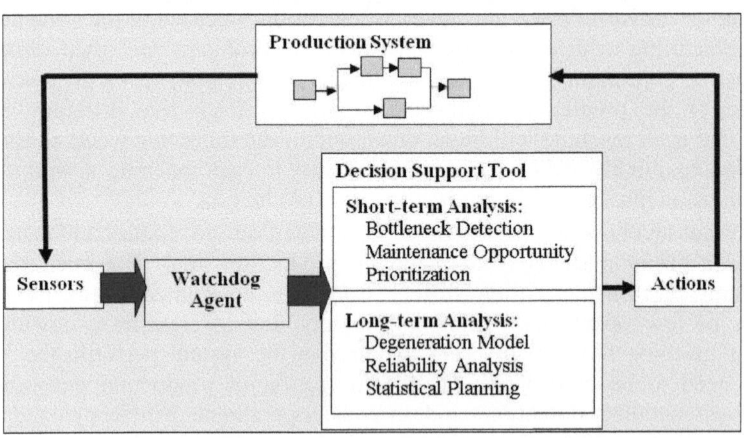

Figure 1.11. Framework for decision support tools

Long-term and short-term time periods are relative definitions. In general, it is difficult to precisely define a period as short term or long term as it depends on the final objective, as well as many other factors. For example, if failures occur frequently, a distribution or pattern may be used to describe the system's performance to study the long-term behaviour. On the contrary, if the failures are rare, then short-term analysis may be a more suitable approach compared to statistical distributions.

The definition of short term may refer to an operational period not long enough for machines' failure behaviours to assume a statistical distribution or for system behaviours to approach a steady state; it could be hours, shifts, or days in a mass-production environment. According to Figure 1.11, the short-term analysis and long-term analysis uses different technologies for the decision-making process. For short-term analysis, real data is used for analysis, and the study focuses on process control. The technologies generally include bottleneck detection, maintenance opportunity calculation and maintenance task prioritisation. On the other hand, a long-term study solves the problem of process planning. After receiving data from sensors, the Watchdog Agent processes and transfers data into useful information. The data may include failure records, maintenance records, blockage time, starvation time, and throughput records. Then decisions based on the long-term information will be made to realise the production demand. Degradation study, reliability analysis, and statistical planning are the common ways for long-term decision making. Although the methods used in short term and long term are often different from each other, both analyses are necessary to improve system performance. Combining long-term and short-term analysis can lead to a smart final decision for system performance improvement.

1.3.4 Implementation of the Informatics Platform

The widespread implementation of large-scale and distributed applications in the e-manufacturing environment raises a key challenge to software engineers in view of diverse, restrictive, and conflicting requirements. A reconfigurable platform is highly demanded for the purpose of information transparency in order to reduce the development costs and time to market and lead to an open architecture supporting software reuse, reconfiguration and scalability, including both stand-alone and distributed applications.

The reconfigurable platform for e-manufacturing should be an easy-to-use platform for decision makers to assess and predict the degradation of asset performance. The platform should integrate both hardware and software platform and utilise autonomic computing technologies to enable remote and embedded data-to-information conversion, including health assessment, performance prediction and diagnostics.

The hardware platform, as shown in Figure 1.12, should have the necessary data acquisition, computation and communication capabilities. Considering the data processing and the reconfigurable requirements, the hardware platform is required to have high computational performance and scalability. A PC/104 platform, which is a popular standardised embedded platform for small computing modules typically used in industrial applications, is selected as the hardware platform. With the

Windows XP embedded system installed, all Win32 compatible software modules are supported. A compact flash card is provided for storing the operating system and the programs. A PCM-3718HG-B DAQ card is chosen as the analogue-to-digital data acquisition hardware that has a 12-bit sampling accuracy and supports various data acquisition methods such as software triggering mode, interrupt mode, and direct memory access (DMA) mode. The DAQ card is connected to the main board via a compatible PC/104 extension slot. A PC/104 wireless board and PC/104 GPRS (General Packet Radio Service) module can also be selected to equip the integrated hardware platform with wireless and GPRS communication capabilities, if necessary. These communication boards are also connected to the main board with compatible PC/104 extension slots.

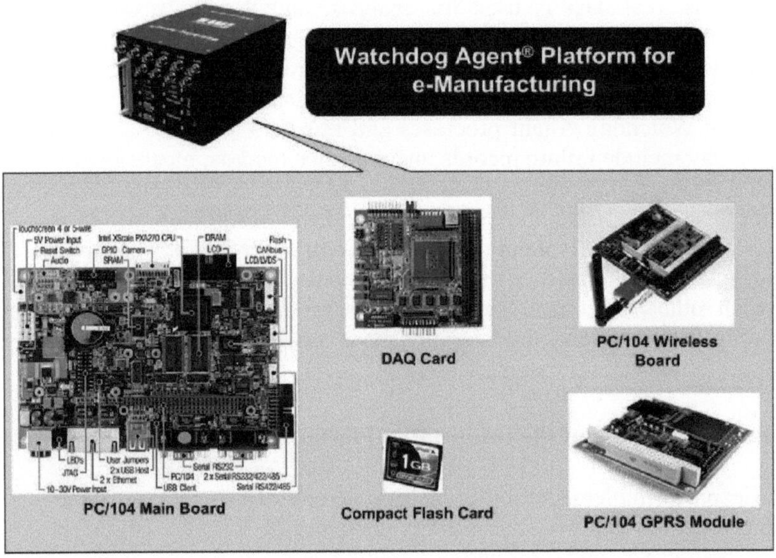

Figure 1.12. Hardware integration for the Watchdog Agent informatics platform

The software architecture is an agent-based reconfigurable architecture, as shown in Figure 1.13. There are three main agents in this reconfigurable architecture: the system agent (SA), knowledge-database agent (KA) and executive agent (EA), which play important roles in the software reconfiguration process. The primary function of SA is to manage both system resources (*e.g.* memory and disk capacity) and device hardware (*e.g.* data acquisition board and wireless board). If a request is received to generate an EA at the initial stage or to modify the EA at the runtime stage, SA creates a vacant agent first. With the interaction with KA, the SA assigns system resources to the agent and executes it autonomously. SA can also communicate with other SAs in the network and receive behaviour requests. KA interacts with the knowledge database to obtain decision making capabilities. It provides components dependencies and model parameters to perform a specific prognostics task. KA also provides coded modules downloaded from the knowledge database to create a functional EA.

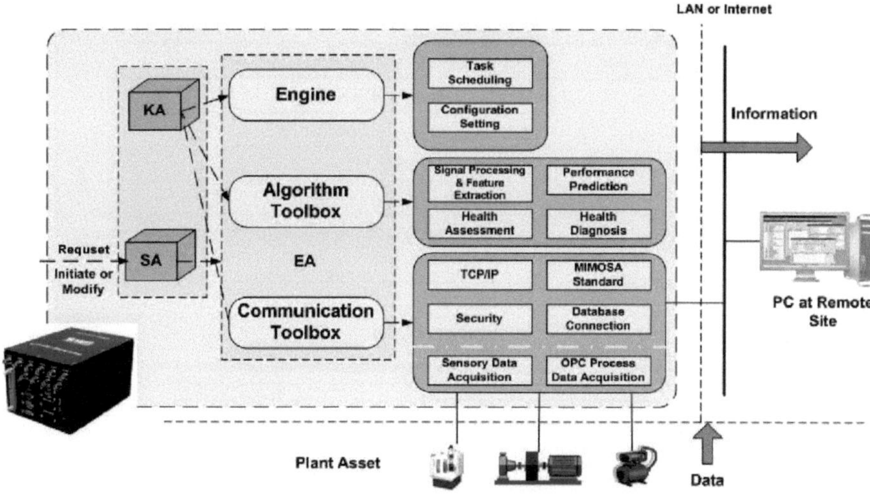

Figure 1.13. Reconfigurable software architecture

1.4 Industrial Case Studies

1.4.1 Case Study 1 – Chiller Predictive Maintenance

For some complex systems (*e.g.* chiller systems), there is no (or rarely) failure mode of the crucial system to follow, which makes it hard to identify a cost-effective threshold for each monitored parameter for preventive maintenance. Therefore, in most of such cases, the replaced components will not be able to perform for their maximum useful life. Even if the thresholds for individual parameters can be set by a maintenance team based on independent analysis of the parameters, there is always the lack of consideration of the interactions among components of the whole system. Therefore, it is critical to evaluate the health condition for the whole system utilising the aforementioned e-manufacturing platform. Case study 1 focuses on illustrating the usage of the e-manufacturing platform, using chiller predictive maintenance as an example.

As shown in Figure 1.14, there are six accelerometers (IMI 623C01) installed on the housing of six bearings (channels 0 to 5) on the chiller. Channel 0 and channel 3 are used for shaft monitoring. Channels 1, 2, 4, and 5 are used to monitor bearings #1 to #4, respectively. The vibration signals are saved in a data logging server that can be accessed by the e-manufacturing platform. OPC (object-linking-and-embedding process control) data including temperature, pressure and flow rate are also obtained from the Johnson Controls OPC server and can be accessed by the e-manufacturing platform. The monitoring objects and the related OPC parameters are listed in Table 1.3. Figures 1.15(a) and 1.15(b) show the raw vibration data for the six channels in normal condition and degradation condition, respectively. The OPC values in normal condition and degradation condition are illustrated in Figures 1.16(a) and 1.16(b), respectively. Obviously, it is hard to tell the health condition of each component by just looking at the raw data.

Table 1.3. OPC parameters and the monitoring objects

Monitoring object	Related OPC parameter
Evaporator	WCC1 return temperature
	WCC1 supply temperature
	WCC1 flow rate
Condenser	WCC1 return temperature
	WCC1 condenser supply temperature
Compressor oil	Oil temperature in separator
	Oil temperature in compressor
Refrigerant circuit	Suction pressure
	Discharge pressure

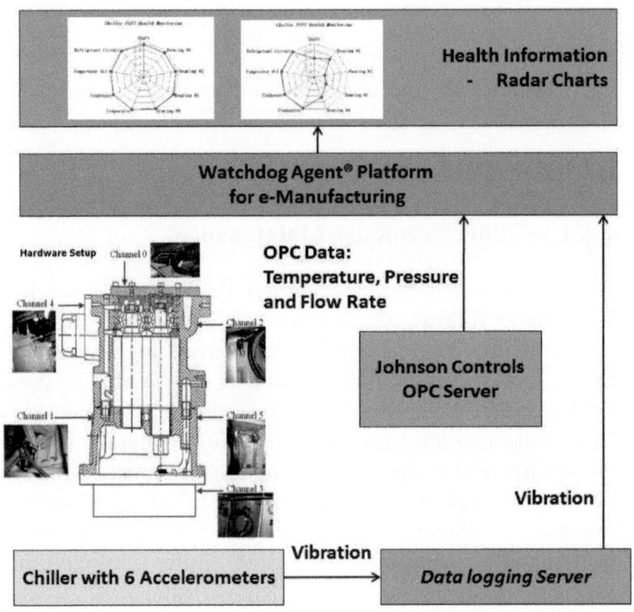

Figure 1.14. System architecture for chiller predictive maintenance

In the experiment, the training and testing datasets of channel 5 (corresponding to bearing #4) in normal condition and degradation condition are used as an example for health assessment. The normal condition data is considered first. Wavelet packet analysis (WPA) is used to extract energy features from the raw vibration data. PCA is then used to find the first two principal components that contain more than 90% variation information. These two principal components are used as the feature space of the baseline for channel 5. The same methods are also applied to the degradation data of channel 5. The next step is to use Gaussian mixture model (GMM) to build mathematical models to approximate the distributions of the baseline feature space and the degraded feature space, in order to determine how far the degraded feature

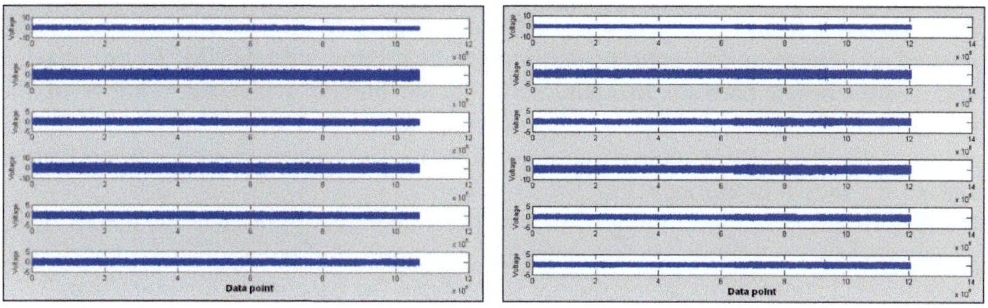

(a) (b)

Figure 1.15. Vibration data (a) in normal condition and (b) in degraded condition

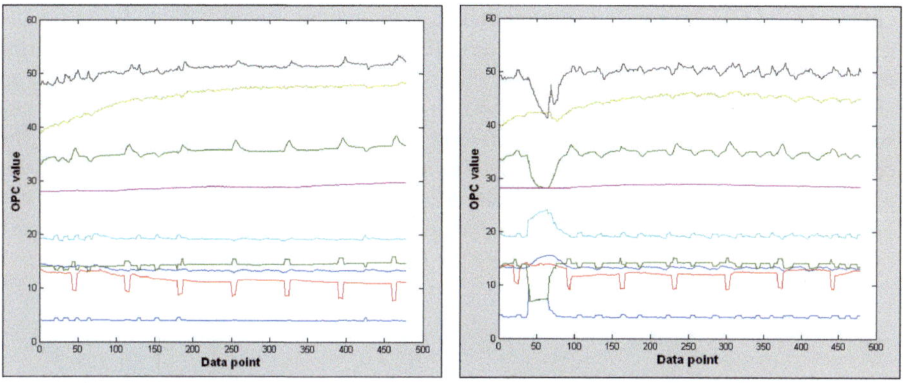

(a) (b)

Figure 1.16. OPC data (a) in normal condition and (b) in degraded condition

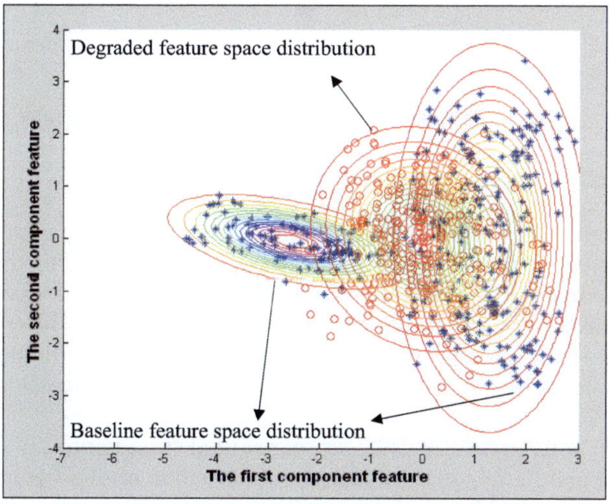

Figure 1.17. GMM approximation results

space deviates from the baseline feature space. Bayesian information criterion (BIC) is used to determine the appropriate number of mixtures for GMM. In this case, 2 and 1 are chosen because the BIC score is the highest when the number of the mixtures is 2 and 1 for the baseline feature space and degraded feature space, respectively. The GMM approximation results for the baseline and degraded feature spaces are shown in Figure 1.17.

A normalised scalar ranging from 0 to 1 is calculated to indicate the health index of the system's performance (0: abnormal, meaning the deviation from baseline is significant; 1: normal, meaning the deviation from the baseline is not significant).

As shown in Figure 1.18, two radar charts are used to show the health assessment results for the monitored components of the chiller system. Each axis on the radar chart indicates the CV of the corresponding component. The components include a shaft, four bearings, evaporator, condenser, compressor oil and refrigerant circuit. If the CV is near 1, it shows that the component is in good condition (in the first radar chart at the left hand side). If the CV is smaller than a predefined threshold (*e.g.* 0.5 in the second radar chart), it indicates that the component is in an abnormal condition. The results of the two radar charts prove that this method can successfully determine the normal and abnormal health conditions of the components on the chiller system. Vibration signals and OPC data (such as temperature, pressure and flow rate) are converted to health information through the informatics e-manufacturing platform, which can guide the decision makers to take further actions to maintain and optimise the uptime of the equipment.

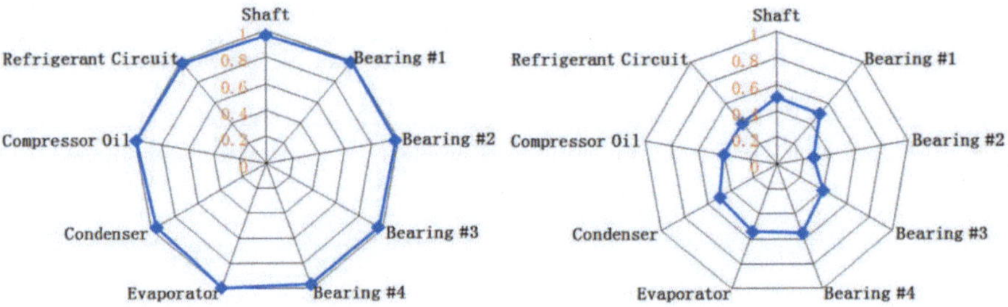

Figure 1.18. Health assessment results for chiller systems

1.4.2 Case Study 2 – Spindle Bearing Health Assessment

Bearings are critical components in machining centres as their failure could cause a sequence of product quality issues and even serious damage to the machines in which they are operating. Health assessment and fault diagnosis have been gaining importance in recent years. Roller bearing failures will cause different patterns of contact forces as the bearing rotates, which cause sinusoidal vibrations. Therefore, vibration signals are taken as the measurements for bearing health assessment and prediction.

In this case, a Rexnord ZA-2115 bearing is used for a run-to-failure experiment. As shown in Figure 1.19, an accelerometer is installed on the vertical direction of the bearing housing. Vibration data is collected every 20 minutes with a sampling rate of 20 kHz. A current transducer is also installed to monitor one phase of the current of the spindle motor. The current signal is used as a time stamp to synchronise the vibration data with the running speed of the shaft of a specific machining process. All the raw data are transmitted through the terminal to the integrated reconfigurable Watchdog Agent platform and then converted to health information locally. This health information is then sent via the Internet, and can be accessed from the workstation at the remote side.

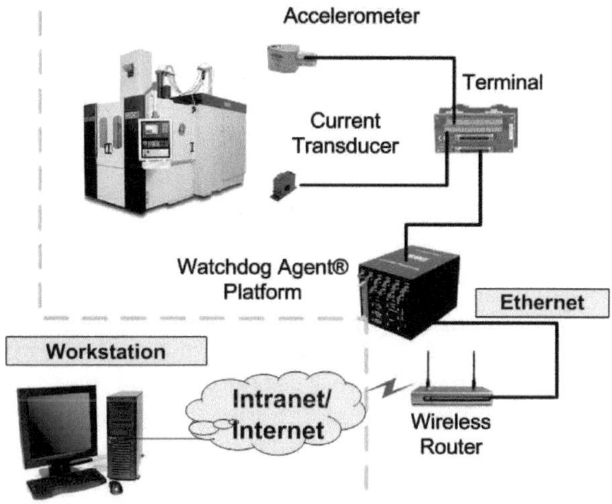

Figure 1.19. System setup for spindle bearing health monitoring

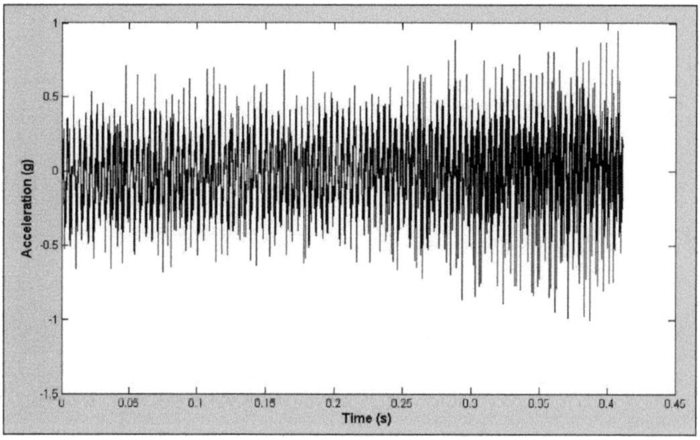

Figure 1.20. Vibration signal for spindle bearing

A magnetic plug is installed in the oil feedback to accumulate debris, which is used as evidence for bearing degradation. At the end of the failure stage, the debris accumulates to a certain level and causes an electrical switch to close to stop the machining centre. In this application, the bearing ultimately developed a roller defect. An example of the vibration signals is shown in Figure 1.20.

FFT was chosen as the appropriate tool for feature extraction because the vibrations can be treated as stationary signals in this case, since the machine is rotating at a constant speed with a constant load. The energy centres around each bearing defect frequency, such as ball passing frequency inner-race (BPFI), ball passing frequency outer-race (BPFO), ball spin frequency (BSF) and foundation train frequency (FTF), and is computed and passed on to the health assessment or diagnosis algorithms in the next step. The equations for calculating those bearing defect frequencies are described in [1.31]. In this case, the BPFI, BPFO and BSF are calculated as 131.73 Hz, 95.2 Hz and 77.44 Hz, respectively.

A task for automatic health assessment is the detection of bearing degradation. Typically, only measurements for normal operating conditions are available. In rare cases there exists historical data of the development of defects in measurements of a complete set of all possible defects. Once a description of the normal machine behaviour is established, anomalies are expected to show up as significant deviations from this description. In this case, the self-organising map (SOM) can be trained only with normal operation data for health assessment purpose. SOM provides a way of representing multidimensional feature space in a one- or two-dimensional space while preserving the topological properties of the input space. For each input feature vector, a best matching unit (BMU) can be found in the SOM. The distance between the input feature vector and the weight vector of the BMU, which can be defined as minimum quantisation error (MQE) [1.32], actually indicates how far away the input feature vector deviates from the normal operating state. Hence, the degradation trend can be visualised by observing the trend of the MQE. As the MQE increases, the extent of the degradation becomes more severe. A threshold can be set as the maximum MQE that can be expected; therefore, the degradation extent can be normalised by converting the MQE into a CV ranging from 0 to 1. After this normalisation, the MQE increases while the CV decreases.

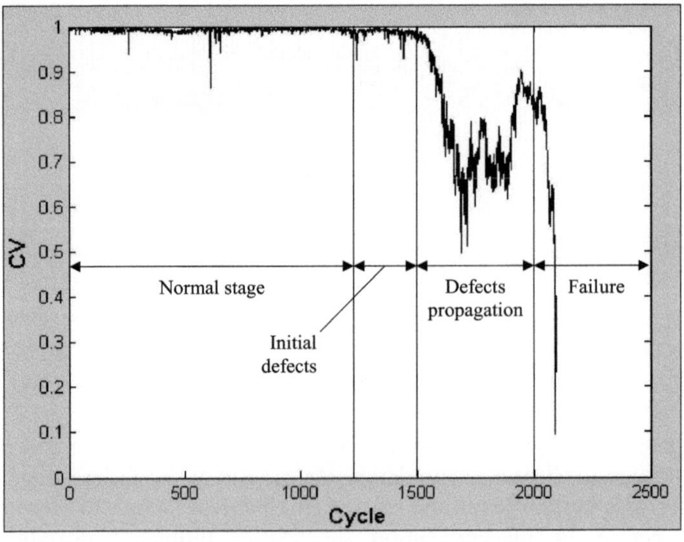

Figure 1.21. CV of the degradation process of the bearing with roller defect

As shown in Figure 1.21, the degradation process of the bearing can be visualised effectively. In the first 1000 cycles, the bearing health is in good condition, as the CVs are near one. From cycle 1250 to cycle 1500, the initial defects appear and the CV begins to decrease. The CV keeps decreasing until it reaches cycle 1700, approximately, which means the defects become more serious. After that and until, approximately, cycle 2000, the CV increases because the propagation of the roller counterbalances the vibration. The CV will decrease sharply after this stage till the bearing fails. When the CV starts to decrease and becomes unstable, after cycle 1300, the amount of debris adhered to the magnetic plug installed in the oil feedback pipe starts to increase. The debris is used as evidence of the bearing degradation. At the end of the failure stage, the debris accumulates to a certain level and it causes an electrical switch to close to stop the machine, which validates the health assessment results.

1.4.3 Case Study 3 – Smart Machine Predictive Maintenance

A smart machine is a piece of equipment having the ability for autonomous data extraction and processing, and decision making. One of its essential components is the health and maintenance technology that is responsible for the overall assessment of the performance of the machine tool including its critical components, such as the spindle, automatic tool changer, motors, *etc.* The overall structure of the health and maintenance of the smart machine is depicted in Figure 1.22. A machine tool system consists of a controller and an assembly of mechanical components, which can be an abundant source of health indicators. Current maintenance packages go as far as extracting raw machine data (vibration, current, pressure, *etc.*) using built-in or add-on sensor assemblies, as well as controller data (status, machine offsets, part programs, *etc.*) utilising proprietary communication protocols. There is a need to transform this voluminous set of data into simple, yet actionable information for the following reasons: a more profound health indicator could be generated if multiple sensor measurements are objectively fused, and providing more sensor and controller readings would just overwhelm an operator, thereby increasing the probability of the wrong decisions being made for the machine. Smart machine health and maintenance employs the Watchdog Agent as a catalyst to reduce data dimension, perform condition assessment, predict impending failures through tracking of component degradation prediction and classify faults in the case that multiple failure modes propagate. The end-goal is an overall assessment value, called the machine tool health indicator, or MTHI, which is a scalar value between 0 and 1 that describes the performance of the equipment; 1 being the peak condition and 0 being an unacceptable working state. Furthermore, it also includes a 'drill-down' capability that will help the operator determine which critical component being monitored is degrading when the MTHI is low. This is exhibited with a component-conscious radar chart that shows the health of the individual components. Finally, the Watchdog Agent provides various visualisation tools that are appropriate for a particular prognostic task.

Figure 1.23 aptly describes how the Watchdog Agent interacts with the demonstration test-bed. The equipment being monitored is a Milltronics horizontal machining centre with a Fanuc controller. Machine data is extracted using sensors

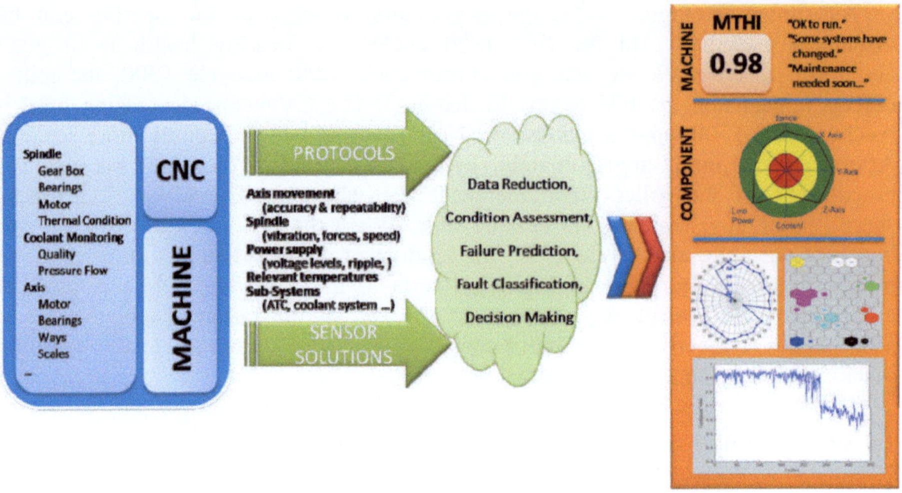

Figure 1.22. Overall structure of the smart machine health and maintenance system

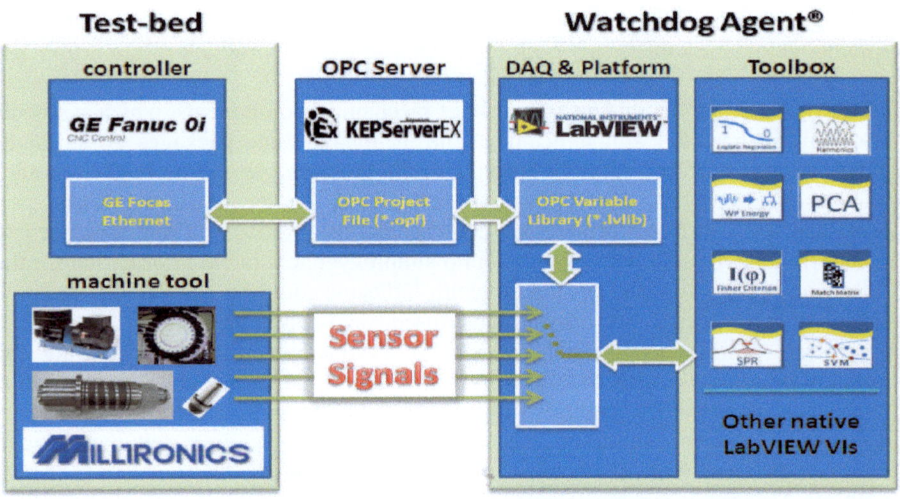

Figure 1.23. Watchdog Agent setup on the demonstration test-bed

added on, and controller data is retrieved through KepServerEX. The Watchdog Agent consists of a data acquisition system and a processing module that uses prognostics tools.

The overall project presented in this example is being conducted in collaboration with TechSolve Inc. (Cincinnati, OH) under the Smart Machine Platform Initiative (SMPI) program. Health and maintenance is one of seven focus areas that the SMPI program has identified. The other technology areas are tool condition monitoring, intelligent machining, on-machine probing, supervisory control, metrology and intelligent network machining. For brevity purposes, the subsequent paragraphs will describe one component of the Watchdog Agent implementation in the smart machine project, *i.e.* tool holder unbalance detection. The problem is briefly described, followed by system setup and a discussion of the results.

An unbalanced tool assembly is detrimental to the quality of the product being produced because it may cause chatter and gouging, and almost inevitably a loss in part accuracy. For severe cases of unbalance, it will also affect the spindle as well as cause accelerated wear on the cutting tool [1.33]. A rotary equipment component experiences unbalance when its mass centreline does not coincide with its geometric centre [1.34]. Most commercially-available-off-the-shelf (COTS) systems that check for unbalance focus on the spindle and the cutting tool. The tool holder is often overlooked while it is almost always tweaked and changed whenever a new cutting tool is required. Furthermore, a dropped tool or a tool crash can also have adverse effects on the geometry of the tool holder.

Experiments were performed on shrink fit tool holders that were free-spun on a horizontal machining centre at a constant spindle speed of 8000 rpm. Three tool holders have had different amounts of material chipped off to induce multiple levels of unbalance. The components to be tested were sent to a third-party company for measurements to verify that indeed the tool holders are at various degrees of unbalance. A new tool holder of the same kind was used as a control sample.

A single-axis accelerometer was connected to the spindle at a location close to the tool assembly. The data acquisition system is triggered by the spindle status that is sent by the machine controller using OPC communications. A time-domain sensor measurement when an unbalanced tool holder was spun is shown in Figure 1.24.

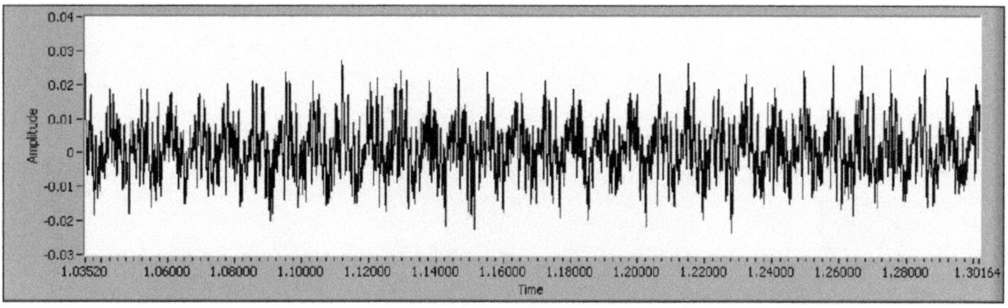

Figure 1.24. Vibration signal from an unbalanced tool holder

The apparent quasi-periodicity in the vibration signal indicates the presence of a strong spectral component. Furthermore, other signal features were also extracted, *e.g.* root mean square (RMS) value, mean value of DC (direct current) and kurtosis. A sample feature plot is given in Figure 1.25 that juxtaposes vibration signals taken from the tool holders in respect to two signal features. Obvious from the plot is the distinct clusters of data from each tool holder. This plot also indicates that the amplitude of the fundamental harmonic alone will be unable to tell when the tool holder experiences a low case of unbalance, as can be seen by the overlap. However, if the RMS value is used in conjunction with the natural frequency, then distinction between the balanced tool and the tool with low unbalance is more apparent. Finally, there seems to be a pattern when the amount of unbalance increases.

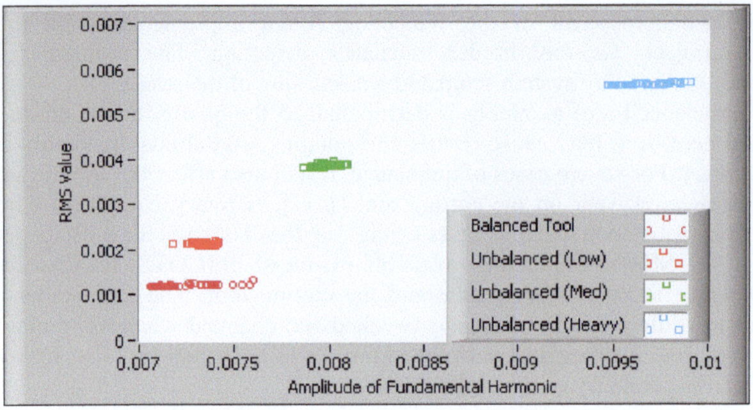

Figure 1.25. Feature space plot

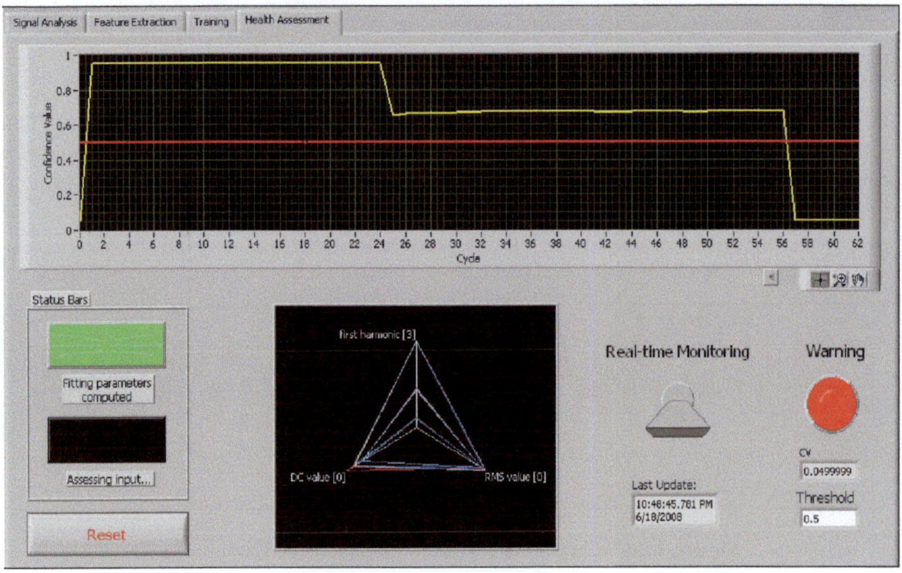

Figure 1.26. Screenshot showing the logistic regression curve

As such, logistic regression is a suitable tool that is able to translate levels of unbalance into confidence values after training the Watchdog Agent with normal data (from a balanced tool) and abnormal data (from an unbalanced tool). The advantage of using logistic regression is that the CV can be customised to reflect tool holder condition based on the requirements of the machine operator. For example, when performing a process with tight tolerances, like milling, then a 'less' unbalanced tool can be used for training data. Meanwhile, for operations that allow a more lenient tolerance, like in drilling, a tool holder with more unbalance can be used for training. Figure 1.26 shows the screenshot of the application interface with the logistic regression curve when the tool holder with the medium unbalance (out of three tool holders) is used for training purposes. As expected, the good tool holder has a high confidence value while the tool holder with medium unbalance has a low confidence value (of around 0.05). On the other hand, the tool holder with light unbalance has a confidence value of around 0.8 and the tool holder with heavy unbalance has a confidence value that is well below 0.05.

1.5 Conclusions and Future Work

This chapter gave an introduction to e-manufacturing system and discussed the enabling tools for its implementation. Following an introduction to the 5S systematic methodology in designing advanced computing tools for data-to-information conversion, an informatics platform, which contains a modularised toolbox and reconfigurable platform for manufacturing applications in the e-manufacturing environment, was described in detail. Three industrial case studies showed the effectiveness of reconfiguring the proposed informatics platform for various real applications.

Future work will be the further development of the Watchdog Agent based informatics platform for e-manufacturing. With respect to the software development, software will be further developed to embed prognostics into products, such as machine tool controllers (Siemens or Fanuc), mobile systems and transportation devices, for proactive maintenance and self-maintenance. For the hardware platform, it is necessary to harvest the developed technologies and standards to enhance interoperability and information security, and to accelerate the deployment of e-manufacturing systems in real-world applications.

Regarding autonomous prognosis design, a signal characterisation mechanism, which can automatically evaluate some mathematical properties of the input signal, is of great interest to facilitate 'plug-n-prognose' with minimum human intervention. The purpose of the signal characterisation mechanism is to cluster identical machine operating conditions that require different prognostics models. For instance, transient and steady states of the motor should be distinguished; different running speed should be distinguished; load/idle condition should be distinguished, too. In some cases, *e.g.* transient and steady states, different features will be extracted and thus different prognostic algorithms will be selected. In other cases, *e.g.* different running speed, the prognostic basis changes and thus separate training procedures are needed for each condition, even if the prognostic algorithms are the same.

Research needs should also be addressed to map the relationship between the machine/process degradation and the economic factors/cost function/loss function to further facilitate decision making and the prioritisation of the actions that should be taken. Advanced maintenance simulation software for maintenance schedule planning and service logistics cost optimisation for transparent decision making is currently under development. Advanced research will be conducted to develop technologies for closed-loop lifecycle design for product reliability and serviceability, as well as to explore research in new frontier areas such as embedded and networked agents for self-maintenance and self-healing, and self-recovery of products and systems.

References

[1.1] Koc, M., Ni, J. and Lee, J., 2002, "Introduction to e-manufacturing," In *Proceedings of the 5th International Conference on Frontiers of Design and Manufacturing.*

[1.2] Kelle, P. and Akbulut, A., 2005, "The role of ERP tools in supply chain information sharing, cooperation, and cost optimisation," *International Journal of Production Economics*, **93–94** (Special Issue), pp. 41–52.

[1.3] Akkermans, H.A., Bogerd, P., Yucesan, E. and Van Wassenhove, L.N., 2003, "The impact of ERP on supply chain management: exploratory findings from a European Delphi study," *European Journal of Operational Research*, **146**(2), pp. 284–301.

[1.4] Da, T. (ed.), 2004, *Supply Chains – A Manager's Guide*, Addison-Wesley, Boston, MA.

[1.5] Andersen, H. and Jacobsen, P., 2000, *Customer Relationship Management: A Strategic Imperative in the World of E-Business*, John Wiley & Sons Canada, Toronto.

[1.6] Cheng, F.T., Shen, E., Deng, J.Y. and Nguyen, K., 1999, "Development of a system framework for the computer-integrated manufacturing execution system: a distributed object-oriented approach," *International Journal of Computer Integrated Manufacturing*, **12**(5), pp. 384–402.

[1.7] Lee, J., 2003, "E-manufacturing – fundamental, tools, and transformation," *Robotics and Computer-Integrated Manufacturing*, **19**(6), pp. 501–507.

[1.8] Koc, M. and Lee, J., 2002, "E-manufacturing and e-maintenance – applications and benefits," In *International Conference on Responsive Manufacturing*, Gaziantep, Turkey.

[1.9] Ge, M., Du, R., Zhang, G.C. and Xu, Y.S., 2004, "Fault diagnosis using support vector machine with an application in sheet metal stamping operations," *Mechanical Systems and Signal Processing*, **18**(1), pp. 143–159.

[1.10] Li, Z.N., Wu, Z.T., He, Y.Y. and Chu, F.L., 2005, "Hidden Markov model-based fault diagnostics method in speed-up and speed-down process for rotating machinery," *Mechanical Systems and Signal Processing*, **19**(2), pp. 329–339.

[1.11] Wu, S.T. and Chow, T.W.S., 2004, "Induction machine fault detection using SOM-based RBF neural networks," *IEEE Transactions on Industrial Electronics*, **51**(1), pp. 183–194.

[1.12] Wang, J.F., Tse, P.W., He, L.S. and Yeung, R.W., 2004, "Remote sensing, diagnosis and collaborative maintenance with web-enabled virtual instruments and mini-servers," *International Journal of Advanced Manufacturing Technology*, **24**(9–10), pp. 764–772.

[1.13] Chen, Z., Lee, J. and Qiu, H., 2005, "Intelligent infotronics system platform for remote monitoring and e-maintenance," *International Journal of Agile Manufacturing*, **8**(1), pp. 3–11.

[1.14] Qu, R., Xu, J., Patankar, R., Yang, D., Zhang, X. and Guo, F., 2006, "An implementation of a remote diagnostic system on rotational machines," *Structural Health Monitoring*, **5**(2), pp. 185–193.

[1.15] Han, T. and Yang, B.S., 2006, "Development of an e-maintenance system integrating advanced techniques," *Computers in Industry*, **57**(6), pp. 569–580.

[1.16] Chau, P.Y.K. and Tam, K.Y., 2000, "Organisational adoption of open systems: a 'technology-push, need-pull' perspective," *Information & Management*, **37**(5), pp. 229–239.

[1.17] Lee, J., 1995, "Machine performance monitoring and proactive maintenance in computer-integrated manufacturing: review and perspective," *International Journal of Computer Integrated Manufacturing*, **8**(5), pp. 370–380.

[1.18] Lee, J., 1996, "Measurement of machine performance degradation using a neural network model," *Computers in Industry*, **30**(3), pp. 193–209.
[1.19] MIMOSA, www.mimosa.org.
[1.20] Djurdjanovic, D., Lee, J. and Ni, J., 2003, "Watchdog Agent – an infotronics-based prognostics approach for product performance degradation assessment and prediction," *Advanced Engineering Informatics*, **17**(3–4), pp. 109–125.
[1.21] Jardine, A.K.S., Lin, D. and Banjevic, D., 2006, "A review on machinery diagnostics and prognostics implementing condition-based maintenance," *Mechanical Systems and Signal Processing*, **20**(7), pp. 1483–1510.
[1.22] Sun, Z. and Chang, C.C., 2002, "Structural damage assessment based on wavelet packet transform," *Journal of Structural Engineering*, **128**(10), pp. 1354–1361.
[1.23] Yan, J. and Lee, J., 2005, "Degradation assessment and fault modes classification using logistic regression," *Transactions of the ASME, Journal of Manufacturing Science and Engineering*, **127**(4), pp. 912–914.
[1.24] Lemm, J.C., 1999, "Mixtures of Gaussian process priors," In *IEEE Conference Publication.*
[1.25] Ocak, H. and Loparo, K.A., 2005, "HMM-based fault detection and diagnosis scheme for rolling element bearings," *Transactions of the ASME, Journal of Vibration and Acoustics*, **127**(4), pp. 299–306.
[1.26] Huang, S.H., Hao, X. and Benjamin, M., 2001, "Automated knowledge acquisition for design and manufacturing: the case of micromachined atomizer," *Journal of Intelligent Manufacturing*, **12**(4), pp. 377–391.
[1.27] Liu, J., Ni, J., Djurdjanovic, D. and Lee, J., 2004, "Performance similarity based method for enhanced prediction of manufacturing process performance," In *American Society of Mechanical Engineers, Manufacturing Engineering Division*, MED.
[1.28] Govers, C.P.M., 1996, "What and how about quality function deployment (QFD)," *International Journal of Production Economics*, **46–47**, pp. 575–585.
[1.29] ReVelle, J. (ed.), 1998, *The QFD Handbook*, John Wiley and Sons.
[1.30] Carnero, M.C., 2005, "Selection of diagnostic techniques and instrumentation in a predictive maintenance program: a case study," *Decision Support Systems*, pp. 539–555.
[1.31] Tse, P.W., Peng, Y.H. and Yam, R., 2001, "Wavelet analysis and envelope detection for rolling element bearing fault diagnosis – their effectiveness and flexibilities," *Journal of Vibration and Acoustics*, **123**(3), pp. 303–310.
[1.32] Qiu, H. and Lee, J., 2004, "Feature fusion and degradation detection using self-organising map," In *Proceedings of the 2004 International Conference on Machine Learning and Applications.*
[1.33] DeJong, C., 2008, "Faster cutting: cut in balance," www.autofieldguide.com.
[1.34] Al-Shurafa, A.M., 2003, "Determination of balancing quality limits," www.plant-maintenance.com.

2

A Framework for Integrated Design of Mechatronic Systems

Kenway Chen, Jonathan Bankston, Jitesh H. Panchal and Dirk Schaefer

The G. W. Woodruff School of Mechanical Engineering
Georgia Institute of Technology, Savannah, GA 31407, USA
Emails: kchen8@gatech.edu, jonathanbankston@gatech.edu, jitesh.panchal@me.gatech.edu, dirk.schaefer@me.gatech.edu

Abstract
Mechatronic systems encompass a wide range of disciplines and hence are collaborative in nature. Currently, the collaborative development of mechatronic systems is inefficient and error-prone because contemporary design environments do not allow sufficient flow of design and manufacturing information across electrical and mechanical domains. Mechatronic systems need to be designed in an integrated fashion allowing designers from both electrical and mechanical engineering domains to receive automated feedback regarding design modifications throughout the design process. Integrated design of mechatronic systems can be facilitated through the integration of mechanical and electrical computer-aided design (CAD) systems. One approach to achieve such integration is through the propagation of constraints. Cross-disciplinary constraints between mechanical and electrical design domains can be classified, represented, modelled, and bi-directionally propagated in order to provide automated feedback to designers of both engineering domains. In this chapter, the authors focus on constraint classification and constraint modelling and provide an example by means of a robot arm. The constraint modelling approach serves as a preliminary concept for the implementation of constraint propagation between mechanical and electrical CAD systems.

2.1 Introduction

Cross-disciplinary integration of mechanical engineering, electrical and electronic engineering as well as recent advances in information engineering are becoming more and more crucial for future collaborative design, manufacture, and maintenance of a wide range of engineering products and processes [2.1]. In order to allow for additional synergy effects in collaborative product creation, designers from all disciplines involved need to adopt new approaches to design, which facilitate concurrent cross-disciplinary collaboration in an integrated fashion. This, in particular, holds true for the concurrent design of mechatronic systems, which is the main focus of this chapter.

Mechatronic systems usually encompass mechanical, electronic, electrical, and software components (Figure 2.1). The design of mechanical components requires a sound understanding of core mechanical engineering subjects, including mechanical devices and engineering mechanics [2.1]. For example, expertise regarding lubricants, heat transfer, vibrations, and fluid mechanics are only a few aspects to be considered for the design of most mechatronic systems. Mechanical devices include simple latches, locks, ratchets, gear drives and wedge devices as well as complex devices such as harmonic drives and crank mechanisms. Engineering mechanics is concerned with the kinematics and dynamics of machine elements. Kinematics determines the position, velocity, and acceleration of machine links. Kinematic analysis is used to find the impact and jerk on a machine element. Dynamic analysis is used to determine torque and force required for the motion of a link in a mechanism. In dynamic analysis, friction and inertia play an important role.

Electronics involves measurement systems, actuators, and power control [2.1]. Measurement systems in general comprise of three elements: sensors, signal conditioners, and display units. A sensor responds to the quantity being measured from the given electrical signal, a signal conditioner takes the signal from the sensor and manipulates it into conditions suitable for display, and in the display unit the output from the signal conditioner is displayed. Actuation systems comprise the elements that are responsible for transforming the output from the control system into the controlling action of a machine or a device. Finally, power electronic devices are important in the control of power-operated devices. The silicon controlled rectifier is an example of a power electronic device that is used to control DC motor drives.

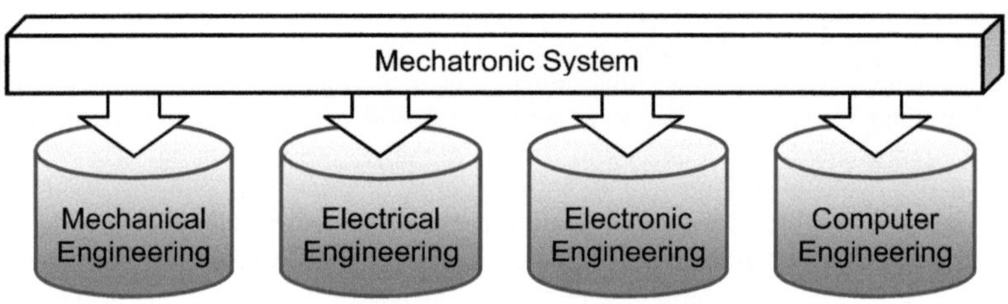

Figure 2.1. The scope of mechatronic system

The electrical aspect of mechatronic systems involves the functional design of electrical plants and control units. This is done through the generation of several types of schematics such as wiring diagrams and ladder diagrams. In addition, programmable logic controllers (PLCs) are widely used as control units for mechatronic systems. PLCs are well adapted to a range of automation tasks. These typically are industrial processes in manufacturing where the cost of developing and maintaining an automation system is relatively high compared to the total cost of the automation, and where changes to the automation system are expected during its operational life.

According to Reeves and Shipman [2.2], discussion about the design must be embedded in the overall design process. The ideal process of concurrent or simultaneous engineering is characterised by parallel work of a potentially distributed community of designers who know about the parallel work of their colleagues and collaborate as necessary. In order to embed the discussion aspect, mechatronic products should be designed in an integrated fashion that allows for designers of both electrical and mechanical engineering domains to automatically receive feedback regarding design modifications made on either side throughout the design process. This means, that if a design modification of a mechanical component of a mechatronic systems will lead to a design modification of an electrical aspect of the mechatronic system or *vice versa*, the engineer working at the counterpart system should be notified as soon as possible.

Obviously, even on the conceptual design level, mechanical and electrical design aspects of mechatronic systems are highly intertwined through a substantial number of constraints existing between their components (see Figure 2.2).

Consequently, in order to integrate mechanical and electrical CAD tools on systems realisation/integration level (see Figure 2.3) into an overarching cross-disciplinary computer-aided engineering (CAE) environment, these constraints have to be identified, understood, modelled, and bi-directionally processed.

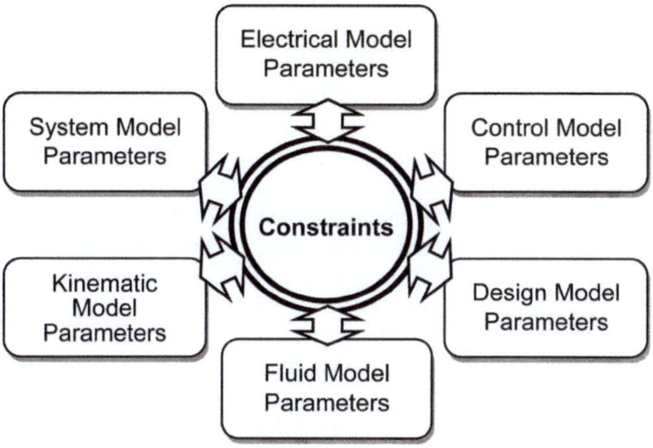

Figure 2.2. Constraints between all domains on the conceptual design level

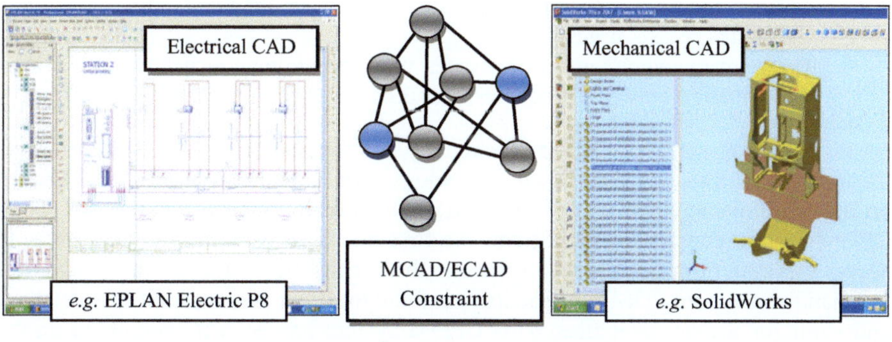

Figure 2.3. Constraints between MCAD and ECAD models on the system realisation level

The organisation of this chapter is as follows: in Section 2.2, an overview of current techniques aimed towards achieving integrated design of mechatronic systems as well as a research gap analysis is provided; in Section 2.3, a constraint-modelling approach towards integrated design of mechatronic systems is proposed, along with a classification of mechanical and electrical constraints; in Section 2.4, the use of the method discussed in Section 2.3 is illustrated by means of a robot arm as an example. Both discipline-specific and cross-disciplinary constraints existing in the robot arm example are identified; in Section 2.5, a framework for integrated mechatronic design approach and the capabilities needed to realise such a framework are discussed; finally, in Section 2.6 closing remarks are provided.

2.2 State of the Art and Research Gaps

In this section, a brief overview of a variety of research activities relevant to the development of approaches towards integrated design of mechatronic systems is presented and the research gaps are identified.

2.2.1 Product Data Management

The amount of engineering-related information required for the design of complex mechatronic products tends to be enormous. Aspects to be considered include geometric shapes of mechanical components, electrical wiring information, information about input and output pins of electronic circuit boards, and so on. In order to keep track of all these product data during the development process, product data management (PDM) systems are used. PDM systems manage product information from design to manufacture, and to end-user support. In terms of capabilities, PDM systems support five basic user functions [2.3]:

1. Data vault and document management that provides storage and retrieval of product information.
2. Workflow and process management that controls procedures for handling product data.
3. Product structure management that handles bills of materials, product configurations, associated design versions, and design variations.
4. Parts management that provides information on standard components and facilitates re-use of designs.
5. Program management that provides work breakdown structures and allows coordination between processes, resource scheduling, and project tracking.

In terms of state of the art technology, contemporary PDM systems have incorporated the use of web-based technology. An example is a component-based product data management system (CPDM) developed by Sung and Park [2.4]. Their CPDM system consists of three tiers: the first tier is focused on allowing users to access the system through a web browser; the second tier is the business logic tier that handles the core PDM functionality; and the third tier is composed of a database and vault for the physical files. This CPDM system has been implemented on the Internet for a local military company that manufactures various mechatronic systems such as power cars, motorised cars, and sensitive electrical equipments. Web-based PDM systems can also be used for similarity search tasks in order to identify existing designs or components of specific shape or manufacturing-related information that may be useful for new designs or design alternatives.

You and Chen [2.5] proposed an algorithm that runs in web-based PDM systems. In their algorithm, a target part is given with characteristic attributes, and similar parts in the database are identified based on their shape or manufacturing features. The results are sorted in the order of similarity. You and Chen's proposed algorithm for similarity evaluation adopts the polar Fourier transform (PFT) method, which is a discrete Fourier transform method.

There are several advantages in utilising web-based PDM systems. One advantage is user-friendliness: the browsers used in the PDM system are the same ones used within the World Wide Web, and hence web-based PDM systems require little training. Another advantage is their great accessibility since these browsers run on different platforms. However, there are several drawbacks as well: first, the information transferring speed is limited compared to the speed of LAN or WAN; second, mistakes relating to acquiring or transferring data can occur if the system is not utilised correctly; and finally, there are major concerns regarding security and exposing a company's trade secrets during the process of information transfer.

2.2.2 Formats for Standardised Data Exchange

PDM systems are tools that allow designers to manage and keep track of the product data throughout the entire design process. However, in order to ensure proper product configuration control, PDM systems must be able to communicate with the CAD systems that the designers use during the design process. In the context of integrated design of mechatronic products, this means communication between CAD/CAE systems of different engineering disciplines, *i.e.* MCAD and ECAD.

For instance, an MCAD model typically contains the following information [2.6]: *features*, which are high-level geometric constructs used during the design process to create shape configurations in the CAD models; *parameters*, which are values of quantities in the CAD model, such as dimensions; *constraints*, which are relationships between geometric elements in the CAD models, such as parallelism, tangency, and symmetry. An MCAD system cannot simply transfer such information to a PDM system or other CAD/CAE system because these systems have significantly different software architectures and data models. One potential approach towards achieving communication between various CAD, CAE, and PDM systems is through the utilisation of neutral file formats, such as, for example, Initial Graphics Exchange Specification (IGES) or the Standard for the Exchange of Product Model Data (STEP).

IGES was created for CAD-CAD information exchange. The fundamental role of IGES was to convert two-dimensional drawing data and three-dimensional shape data into a fixed file format in electronic form and pass the data to other CAD systems [2.7]. Major limitations of IGES include large file size, long processing time, and most importantly, the restriction of information exchange to shape data only [2.7]. Despite these limitations, IGES is still supported by most CAD systems and widely used for CAD information exchange.

Another important neutral file format for representation of product information is STEP, also known as ISO 10303. STEP can be viewed as consisting of several layers [2.7], the top layer being a set of application protocols (APs) that address specific product classes and lifecycle stages. These APs specify the actual information exchange and are constructed from a set of modules at lower layers, called integrated resources, which are common for all disciplines. The APs relevant to electro-mechanical systems integration are AP-203, AP-210, AP-212, and AP-214. AP-203 protocol defines the information exchange of geometric entities and configuration control of products. This protocol can capture common modern MCAD representations including 2D drawing, 3D wireframes, surface models, and solid models. AP-212 is concerned with electro-mechanical design and installation.

Currently, there is an ongoing effort in making STEP information models available in a universal format to business application developers. Lubell *et al.* [2.8] have presented a roadmap for possible future integration of STEP models with widely accepted and supported standard software modelling languages such as UML and XML. STEP provides standardised and rigorously-defined technical concepts and hence shows greater quality than other data exchange standards, but the traditional description and implementation method for STEP has failed to achieve the popularity of XML and UML [2.8]. Thus, emerging XML and UML-based STEP implementation technology shows promise for better information exchange ability.

2.2.3 The NIST Core Product Model

Most PDM systems and the exchange standards used for communication between CAD/CAE/PDM systems focus mainly on product geometry information. However, more attention is needed for developing standard representations that specify design information and product knowledge in a full range instead of solely geometry-oriented. At the US National Institute for Standards and Technology (NIST), an information modelling framework intended to address this issue of expanding the standard representations to a full range has been under development [2.9]. This conceptual product information modelling framework has the following key attributes [2.9]:

1. It is based on formal semantics and will eventually be supported by an appropriate ontology to permit automated reasoning.
2. It deals with conceptual entities such as artefacts and features and not specific artefacts such as motor, pumps, or gears.
3. It is to serve as information repository about products, and such information includes product description that are not at the time incorporated.
4. It is intended to foster the development of applications and processes that are not feasible in less information-rich environments.

One major component of this information modelling framework is the core product model (CPM). The CPM is developed as a basis for future CAD/CAE information exchange support system [2.7]. CPM is composed of three main components [2.7]:

(a) *Function*: models what the product is supposed to do.
(b) *Form*: models the proposed design solution for the design problem specified by the function, usually the product's physical characteristics are modelled in terms of its geometry and material properties.
(c) *Behaviour*: models how the product implements its function in terms of the engineering principles incorporated into a behavioural or casual model.

CPM was further extended to components with continuously varying material properties [2.10]. The concept for modelling continuously varying material properties is that of distance fields associated with a set of material features, where values and rate of material properties are specified [2.10]. This extension of CPM uses UML to represent scalar-valued material properties as well as vector- and tensor-valued material properties.

2.2.4 Multi-representation Architecture

Another approach that can be used to support the integrated design of mechatronic systems is the multi-representation architecture (MRA) proposed by Peak *et al.* [2.11]. It is a "design analysis integration strategy that views CAD-CAE integration as an information-intensive mapping between design models and analysis models" [2.11]. In other words, the gap that exists in CAD/CAE between design models and analysis models is considered too large to be covered by a "single general integration bridge"; hence MRA addresses the CAD-CAE integration problem by placing four information representations as "stepping stones" between design and analysis tools in the CAD/CAE domains. The four information representations are: solution method models, analysis building blocks, product models, and product model-based analysis models.

Solution method models represent analysis models in low-level, solution method-specific form. They combine solution tool inputs, outputs, and control into a single information entity to facilitate automated solution tool access and results retrieval. Analysis building blocks represent engineering analysis concept and are largely independent of product application and solution method. They represent analysis concepts using object and constraint graph techniques and have a defined information structure with graphical views to aid development, implementation, and documentation. Product models represent detailed, design-oriented product information. A product model is considered the master description of a product that supplies information to other product lifecycle tasks. It represents design aspects of products, enables connections with design tools, and supports idealisation usable in numerous analysis models. And finally, product model-based analysis models contain linkages that represent design-analysis associativity between product models and analysis building blocks.

The MRA can be used to support integrated design of mechatronic systems because it has the flexibility to support different solution tools and design tools and also accommodating analysis models from diverse disciplines. Object and constraint graph techniques used in the MRA provide modularity and semantics for clearer representation of design and analysis models. Peak *et al.* [2.11] have evaluated the MRA using printed wiring assembly solder joint fatigue as a case study. Their results show that specialised product model-based analysis models "enable highly automated routine analysis and uniformly represent analysis models containing a mixture of both formula-based and finite element-based relations" [2.11]. With the capability of analysing formula-based and finite element-based models, the MRA can be used to create specialised CAE tools that utilise both design information and general purpose solution tools.

2.2.5 Constraint-based Techniques

Many mechanical engineering CAD systems provide parametric and feature-based modelling methods and support frequent model changes. In order to define and analyse various different attributes of a product, a variety of models including design models, kinematic models, hydraulic models, electrical models, and system models are needed. Except for geometry-based data transfers, there is neither exchange nor integration of data for interdisciplinary product development available. Kleiner *et al.* [2.12] proposed an approach that links product models using constraints between parameters. The integration concept is based on parametric product models, which share their properties through the utilisation of constraints. In their context, a virtual product is represented by partial models from different engineering disciplines and associated constraint models.

The fundamentals for the development of neutral, parametric information structures for the integration of product models are provided by existing data models from ongoing development as well as concepts from constraint logic programming [2.13]. The parametric information model that Kleiner *et al.* [2.12] developed is based on the Unified Modelling Language (UML). The model contains the class *Item*, which represents real or virtual objects such as parts, assemblies, and models. Every object *Item* has a version (class *ItemVersion*) and specific views (class *DesignDisciplineItem Definition*). A view is relevant for the requirements of one or more lifecycle stages and application domains and collects product data of the *Item* and *ItemVersion* object. The extension of STEP product data models includes general product characteristics (class *Property*), attributes (class *Parameter*) and restricted relationships (class *Constraint*). The information model Kleiner *et al.* [2.12] developed is based on the integration of independent CAx models using their parameters. The links between CAx models are implemented using the class *Constraint*, which can set parameters of different product models in relationship to each other. On one hand, a constraint restricts at least one parameter and on the other hand, a parameter may be restricted by several constraints, which are building a constraint net. Different types of constraints are implemented in subclasses in order to characterise the relationship between parameters in detail.

The constraint-based parametric integration offers an alternative solution compared to unidirectional process chains or file-based data exchange procedures using neutral data formats (*e.g.* IGES and STEP). Model structures and properties could be imported, analysed, and exported by linking different CAx models. Kleiner *et al.* [2.12] developed a Java-based software system that supports product data integration for the collaborative design of mechatronic products. The software system, called Constraint Linking Bridge (Colibri), is developed based on the constraint-based integration concept and the described information model. It sets up connections to CAx systems in order to analyse model structures and parameters. Different interfaces to CAD/CAE systems allow the transfer of appropriate model structures and parameters as well as parameter transformation activities. Browsers for models (model viewer) and constraints (constraint viewer) are used as graphical user interfaces to analyse, specify, check, and solve constraints. The current version of Colibri contains implemented interfaces to the CAD system Pro/ENGINEER, the simulation application system MATRIXx/SystemBUILD as well as to the multi-body simulation software system AUTOLEV. Colibri is integrated into a distributed product development environment, called *Distributed Computing Environment/ Distributed File System* (DCE/DFS). This infrastructure offers platform-independent security and directory services for users and groups, policy and access control management as well as secure file services storing objects in electronic vaults.

This software Colibri, though is just a prototype, it offers an alternative technology for sharing product data and illustrates the possibility of integrating difference CAx models by linking them and using constraints to specify their product development relationships. This work done by Kleiner *et al.* opens up opportunities for further development that supports multi-disciplinary modelling and simulation of mechatronic systems and offers user functions for data sharing.

2.2.6 Active Semantic Networks

In the area of electrical engineering CAD, Schaefer *et al.* [2.14] proposed a shared knowledge base for interdisciplinary parametric product data models in CAD. This approach is based on a so-called Active Semantic Network (ASN). In ASNs, constraints can be used to model dependencies between interdisciplinary product models and co-operation, and rules using these constraints can be created to help designers to collaborate and integrate their results to a common solution. With this design approach, designers have the ability to visualise the consequences of design decisions across disciplines. This visualisation allows designers to integrate their design results and to improve the efficiency of the overall design process.

A semantic network is a graphic notation for representing knowledge in patterns of interconnected nodes and arcs. It is a graph that consists of vertices, which represent concepts, and arcs, which represent relations between concepts. An ASN can be realised as an active, distributed, and object-oriented DBMS (database management system) [2.15]. A DBMS is computer software designed for the purpose of managing a database. It is a complex set of software programs that controls the organisation, storage and retrieval of data in a database. An active DBMS is a DBMS that allows users to specify actions to be taken automatically when certain conditions arise.

Constraints defined in product models can be specified by rules. When a constraint is violated, possible actions include extensive inferences on the product data and notifications to the responsible designers to inform them about the violated constraints. There exists constraint propagation within the ASN. Constraints are mainly used in CAD to model dependencies between geometric objects. In the ASN, constraints are used to model any kind of dependency between product data [2.15].

A database object in the ASN consists of the data itself, a set of associated rules, and co-operative locks. This means that event-condition-action (ECA) rules can be connected to database objects to specify their active behaviour. Constraints are modelled as normal database objects that are also subject to user modifications. ECA rules linked to a constraint object specify the active behaviour of the constraint and are evaluated at run time. The ASN uses a rule-based evaluation method for constraint propagation.

2.2.7 Summary and Research Gap Analysis

Product data management systems manage product information from design to manufacture to end-user support. They ensure that "right information in the right form is available to the right person at the right time" [2.3]. PDM technology, if it can be implemented to all CAD/CAE systems, can significantly improve the design process of mechatronic systems because its infrastructure is user-friendly and has great accessibility. For example, Windchill developed by PTC has a browser-based user interface that uses standard HTML for bi-directional communication of form-based information and Java applets to deliver interactive application capabilities [2.3]. Rolls-Royce AeroEngines designs and manufactures mid-range aircraft engines. They use PTC Windchill for their integrated product development environments, allowing them to maintain a quicker product delivery with increasing costs. Modern PDM systems, such as PTC Windchill, though possessing the capability of communicating with several MCAD systems, lack the ability to communicate with ECAD systems and other CAD/CAE systems that may be used during the design of mechatronic systems.

Standardised data exchange formats, such as ISO 10303 (STEP), provide information exchange for parameterised feature-based models between different CAD systems. They provide communication between CAD systems through system-independent file formats that are in computer-interpretable forms for data transmission. These data exchange formats cover a wide range of application areas: aerospace, architecture, automotives, electronic, electro-mechanical, process plant, ship building, and the list can go on and on. However, as with any standard-based exchange of information between dissimilar systems, it is impossible to convey certain elements defined in some particular CAD systems but having no counterparts in others [2.6]. Furthermore, past testing experience has shown that differences in the internal accuracy criteria of CAD systems can lead to problems of accuracy mismatch, which has caused many translation failures. Also, the provision of explicit geometric constraints adds possibilities for redundancy in shape models, and such geometric redundancy implies more possibility of accuracy mismatch. Hence, despite the ability to cover a wide range of applications areas, data exchange formats also have numerous problems to be solved.

NIST CPM and its extensions are abstract models with general semantics with specific semantics about a particular domain embedded within the usage of the models for that domain. CPM represents a product's form, function, and behaviour, as well as its physical and functional decompositions, and the relationships among these concepts. CPM is intended to capture product and design rationale, assembly, and tolerance information from the earliest conceptual design stage to the full lifecycle, and also facilitates the semantic interpretability of CAD/CAE/CAM systems. The current model also supports material model construction, material-related queries, data transfer, and model comparison. The construction process involves the definition of material features and choosing properties for distance field. Further research is needed to develop an API (application programming interface) between CPM and PLM systems, and to identify or develop standards for the information interchange [2.9].

Multi-representation architecture is aimed at satisfying the needs in the links between CAE and CAD. These needs include [2.11]:

1. Automation of ubiquitous analysis
2. Representation of design and analysis associativity and of the relationships among models
3. Provision of various analysis models throughout the lifecycle of the product

The initial focus of the MRA is on ubiquitous analyses, which are analyses that are regularly used to support the design of a product [2.16]. The MRA supports capturing knowledge and expertise for routine analysis through semantic-rich information models and the explicit associations between design and analysis models. While the MRA captures routine analysis and the mapping between design parameters and analysis parameters, there is still the opportunity for model abuse [2.17]. The MRA enables reuse of the analysis templates in product development. The behaviour model creators (such as the analysts) and behaviour model users (such as the designers) often do not have the same level of understanding of the model and thus limit the reuse of a model [2.17]. The gap between designers and analysts is decreased by providing engineering designers with increased knowledge and understanding about behavioural simulation. Plans for future work regarding the MRA include [2.17]: further instantiation of the behavioural model repository, refinement of knowledge representation using ontology languages, and implementation to support instantiation with design parameters for execution.

The collaborative design system Colibri developed by Kleiner *et al.* [2.12] offers a new approach for exchanging information across the disciplinary divide as compared to unidirectional process chains or file-based data exchange procedures using neutral data formats (*e.g.* IGES and STEP); however, it focuses on linking various CAx models and does not cover the information gap between mechanical CAD and electrical CAD systems.

In designing a mechatronic product, there are many situations that require the exchange of information between MCAD models and ECAD models. Modifications made on MCAD site may lead to significant design modifications to be made on ECAD site and *vice versa*. Obviously, there exist a huge number of constraints between a mechanical part of a mechatronic product and its electrical counterpart that have to be fulfilled to have a valid design configuration. As yet, such interdisciplinary constraints between models of different engineering design domains cannot be handled in a multi-disciplinary CAE environment due to the lack of appropriate multi-disciplinary data models and appropriate propagation method. Table 2.1 summarises the research gaps in the aforementioned integration approaches. The framework for integrated design of mechatronic systems proposed in the following sections focuses on integrating the mechanical and the electrical domains. The framework is intended to support the information exchange of design modification during the design process.

Table 2.1. Research gaps in the integration approaches

Approach	Description	Research gap
PDM	Manages product information allowing multiple designers to work on a shared repository of design information.	Preserves only file-level dependencies between information from multiple domains. Does not capture fine-grained information dependencies such as at a parameter level.
Standardised Data Exchange	Supports information exchange between CAD/CAE/PDM systems.	The emphasis of data exchange standards is on information flow across systems in a given domain, not on cross-discipline integration.
Multi-representation Architecture	Supports CAD-CAE integration through the usage of four models each supporting different levels of product information.	The focus in MRA is on integrating geometric and analysis information. It does not address the information link between mechanical and electrical CAD systems.
Colibri	Shares data across design teams of different domains through constraints and parametric relations in CAx.	Does not provide information exchange between mechanical CAD and electrical CAD.

2.3 An Approach to Integrated Design of Mechatronic Products

2.3.1 Modelling Mechatronic Systems

The constraint modelling-based approach proposed in this chapter (Figure 2.4) is similar to the semantic network approach briefly described before in that constraints are being modelled as nodes and relationships are drawn between nodes. The components of mechatronic systems are modelled as objects with attributes, and constraints between these attributes are identified and modelled. The procedures of the proposed constraint modelling approach are as follows:

STEP 1: List all components of the mechatronic system and their attributes and classify the components in either the mechanical domain or the electrical domain.

STEP 2: Based on the attributes of the component, draw the constraint relationship between the components in the domain and appropriately label the constraint by the constraint categories.

STEP 3: Based on the attributes of the component, draw the constraint relationship between the components across the domains and appropriately label the constraint by the constraint categories.

STEP 4: Construct a table of constraints for the particular mechatronic system. The table contains a complete list of the every component of the mechatronic system, the table is to indicate that, when a particular attribute of the component is being modified, which attribute of which component (both within the domain and across the domain) would be affected.

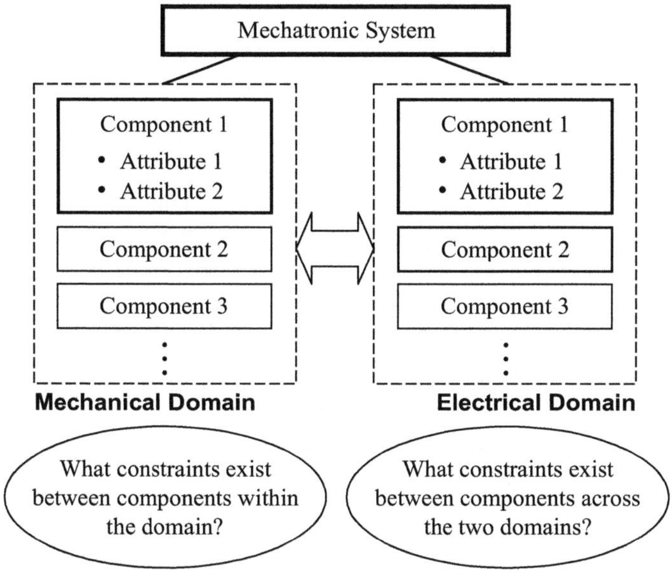

Figure 2.4. A graphical view of the constraint modelling approach

2.3.2 Constraint Classification in Mechanical Domain

Geometric Constraints

Most CAD systems allow the creation of variational models with parameterisation, constraints and features. The set of common geometric constraint types is listed as follows [2.18]:

- *Parallelism* – this has an undirected form and a directed form with one reference element. There is also a dimensional subtype, in which a constrained distance can be specified.
- *Point-distance* – in the directed case, the reference element may be either point, line, or plane. Multiple points may be constrained. In the undirected case, the number of constrained points is limited to two, and a dimensional value is required.
- *Radius* – has a dimensionless form, for example, "the radii of all these arcs are the same", and a dimensional form, for example, "the radii of all the constrained arcs have the same specified value".
- *Curve-length* – asserts that the lengths of all members of a set of trimmed curves are equal. There is a dimensional form allowing the value of the length to be specified.
- *Angle* – constraints a set of lines or planes to make the same angle with a reference element, or in the undirected case specifies the angle between precisely two such elements.
- *Direction* – a vector-valued constraint used for constraining the directional attributes of linear elements such as lines or planes.

- *Perpendicularity* – there may be either one or two reference elements (lines or planes), and all the constrained elements are required to be perpendicular to them. There is also an undirected form in which two or three elements are required to be mutually perpendicular.
- *Incidence* – in its simplest form, it simply asserts that one or more constrained entities are contained within that reference element.
- *Tangency* – may be used to specify multiple tangencies between a set of reference elements, and as set of constrained elements.
- *Coaxial* – constrains a set of rotational elements to share the same axis or to share a specified reference axis.
- *Symmetry* – constrains two ordered sets of elements to be pair-wise symmetric with respect to a given line or plane.
- *Fixed* – used to fix points and directions in absolute terms for anchoring local coordinated systems in global space.

Kinematics Constraints

Kinematics is a branch of mechanics that describes the motion of objects without the consideration of the masses and forces that bring about the motion [2.1]. Kinematics is the study of the position of an object changes with time. Position is measured with respect to a set of coordinates. Velocity is the rate of change of position. Acceleration is the rate of change of velocity. In designing mechatronic systems, the kinematics analysis of machine elements is very important. Kinematics determines the position, velocity, and acceleration of machine links. Kinematics analysis helps to find the impact and jerk on a machine element.

Force Constraints

In mechanical engineering, and in particular, in dealing with "machines in mechatronics", it often involves the study of relative motion between the various parts of a machine as well as the forces acting on them, hence the knowledge in this subject of "forces" is very essential for an engineer to design the various parts of mechatronic systems [2.1]. Force is an important factor as an agent that produces or tends to produce, destroys or tends to destroy motion. When a body does not move or tend to move, the body does not have any friction force. Whenever a body moves or tends to move tangentially with respect to the surface on which it rests, the interlocking properties of the minutely projected particles due to the surface roughness oppose the motion. This opposition force that acts in the opposite direction to the movement of the body is the force of friction. Both force and friction play an important role in mechatronic systems.

In considering the force constraint in mechanical systems, there are three major parameters that can affect the mechanical systems: the stiffness of the system, the forces opposing motion (such as frictional or damping effects), and the inertia or resistance to acceleration [2.19]. The stiffness of a system is described by the relationship between the forces used to extend or compress a spring and the resulting extension or compression. The inertia or resistance to acceleration exhibits the property that the bigger the inertia (mass) the greater the force required to give it a specific acceleration.

Energy Constraints

Energy is a scalar physical quantity, which is a property of objects and systems that is conserved by nature. Energy can be converted in a variety of ways. An electric motor converts electrical into mechanical and thermal energy, a combustion engine converts chemical into mechanical and thermal energy, and so on. In physics, mechanical energy describes the potential energy and kinetic energy present in the components of a mechanical system. If the system is subject only to conservative forces, such as only to gravitational force, by the principle of conservation of mechanical energy the total mechanical energy of the system remains constant.

Material Constraints

The various machine parts of the mechatronic system often experience different loading conditions. If a change of motion of the rigid body (the machine parts) is prevented, the force applied will cause a deformation or change in the shape of the body. Strain is the change in dimension that takes place in the material due to an externally applied force. Linear strain is the ratio of change in length when a tensile or compressive force is applied. Shear strain is measured by the angular distortion caused by an external force. The load per unit deflection in a body is the stiffness. Deflection per unit load is the compliance. If deformation per unit load at a point on the body is different from that at the point of application of the load then compliance at that point is called cross-compliance. In a machine structure, cross-compliance is an important parameter for stability analysis during machining.

The strength of a material is expressed as the stress required causing it to fracture. The maximum force required to break a material divided by the original cross-sectional area at the point of fracture is the ultimate tensile strength of the material. It is obvious that the stress allowed in any component of a machine must be less than the stress that would cause permanent deformation. A safe working stress is chosen with regard to the conditions under which the material is to work. The ratio of the yield stress to allowable stress is the factor of safety.

Tolerance Constraints

The relationship resulting from the difference between the sizes of two features is the fit. Fits have a common basic size. They are broadly classified as clearance fit, transition fit, and interference fit. A clearance fit is one that always provides a clearance between the hole and shaft when they are assembled. A transition fit is one that provides either a clearance or interference between the hole and the shaft when they are assembled. An interference fit is one that provides interference all along between the hole and the shaft when they are assembled.

The production of a part with exact dimensions repetitively is usually difficult. Hence, it is sufficient to produce parts with dimensions accurate within two permissible limits of size, which is the tolerance. Tolerance can be provided on both sides of the basic size (bilateral tolerance) or on one side of the basic size (unilateral tolerance). The ISO systems of tolerance provides for a total of 20 standard tolerance grades.

In any engineering industry, the components manufactured should also satisfy the geometrical tolerances in addition to the common dimensional tolerances. Geometrical tolerances are classified as:

- *Form tolerance* – straightness, flatness, circularity, cylindricity, profile of any line, and profile of any surface
- *Orientation tolerance* – parallelism, perpendicularity, and angularity
- *Location tolerance* – position, concentricity, co-axiality, and symmetry
- *Run-out tolerance* – circular run-out and axial run-out

2.3.3 Constraint Classification in Electrical Domain

Electrical Resistance

Electrical resistance is a measure of the degree to which an object opposes the electric current through it. The electrical resistance of an object is a function of its physical geometry. The resistance is proportional to the length of the object and inversely proportional to the cross-section area of the object. All resistors possess some degree of resistance. The resistances of some resistors are indicated by numbers. This method is used for low value resistors. Most resistors are coded using colour bands, and the way to decode these bands is as follows: the first band gives the resistance value of the resistor in ohms. The fourth band indicates the accuracy of the value. Red in this region indicates 2%, gold indicates 5%, and silver indicates 10% accuracy.

Electrical Capacitance

Electrical capacitance is a measure of the amount of electric charge stored for a given electric potential. The most common form of charge storage device is a two-plate capacitor, which consists of two conducting surfaces separated by a layer of insulating medium called dielectric. The dielectric is an insulating medium through which an electrostatic field can pass. The main purpose of the capacitor is to store electrical energy.

Electrical Inductance

A wire wound in the form of a coil makes an inductor. The property of an inductor is that it always tries to maintain a steady flow of current and opposes any fluctuation in it. When a current flows through a conductor, it produces a magnetic field around it in a plane perpendicular to the conductor. When a conductor moves in a magnetic field, an electromagnetic force is induced in the conductor. The property of the inductor due to which it opposes any increase or decrease in current by the production of a counter emf (electromotive force) is known as self-inductance. The emf force developed in an inductor is proportional to the rate of current through the inductor, and mathematically it depends on the amount of current, the voltage developed, and a proportionality constant that represents the self-inductance of the coil.

Motor Torque

The torque generation in any electric motor is essentially the conversion process of converting electrical energy into mechanical energy. It can be viewed as a result of the interaction of two magnetic flux density vectors: one generated by the stator and one generated by the rotor. In different motor types, the way these vectors generated is different [2.20]. For instance, in a permanent brushless motor the magnetic flux vector is generated by the current in the windings. In the case of an AC induction motor, the stator magnetic flux vector is generated by the current in the stator winding, and the rotor magnetic flux vector is generated by induced voltages on the rotor conductors by the stator field and resulting current in the rotor conductors. The torque production in an electric motor is proportional to the strength of the two magnetic flux vectors (stator's and rotor's) and the sine of the angle between the two vectors.

System Control

The control system provides a logical sequence for the operating program of the mechatronic system. It provides the theoretical values required for each program step, it continuously measures the actual position during motion, and it processes the theoretical versus actual difference [2.21]. In controlling a robot, for example, there are two types of control techniques: point-to-point and continuous path. The point-to-point control involves the specification of the starting point and end point of the robot motion and requiring a control system to render feedbacks at those points. The continuous path control requires the robot end-effector to follow a stated path from the starting point to the end point.

2.4 Illustrative Example: a Robot Arm

2.4.1 Overview of the Robot Arm

A robot is a mechatronic system capable of replacing or assisting the human operator in carrying out a variety of physical tasks. The interaction with the surrounding environment is achieved through sensors and transducers, and the computer-controlled interaction systems emulate human capabilities. The example investigated is the SG5-UT robot arm designed by Alex Dirks of the CrustCrawler team [2.22] (see Figure 2.5).

List of major mechanical components:

- Base and wheel plates
- Links: bicep, forearm, wrist
- Gripper
- Joints: shoulder, elbow, and wrist
- Hitec HS-475HB servos (base, wrist and gripper)
- Hitec HS-645MG servos (elbow bend)
- Hitec HS-805BB servos (shoulder bend)

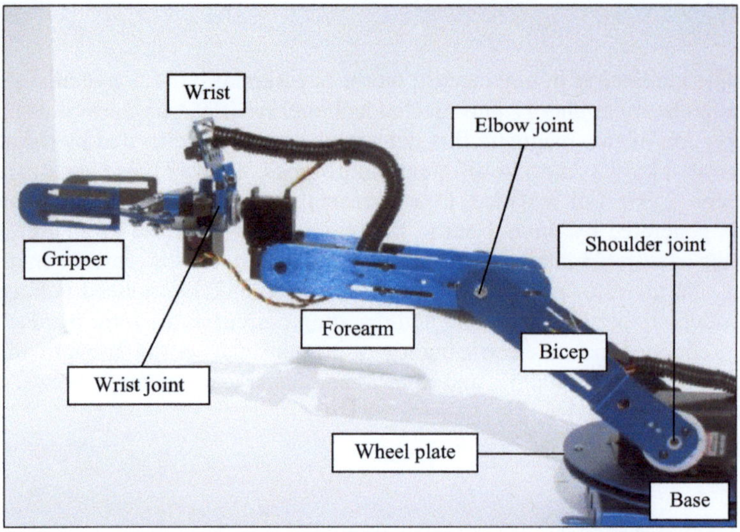

Figure 2.5. The CrustCrawler SG5-UT robot arm

(a) Gripper

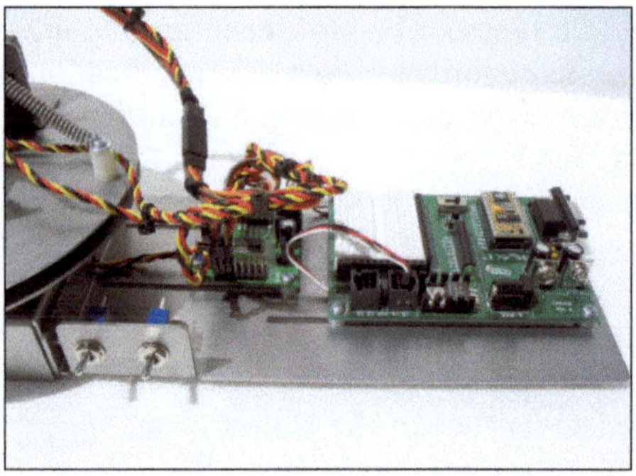

(b) Microcontroller

Figure 2.6. Major components of the robot arm

The most critical aspect of any robot arm is in the design of the gripper [2.22]. The usefulness and functionality of a robot arm is directly related to the ability to sense and successfully manipulate its immediate environment. The gripper drive system, as shown in Figure 2.6(a), consists of a resin gear train driven by an HS-475HB servo. The servos are needed to provide motion to the various mechanical links as well as the gripper. The mounting site of the servos and the power routing to servos and supporting electronics are some of the important aspects to be considered in the design of this robot arm. The microcontroller board (Figure 2.6(b)) is essential for communication between the robot and a PC, providing users the ability to manipulate the robot. It is important to have accurate information on the pin connections and the corresponding components that are being controlled.

2.4.2 Modelling Constraints for SG5-UT

As mentioned in Section 2.3.1, a four-step procedure is used for modelling the SG5-UT robot arm.

STEP 1: List all components of the SG5-UT robot arm and their attributes and classify the components in either the mechanical domain or the electrical domain. These components are listed in Figure 2.7.

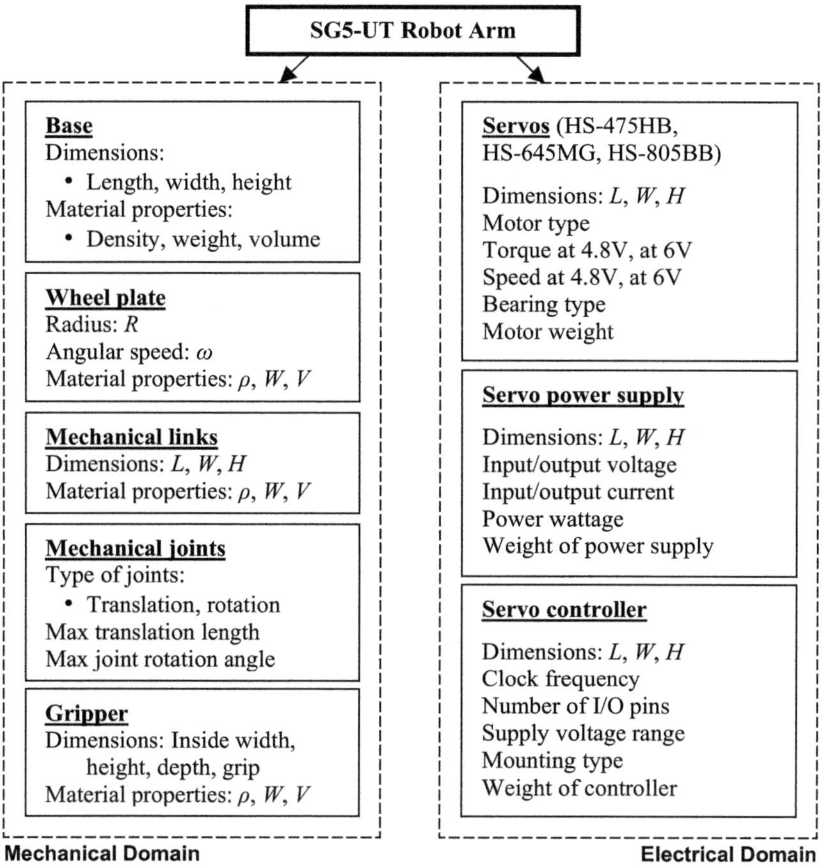

Figure 2.7. A list of major mechanical and electrical components and attributes

STEP 2: Based on the attributes of the components, draw the constraint relationship between the components in the domain and appropriately label the constraint by the constraint categories, as presented below in Figure 2.8 and in Tables 2.2 and 2.3.

For the purpose of brevity, only geometric constraints are described as constraints in the mechanical domain. However, there are many constraints that exist

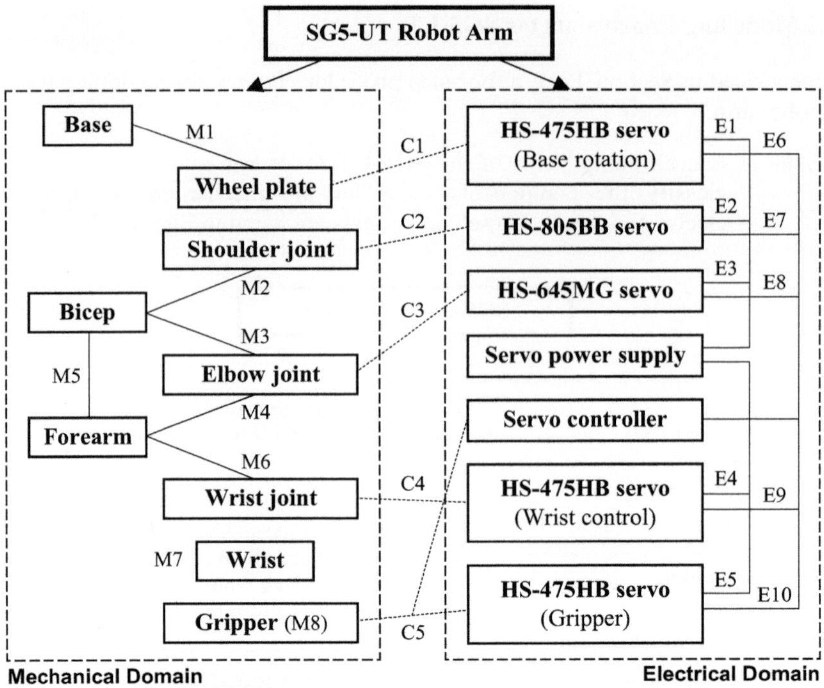

Figure 2.8. Constraints model of SG5-UT robot arm (solid line represents constraints within the domain, M is mechanical and E is electrical; dashed line represents cross-disciplinary, C)

Table 2.2. Mechanical constraints for the SG5-UT robot arm

	Constraint type	Constraint description
M1	*Geometric*: *fixed* Between base and wheel plate	The co-ordinate of base contact point with wheel plate is the co-ordinate of wheel plate centre.
M2	*Geometric*: *coaxial* Between bicep and shoulder joint	The axis of shoulder joint is coaxial with the axis of bicep rotation.
M3	*Geometric*: *fixed* Between bicep and elbow joint	The co-ordinate of the bicep contact point with the forearm is the co-ordinate of elbow joint.
M4	*Geometric*: *coaxial* Between elbow joint and forearm	The axis of elbow joint is coaxial with the axis of forearm rotation.
M5	*Geometric*: *angle* Between bicep and forearm	The angle between bicep and forearm is between 90 and 270 degrees. The forearm is not permitted to crush into the bicep.
M6	*Geometric*: *fixed* Between forearm and wrist joint	The co-ordinate of wrist contact point with forearm is the co-ordinate of wrist joint.
M7	*Geometric*: *coaxial* Between wrist joint and gripper	The axis of wrist joint is coaxial with the axis of gripper rotation.
M8	*Geometric*: *symmetry* Gripper	The left half of gripper and the right half of gripper are symmetric.

Table 2.3. Electrical constraints for the SG5-UT robot arm

	Constraint type	Constraint description
E1 ~ E5	*Maximum torque* Between the five servos and power supply	There is a maximum torque for a particular given voltage from the power supply.
E6 ~ E10	*System control* Between the five servos and servo controller	The servo controller determines the appropriate drive signal to move the actuator towards its desired position.

under other categories, such as force constraints, material constraints. For example, if the density of the material of the robot arm component is uniform, then the weight of that component would be the product of density of the material, volume of that component, and the gravitational acceleration.

STEP 3: Based on the attributes of the components, draw the constraint relationship between the components across the domains and appropriately label the constraint by the constraint categories, as presented below in Table 2.4.

Table 2.4. Cross-disciplinary constraints for the SG5-UT robot arm

	Constraint type	Constraint description
C1	*Kinematics–force–motor torque* Between the wheel plate and base rotation servo	The rotation speed of the wheel plate is dependent on the weight of the entire arm structure and the torque provided by the base rotation servo.
C2 | C4	*Geometry–force–motor torque* Between the servos and mechanical links	The torque required about each joint is the multiple of downward forces (weight) and the linkage lengths. This constraint exists in each lifting actuators.
C5	*System control–kinematics–force* Between the gripper, gripper servo, and servo controller	The servo controller determines the current state of the gripper given the current state of the actuators (position and velocity). The controller also adjusts the servo operation given the knowledge of the loads on the arm.

STEP 4: Construct a table of constraints for the particular mechatronic system. The table contains a complete list of the every component of the mechatronic system; the table is to indicate that, when a particular attribute of the component is being modified, which attribute of which component (both within the domain and across the domain) would be affected.

An example of the cross-disciplinary constraints would be the force relationship that is needed for motor selection. The motor that is chosen for the robot arm must not only support the weight of the robot arm but also support what the robot will be carrying. To perform the force calculation for each joint, the downward force (weight) of the components that has effect on the moment arm of the joint is multiplied by the linkage length and all the forces are summed to provide the torque required about each joint. This calculation needs to be done for each lifting actuator. For each degree of freedom added to the robot arm, the mathematical calculation gets more complicated, and the joint weights become heavier. Figure 2.9 illustrates the force calculation for a simple robot arm that has two degrees of freedom. And in Table 2.5, the cross-disciplinary constraints for the SG5-UT robot arm are identified and listed.

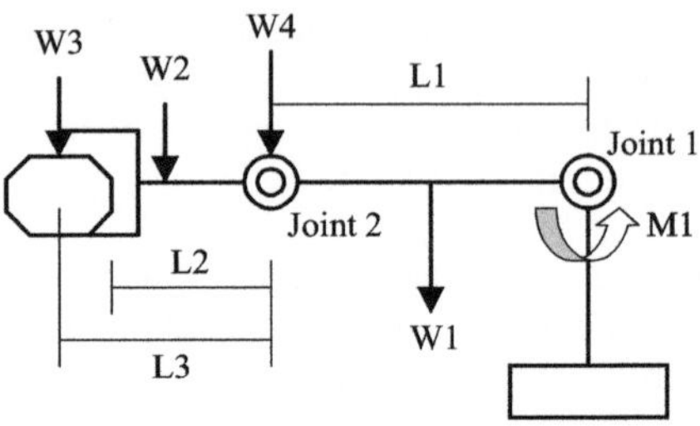

Figure 2.9. Force body diagram of a robot arm stretched out to its maximum length

Table 2.5. Table of constraints for the SG5-UT robot arm

Component (attribute)	Constrained attribute within the domain	Constrained attribute across the domain
Bicep (*L*, *W*, *H*)	Bicep (weight, volume)	Shoulder servo (torque)
Shoulder joint (location)	Bicep (location of axis of rotation)	
Forearm (*L*, *W*, *H*)	Forearm (weight, volume)	Shoulder servo, elbow servo (torque)
Elbow joint (location)	Forearm (location of axis of rotation)	
Wrist (*L*, *W*, *H*)	Wrist (weight, volume)	Shoulder servo, elbow servo, wrist servo (torque)
Base servo (torque)		Wheel plate (rotation speed)
Servo power supply	All servos (torque)	
Servo controller	All servos (control)	Gripper (location, velocity)

Torque about joint 1:

$$M1 = \frac{L1}{2} * W1 + L1 * W4 + (L1 + \frac{L2}{2}) * W2 + (L1 + L3) * W3 \quad (2.1)$$

Torque about joint 2:

$$M2 = \frac{L2}{2} * W2 + L3 * W3 \tag{2.2}$$

2.5 Requirements for a Computational Framework for Integrated Mechatronic Systems Design

Currently, the collaborative development of mechatronic systems occurs in an overlapping three-phase process. First, the mechanical design is synthesised based on customer requirements and technical constraints. Second, the technical specifications from the mechanical design are communicated to the electrical engineering team, which designs the electrical system to support the mechanical design. After an iterative process between the mechanical and electrical teams, the electrical engineers begin to finalise the electrical design as the software engineers start the third phase of supporting software design. Another iteration phase between the electrical and software engineers leads to the finalisation of the software design. The collaborative development of mechatronic systems can be considered from the point of view of the electrical, software, mechanical, and electronic engineers, each discipline has its design requirements to be fulfilled. The following are descriptions of the various design requirements in each discipline.

2.5.1 Electrical Design

Table 2.6 lists the design requirements from electrical engineering point-of-view.

2.5.1.1 Basic Requirements

Defined technical standard

Electrical engineering design standards are not identical worldwide. Many companies in the USA still use an iteration of older Joint Industrial Council (JIC) standards. These standards, developed in the early 1950s, were taken over by the National Fire Protection Association (electrical standards) and the National Fluid Power Association (hydraulic and pneumatic standards). Therefore, the National Fire Protection Association issued the current US electrical standards in NFPA 79 Electrical Standard for Industrial Machinery [2.1]. Most countries follow the electrical engineering guidelines outlined by the International Electro-technical Commission (IEC). The IEC was created in 1906 and has unified electrical design standards for most of the world. The IEC guides the design of electrical systems from metric definition to software/diagram specifications.

All designers on a project must agree upon the standard to be used. This will simplify data exchange throughout the design and avoid complications in unit conversion between groups in various corporate divisions.

Chosen software package

There are several software packages to facilitate electrical design today. These range from database programs to 2D CAD with electrical add-ons to fully developed electrical design studio packages. As most of these products still rely on proprietary data formats, designers (and their respective companies) must agree upon a software system or series of compatible software systems. This will ease restrictions in drawing and component information sharing during the design process.

Table 2.6. Requirements from an electrical engineering point of view

	Requirements list for current design of mechatronic systems
	A. *Initial requirements*
D	Defined technical standard (JIC, IEC)
D	Chosen software package, *i.e.* defined data representation/storage methods
	B. *Collaboration requirements*
D	Communication between mechanical, electrical, software engineers
W	Shared access to design data of all domains
D	Defined data storage procedures, organisation, and file formats
	C. *Energy requirements*
D	Load requirements for power consuming devices
D	Voltage/current entering the system
D	Space allocations for components and cabinets (geometry of mechanical system)
W	Cabinet locations to enable arrangement of terminal strips
	D. *Control requirements*
D	Type of control desired (microcontroller *vs.* PLC)
D	Required controls for each component (degrees of freedom, bounds, *etc.*)
D	What type of devices to control (motor, actuator, *etc.*)
D	Locations and types of inputs from sensors and switches
D	Necessary indicator lights
	E. *Installation concerns*
W	Company-preferred part vendors
D	Clear, intelligent diagrams for construction/installation
D	Complete set of matching connection point designations
D	Parts list
D	Checks for OSHA, UL, IEEE, *etc.* for safety compliance
W	Wire size/type of current installation (for system upgrades/product revisions)

D = Demands, must be satisfied during the design process
W = Wishes, would be ideal to satisfy during the design process, but not required

2.5.1.2 Collaboration Requirements

Communication between mechanical, electrical, software engineers

In order to design a mechatronic product, there must be constant facilitated communication between all involved disciplines. Currently, this communication exists as anything from yelling over cubicle walls to utilisation of cutting edge collaboration software, depending on corporate policy and available tools.

Shared access to design data of all domains

To efficiently perform a cross-discipline design, applicable data from all domains must be accessible to everyone involved in the design process. Currently, this data is rarely centrally located. There is a need for designers to access the updated information in a reasonable time frame. This may involve direct physical sharing of paper copies of design sketches, models, and function information or a network data storage solution.

Defined data storage procedures, organisation, and file formats

To begin an electrical design, the design team must agree upon how project data will be stored. Everyone on the team must know which office or person to send design document change requests. There must be clear organisation of design materials, be it through a product data management system or a common file naming convention. The team must also decide on one file format for design information storage from each design software system.

2.5.1.3 Energy Requirements

Load requirements for power consuming devices

The main connection between the mechanical system and the electrical system is when energy is converted from electrical power to mechanical motion. This conversion occurs in several power consuming devices including motors, heating elements, and lights.

Voltage/current entering the system

To select and arrange the proper components for an electrical design, the designer must know the voltage and current coming into the system. If necessary, engineers can use transformers to achieve the desired voltage or modify component selection to a more appropriate part number.

Space allocations for components and cabinets (geometry of mechanical system)

Using information from the mechanical engineering team, designers can plan the size of control cabinets based on the available space around the mechanical system. This enables designers to arrange the components inside the cabinet, an integral part of electrical design.

Cabinet locations to enable arrangement of terminal strips

Knowing the location of mechanical components, electrical engineers can plan the locations and arrangement of supporting electrical control cabinets. This allows them to logically order the wiring of control cabinets and accurately label terminal strips and wiring harnesses. Logical, accurate terminal strip numbers leads to accurate electrical wiring diagrams and more logical systems for the manufacturing team.

2.5.1.4 Control Requirements

Type of control desired (microcontroller vs. programmable logic controller)

Depending on the expected production volume, engineers must decide how to control mechatronic systems. Microcontrollers are meant for high volume applications, where a cheap, integrated chip can be specifically programmed for a certain function. For lower volume, more specialised applications, PLCs provide more versatility to the manufacturer and end user. PLCs are "well-adapted to a range of automation tasks. These are typically industrial processes in manufacturing where the cost of developing and maintaining the automation system is high relative to the total cost of the automation, and where changes to the system would be expected during its operational life" [2.2]. A designer must have intimate knowledge of these controllers and the target production figures for the design.

Required controls for each component (degrees of freedom, bounds, etc.)

To plan the control setup for an electrical system, some mechanical constraints are required. For example, in designing a robot arm, the design team needs to know how many degrees of freedom and how many motors to use in the robot arm. Furthermore, the limits of each of those degrees of freedom will translate into stopping points for the arm's motion.

What type of devices to control (motor, actuator, etc.)

To ensure that control cabling and signals are routed to the right components, electrical engineers must know what type of devices – from connection point designations to part numbers – are part of the control network.

Locations and types of inputs from sensors and switches

To provide the proper interface setup, electrical engineers must know what type and range of signals to expect from sensors and switches. They must also know the physical location of these objects. This helps them include the right parts based on reliability needs and control requirements (pushbutton *vs.* toggle *vs.* variable switch). This also serves to counteract miscommunication between electrical and mechanical engineers.

Necessary indicator lights

Finally, control design usually ends with the inclusion of indicator lights for operators. These lights indicate powered systems, tripped circuits, automated safety systems operation, and more. The control engineers specify the need for indication lights while the mechanical engineer indicates light locations on the instrument panel or equipment. It is the electrical engineering team's duty to provide power and signal feedback to these indicators, as well as communicate their existence to the software engineer.

2.5.1.5 Installation Concerns

Company-preferred part vendors

It is no secret that companies prefer to reuse vendors who provided reliable parts and trustworthy service. This helps the company establish strong vendor-user relationships that can reduce costs. This also creates a more uniform system environment, with a warehouse supply of redundant parts. Most engineers on the team will be comfortable with the vendors used in the past. New team members should be given a list of preferred vendors before the design begins.

Clear, intelligent diagrams for construction/installation

When electricians install electrical components in cabinets, they connect components to one another and to terminal strips based on connection point numbers. The connections between components are mapped out on wiring diagrams and power/control cabinet layouts.

The quality and accuracy of an electrical cabinet's wiring diagrams and component layout views decide the lifelong reliability of the cabinet. Ideally, the wiring diagrams will feature the same connection point numbers as the real part, label the colour of each wire in the bundle, and clearly show the sequential routing of wires from one component to the next.

Table 2.7. Requirements from a mechanical engineering point of view

	Requirements list for current design of mechatronic systems
	A. Kinematics requirements
D	Precise positioning of mechanical parts
D	Accurate measurement of velocity and acceleration
D	Control of type of motion and direction of motion
	B. Forces requirements
D	Knowledge of force, weight, load and deformation
D	Knowledge of the existing friction forces and their effects
D	Appropriate application of lubrication
W	Maximum reduction of wear and tear due to friction
	C. Energy requirements
D	Sufficient supply of energy sources (electricity, fuel, *etc.*)
D	Secure check of heating, cooling, and ventilation
W	High efficiency with low power consumption
	D. Material properties requirements
D	Solid understanding of the materials' stress-strain behaviour
D	The material does not fracture, fatigue, and display other types of failure during the lifecycle of the product
D	Appropriate choosing of factor of safety in stress analysis
	E. Material selection requirements
D	The materials selected satisfy the functional requirement
D	The materials are easily fabricated and readily available
D	The materials satisfy thermal and heat treating specifications
W	The materials allow the system to operate at higher speed/feeds
	F. Geometric constraint requirements
D	Accurate description of the relationship between features
D	Parts with dimensions accurate within the permissible tolerance
D	Parts satisfy the geometrical tolerances (*i.e.* parallelism)
W	Maximum balance between tolerance level and cost required
	G. Manufacturability requirements
D	Appropriate overall layout design
D	Appropriate form design of components
D	Appropriate use of standards and bought-out components
D	Appropriate documentation

D = Demands, must be satisfied during the design process
W = Wishes, would be ideal to satisfy during the design process, but not required

Complete set of matching connection point designations

When electrical parts are manufactured, each wire receptacle is given a specific connection point designation. The number of that receptacle will be the same for every part of that design manufactured. Electrical designers must know how the connection points on devices they use are numbered. Designers can then use those numbers on all wiring diagrams to reduce confusion during installation.

Parts list

To aid in billing, supply chain, and installation prep, electrical engineers should include a parts list in their design package, also known as a Bill of Materials. Among the benefits are checks from other designers and suppliers against incorrectly selected parts, misprinted part numbers, and supply availability.

Checks for OSHA, UL, IEEE, etc. for safety compliance

Safety is the utmost concern for all engineers. This is no different for electrical engineers, whose designs can be operated by customers with little or no technical experience or training. To help protect against shock, fire hazards, and general safety concerns, several organisations exist to certify that electrical designs meet stringent safety standards. Engineers must be comfortable with these standards and check for compliance during and after completion of electrical designs.

Wire size/type of current installation (for system upgrades/product revisions)

To protect continuity in existing products and systems, engineers must know the pre-design specifications. This includes the type and size of wire used.

2.5.2 Mechanical and Electronic Design

Tables 2.7 and 2.8 list the design requirements from mechanical and electronic engineering points of views, respectively.

2.5.3 Integrated Design

Table 2.9 lists the design requirements for integrated mechatronic design approach.

2.5.3.1 Mechanical-electronic Requirements

Even though many MCAD systems are able to generate output files that can be understood by CAD layout tools, in many cases this interface uses standard formatted files that usually do not contain the information listed above and only provide basic board outline as a set of lines that the electronic and mechanical designers has to interpret in later stages of design process.

The procedure for transferring PWBA data between electronic design and mechanical design during the design process is as follows:

1. In MCAD, define PWB outlines; define keep-in/keep-out areas, holes, cut-outs, *etc.*; pre-place ICs, connectors, switches and other fixed components.

Table 2.8. Requirements from electronic engineering point of view

	Requirements list for current design of mechatronic systems
	A. Electronic device requirements
D	Accurate specification and labelling of electronic features (resistance, capacitance, inductance, *etc.*)
D	Right amount of impurities are added to the semiconductor materials to provide desired conductivity
W	The device can respond to very small control signal
W	The device can respond at a speed far beyond the speed at which mechanical or electrical devices can
	B. Circuitry requirements
D	The printed circuit board layout satisfies the mechanical and space constraints imposed by the amount of area required to physically locate all the components on the board
D	Specify the keep-in and keep-out areas on the board
D	Specify the height restrictions for component placement
D	Specify specific features such as holes and cut-outs
D	Specify the pre-placement of components that are linked to mechanical constraints (ICs, connectors, switches, *etc.*)
	C. Signal requirements
D	Specify input and output
D	Appropriate form, display, and control impulse
D	If conversion required, specify whether it is digital-to-analogue or analogue-to-digital
D	For A/D conversion, choose the appropriate schemes and the appropriate variation within each scheme
W	Maximum reduction of ambient noise
	D. Controller requirements
D	Designed to withstand vibration, change in temperature and humidity, and noise
D	The interfacing circuit for input and output is inside the controller
D	Specify the required processing information (number of input and output pins, memory and processing speed required, *etc.*)
W	Easily programmed and have an easy programming language

D = Demands, must be satisfied during the design process
W = Wishes, would be ideal to satisfy during the design process, but not required

2. Convert the above MCAD information to IDF (or other standard format such as DXF and STEP) and transfer the file to ECAD.
3. In ECAD, read the IDF file, write ECAD model where the board structure is defined, all components are placed and all interconnect are routed. This will create an ECAD file.
4. Convert the above ECAD file to IDF (or other standard format) and transfer the file to MCAD.
5. In MCAD, the PWB is imported as an assembly file that has component information, including location and properties. Based on the information imported, perform component height analysis, thermal analysis, structural analysis, *etc.* using MCAE tools. It is important for the MCAD to have a

fully populated library of MCAE components containing thermal and mechanical models of all parts present in the electrical layout.

Table 2.9. Requirements for integrated mechatronic design

	Requirements list for an integrated mechatronic design framework
	A. *General collaboration requirements*
D	Corporate specific technical standards and software package selections
D	Centralised database for all product information with project-wide data storage procedures
D	Enhanced instant messaging/data sharing software integrated into design tools (ECAD/MCAD/Office)
	B. *Energy requirements*
D	Automated optimisation of wire specifications based on mechanical system energy needs
D	Automated synthesis of protection devices based on selection of power consuming devices
D	Real-time calculation of voltage/current entering and running through the system
D	Automated development of electrical system based on demands from electrical side and corporate-preferred parts database
	C. *Control requirements*
D	Automated generation of graphical/logical control system mapping with electrical design, enhanced with sensor, user input, output, and kinematic data
D	Software code classes for common I/O needs (temp sensors, motor control, switchboard, *etc.*)
D	Automated generation of computer simulation of completed control system for testing
	D. *Installation concerns*
D	Automated synthesis of necessary power/control cabinets followed by optimisation of control cabinet location and automated terminal strip connection point numbering (based on component selection)
D	Automated generation of clear, intelligent diagrams for construction/installation, based on system synthesis
D	Automated parts list generation and communication to supply chain
D	Automated system checks for OSHA, UL, IEEE, *etc.*, for safety compliance
	E. *Maintenance concerns*
D	Automated allowance for maintenance operations and upgrade modules
W	Automated user manual generation based on component modules
	F. *Mechanical-electronic requirements*
D	Electronic assembly layout and the PCB board design must allow for the physical style and functionality of mechanical design (*i.e.* satisfy the imposed geometrical constraints)
D	Mechanical material selection and manufacturing accounts for the physical aspects of the internal electronic such that there is no device malfunctioning
D	Electronic CAD systems must support 3D modelling at components level and have facilities to export accurate 3D design data
D	Bi-directional flow of complete design information between mechanical CAE environment and electronic CAE environment

Table 2.9. Requirements for integrated mechatronic design (continued)

	G. Product lifecycle requirements
D	Frequent communication with the customers
D	Ability to carry out customisation requirement quickly and correctly
D	Robustness of the product: long mean time between failure and requires little maintenance
D	Enables standardisation and reusability
D	Effective end-of-lifecycle processing: recycling or reuse

D = Demands, must be satisfied during the design process
W = Wishes, would be ideal to satisfy during the design process, but not required

Efficiently bridging the gap between the mechanical and electronic design processes is, therefore, the key towards collaborative and successful product development. Rather than simply passing raw dimensioning and positional data from the ECAD to MCAD environment, it would be far more beneficial for the design tools to allow a bi-directional flow of comprehensive data between the two CAD environments. In other words, the ECAD must possess the ability to import and seamlessly integrate 3D component data from an MCAD environment, and then pass a full and accurate 3D representation of the board assembly back to the MCAD. To harness this potential, the electronics design system must support 3D modelling at the component level. This ability plus facilities to export accurate 3D design data would support the necessary interaction between the mechanical and electrical environments.

2.5.3.2 Product Lifecycle Requirements

Since industry networks can only get more complex and since information flow can only grow and not shrink, the following problems can be expected to exist in designing mechatronic systems:

- Material management: information maintenance is difficult, and so is maintaining the internal and external communication of the company regarding product data and the changes that have taken place in it; hence the following problems can occur.
- The item management of the company is not in order.
- The company's own component design and manufacturing is inefficient.
- Company buys the same type of components from different suppliers and ends up handling and storing these same types of components as separate items.
- Company makes overly fast and uncontrolled changes in the design of the product.

In order to resolve these problems listed above, the lifecycle of the product must be analysed. The useful life of a product can be measured by the length of its lifecycle. At the end of this lifecycle, the efficiency of the product is so low as to warrant the purchase of a newer one. The product can no longer be upgraded due to financial feasibility or lack of adequate updates. Lengthening this lifecycle will provide customers with a more useful product and will help reduce the strain on the environment from both the supply and the waste aspects. As the business world continues to grow more consumer-friendly, more products will have to become instantly customisable. To measure this mass customisation, a count of the number of variants into which a product can be customised gives a quick estimate of the abilities of its design method.

2.6 Conclusions

Mechatronic systems comprise of, among others, mechanical and electrical engineering components. An example is a robot arm that consists of mechanical links and electrical servos. Mechanical design changes lead to design modifications on the electrical side and *vice versa.* Unfortunately, contemporary CAE environments do not provide a sufficient degree of integration in order to allow for bi-directional information flow between both CAD domains.

There are several approaches to support the information exchange across different engineering domains. PDM systems manage product information from design to manufacture to end-user support. Standard data exchange formats are developed to achieving communication between various CAD, CAE, and PDM systems.

The approach to achieving integration of mechanical and electrical CAD systems proposed in this chapter is based on cross-disciplinary constraint modelling and propagation. Cross-disciplinary constraints between mechanical and electrical design domains can be classified, represented, modelled, and bi-directionally propagated in order to provide immediate feedback to designers of both engineering domains. In constraint classification, a selected mechatronic system is being analysed to identify and classify discipline-specific constraints based on associated functions, physical forms, system behaviour, and other design requirements. In constraint modelling, the mechatronic system is modelled in block-diagram form and relationships between domain-specific constraints are identified and categorised.

Most development activities for mechatronic products are of a collaborative nature. In the past, electrical engineers and mechanical engineers had to have personal interactions with each other to collaborate. Today, they collaborate through information systems. An integrated framework for designing mechatronic product helps to improve collaborative design activities. This collaboration can happen everywhere from within one office, one company, to world-wide distributed parties in a virtual environment.

References

[2.1] Appuu Kuttan, K.K., 2007, *Introduction to Mechatronics*, Oxford University Press, pp. 1–11.

[2.2] Reeves, B. and Shipman, F., 1992, "Supporting communication between designers with artifacts-centered evolving information spaces," In *Proceedings of the 1992 ACM Conference on Computer-supported Cooperative Work*, pp. 394–401.

[2.3] Liu, D.T. and Xu, X.W., 2001, "A review of web-based product data management systems," *Computers in Industry*, **44**, pp. 251–262.

[2.4] Sung, C.S. and Park, S.J., 2007, "A component-based product data management system," *International Journal of Advanced Manufacturing Technology*, **33**, pp. 614–626.

[2.5] You, C.F. and Chen, T.P., 2007, "3D part retrieval in product data management system," *Computer-Aided Design and Applications*, **3**(1–4), pp. 117–125.

[2.6] Pratt, M.J., Anderson, B.D. and Ranger, T., 2005, "Towards the standardised exchange of parameterised featured-based CAD models," *Computer-Aided Design*, **37**(12), pp. 1251–1265.

[2.7] Fenves, S.J., Sriram, R.D., Subrahmanian, E. and Rachuri, S., 2005, "Product information exchange: practices and standards," *Journal of Computing and Information Science in Engineering*, **5**(3), pp. 238–246.

[2.8] Lubell, J., Peak, R., Srinivasan, V. and Waterbury, S., 2004, "STEP, XML, and UML: complementary technologies," In *Proceedings of ASME 2004 Design Engineering Technical Conference and Computers and Information in Engineering Conference.*

[2.9] Sudarsan, R., Fenves, S.J., Sriram, R.D. and Wang, F., 2005, "A product information modelling framework for product lifecycle management," *Computer-Aided Design*, **37**(13), pp. 1399–1411.

[2.10] Biswas, A., Fenves, S.J., Shapiro, V. and Sriram, R.D., 2008, "Representation of heterogeneous material properties in the core product model," *Engineering with Computers*, **24**(1), pp. 43–58.

[2.11] Peak, R.S., Fulton, R.E., Nishigaki, I. and Okamoto, N., 1998, "Integrating engineering design and analysis using multi-representation architecture," *Engineering with Computers*, **14**(2), pp. 93–114.

[2.12] Kleiner, S., Anderl, R. and Grab, R., 2003, "A collaborative design system for product data integration," *Journal of Engineering Design*, **14**(4), pp. 421–428.

[2.13] ISO, 2001, "Product data representation and exchange. Integrated application resource: parameterisation and constraints for explicit geometric product models," *ISO Committee Draft (CD) 10303-108*, ISO TC 184/SC4/WG12/N940, Geneva, Switzerland.

[2.14] Schaefer, D., Eck, O. and Roller, D., 1999, "A shared knowledge base for interdisciplinary parametric product data models in CAD," In *Proceedings of the 12th International Conference on Engineering Design*, pp. 1593–1598.

[2.15] Roller, D., Eck, O. and Dalakakis, S., 2002, "Advanced database approach for cooperative product design," *Journal of Engineering Design*, **13**(1), pp. 49–61.

[2.16] Fulton, R.E., Ume, C., Peak, R.S., Scholand, A.J., Stiteler, M., Tamburini, D.R., Tsang, F. and Zhou, W., 1994, "Rapid thermomechancial design of electronic products in a flexible integrated enterprise," *Interim Report, Manufacturing Research Centre*, Georgia Institute of Technology, Atlanta.

[2.17] Mocko, G., Malak, R., Paredis, C. and Peak, R., 2004, "A knowledge repository for behaviour models in engineering design," In *Proceedings of ASME 2004 Design Engineering Technical Conference and Computers and Information in Engineering Conference.*

[2.18] Klein, R., 1998, "The role of constraints in geometric modelling," In *Geometric Constraint Solving and Applications*, Bruderlin, B. and Roller, D (eds), Springer-Verlag, pp. 3–23.

[2.19] Bolton, W., 2003, *Mechatronics: Electronic Control Systems in Mechanical and Electrical Engineering*, Pearson Education Limited, pp. 185–197.

[2.20] Centikunt, S., 2007, *Mechatronics*, John Wiley and Sons, pp. 393–395.

[2.21] Rembold, U., Nnaji, B.O. and Storr, A., 1993, *Computer Integrated Manufacturing and Engineering*, 3rd edn, Addison-Wesley, pp. 449–464.
[2.22] http://www.crustcrawler.com/product/arm5.php?prod=0

3

Fine Grain Feature Associations in Collaborative Design and Manufacturing – A Unified Approach

Y.-S. Ma[1], G. Chen[2] and G. Thimm[2]

[1] Department of Mechanical Engineering, Faculty of Engineering
University of Alberta, Edmonton, Alberta T6G 2E1, Canada
Email: yongsheng.ma@ualberta.ca

[2] School of Mechanical and Aerospace Engineering
Nanyang Technological University, Singapore
Emails: chengang66@pmail.ntu.edu.sg, mgeorg@ntu.edu.sg

Abstract
In the context of concurrent and collaborative engineering, the validity and consistency of product information become important. However, it is difficult for the current computer-aided systems to check the information validity and consistency because the engineers' intent is not fully represented in a consistent product model. This chapter consolidates a theoretic unified product modelling scheme with fine grain feature-based methods for the integration of computer-aided applications. The scheme extends the traditional feature concept to a flexible and enriched data type, unified feature, which can be used to support the validity maintenance of product models. The novelty of this research is that the developed unified feature scheme is able to support entity associations and propagation of modifications across product lifecycle stages.

3.1 Introduction

Product development comprises several lifecycle stages, such as conceptual design, detailed design, process planning, machining, assembly, *etc*. Commonly, computer-aided tools (called 'CAx systems' hereafter) are used to support activities associated to these stages. Traditionally, stand-alone CAx systems for individual stages produce separate models, such as a product design or a process plan. The existing CAx technologies have difficulties in maintaining the integrity of the comprehensive product model as inter-stage data transfer or sharing is insufficiently supported, especially for non-geometric data. Furthermore, validity checking of product models is difficult as the engineering knowledge applied in product designs or process plans is usually not stored with the product model as the existing technology does not allow for this. Recently, due to the drive for industrial globalisation and mass customisation, the trend of concurrent and collaborative engineering has led to tight integration of product and process domains as well as CAx systems [3.1].

This research accommodates product model validity and consistency by proposing a comprehensive product model consisting of linked geometric and non-geometric data throughout all product lifecycle stages based on feature technology with consideration of knowledge engineering, system integration, and collaboration. The goal of this research is to establish a paradigm in product modelling across multiple lifecycle stages. The multiple aspects of product modelling are integrated in a systematic and scalable manner. The paradigm is expected to allow multiple applications to share a consistent product model with supporting mechanisms and to maintain its integrity and validity.

3.2 Literature Review

Traditional application integration approaches focus on geometric data sharing. For example, system integration between design and reverse engineering, rapid prototyping, co-ordinate measuring machine, mesh generation for CAE, and virtual reality has been widely studied [3.2–3.7]. The most common approach to support application integration is using geometric data file exchange via a set of neutral formats, such as the Initial Graphics Exchange Specification (IGES) or the STandard for the Exchange of Product model data (STEP) [3.8]. This situation is no longer satisfactory to support modern product lifecycle management [3.1]. To support application integration fully, more comprehensive data sharing is needed than provided by the existing IGES or STEP standards.

Features combine geometric and non-geometric entities. Therefore, compared with geometric models, more complex relations exist in feature models. Managing these relations, especially the non-geometric ones, is essential for the validity of a product model. Relations in a feature-based product model can be classified as shown in Table 3.1.

Table 3.1. Summary of research on relations in a feature-based application

Relation	Related entity	Representation	Source
Geometric relations	Between geometric entities	Geometric constraints	[3.9, 3.10]
	Between features	Interaction constraints	[3.11, 3.12]
Non-geometric relations	Between features and the corresponding geometric entities	Features referred to the corresponding geometric entities	[3.13–3.15]
	Between features and other non-geometric entities, such as functions, behaviours, assembly methods, machines, cutting tools	Tables, graph, rules, *etc.*	[3.16–3.25]

3.2.1 Geometric Relations

Many publications focus on geometric relations in a feature model [3.9]. All these relations are explicitly declared and represented as geometric constraints, which maintain the geometric integrity of features. However, unintentional feature interactions, may also affect the validity of features [3.11, 3.12]. These interactions usually cannot be prevented by geometric or algebraic constraints. This work will show that the geometric feature interactions can only be managed through the associations between the feature model and the geometric model.

3.2.2 Non-geometric Relations

Non-geometric relations refer to dependency relations involving non-geometric entities. For example, in process planning, the clamping faces or accessing faces are required and are to be preserved when machining a feature and they are associated to the machining processes and sequence used. Furthermore, two features, which do not spatially overlap, even belong to different product lifecycle stages, may interact with each other. How to represent these non-geometric feature relations has not been fully investigated.

Non-geometric relations also exist among features and non-geometric entities. For example in functional design stage, functional-form matrixes, bipartite function-feature graphs, design flow chain and key characteristics, and mapping hierarchy are used to link features to product functions [3.17, 3.20, 3.21, 3.24, 3.26]. In the process planning stage, features are also related to non-geometric entities, such as machines, cutting tools, and machining processes [3.22]. The methods of using non-geometric relations to validate product models have not been developed.

A product model has to be constructed or analysed iteratively using engineering knowledge from different aspects of expertise to fulfil requirements, such as functional or manufacturing requirements. In addition, lifecycle stages are inter-related and mutually constraining. Any modification in one stage may provoke a chain of subsequent modifications to entities of the same or other stages. This propagation of changes requires the management of inherent relations within and among these stages. In other words, a product model must have a sound mechanism to check its validity. Compared to the strict validity maintenance mechanisms of B-rep or CSG, current feature-based modelling schemes are weak in this aspect.

Laakko and Mantyla [3.14], and Rossignac [3.27] suggested that a feature's validity should be defined in terms of the referenced geometric entities and of their existence, shape, and relations to other geometric elements of the model. A feature model is valid if the geometric and algebraic constraints specified on features are satisfied. However, with the introduction of *associative features* [3.28], the validity of features must be checked in more complex scenarios. The associative feature concept expands feature definitions of specific application-related shapes into a set of well-constrained geometric entities. By using an object-oriented approach, a feature type can be modelled in a declarative manner that basically consists of the properties and behaviours. Feature properties define the geometric entities whose behaviours define the related constraints and logics in functioning methods throughout the lifecycle of any feature instance. With the built-in object polymorphism capability, a systematic modelling scheme for a generic and abstractive parent feature class, with levels of specification as per application domain requirements, can be developed. Such a generic feature definition scheme unifies many traditionally defined, application-oriented feature definitions and supports XML representation and fine grain database repository. Under the associative feature concept, where the associative constraints across multiple phases of applications of a product lifecycle, complicated engineering features (patterns) and engineering intent can be implemented. An example associative feature, cooling channel pattern in plastic injection mould design, was given in [3.28]. An initial sketch-based conceptual pattern in the early mould design stage is implemented and its downstream cooling *hole* features are derived from the pattern; and then the related assembly interfacing features and associated standard components at the manufacturing and assembly stages are associatively generated and managed via a well-defined feature class model.

Feature validity is concerned with a feature's internal semantic characteristic properties, logics, constraints and attributes. This validity aspect is largely categorised as the constraint satisfaction problem, which has been partly addressed to a wide extent.

Feature consistency refers to the tally relations between related features or more abstracted semantic entities. Feature consistency is related to the semantic relations. The consistency requirement can have different types. Some researchers suggest that feature consistency means that the feature concerned is agreeable to the engineering intent [3.29]. In their publications, engineering intent must be transformed into a set of geometric, algebraic or preliminary semantic constraints, such as the boundary or interaction constraints [3.15]. However, during the transformation process, engineering intent may be lost because it has not been modelled explicitly so far. Others emphasise that non-geometric constraints, such as a dependency constraint, specified on the features have to be satisfied. For example, the presence of features, or the values of feature parameters, may be determined by functional requirements [3.18]. For another example, different machining sequences may influence the presence, form, volume, and validity of machining features. Hence, the presence of a machining feature is coupled with a machining process. Currently the representation, checking and maintenance methods of inter-feature non-geometric constraints are immature. Few researchers have touched on the feature consistency aspect although they are equally important for product modelling. A more detailed literature review by the authors is available [3.30]. This work introduces a solution framework that entails major class definitions, association structures, as well as integration and reasoning mechanisms based on a unified feature concept.

3.3 Unified Feature

Unified feature is a feature class definition that can generically represent the common properties as well as the required methods throughout product lifecycle stages. A unified feature is defined as a set of constrained associations among a group of geometric and non-geometric entities. The commonalities of application features, such as conceptual design features, detailed design features, and process planning features, are defined in the unified feature class as generic fields and methods. A brief publication can be found in [3.31]. Table 3.2 gives the major fields and methods defined in the unified feature class.

Figure 3.1 gives the generic definition using a UML diagram [3.32]. The UML symbols used in the figure are explained here. Rectangles represent classes, such as

Table 3.2. Major fields and methods of the unified feature class

	Name		Description
Fields	Attributes	Association attributes	Identities of the associated objects, such as functions and behaviours in a conceptual design, machines and cutters in a process plan, other features, *etc.*
		Self-describing attributes	Material, surface finish, belonging application, *etc.*
	Parameters		Variables used as input to geometry creation methods
	Constraints	Geometric constraints	Identities of geometric constraints that the feature's topological entities participate in
		Algebraic constraints	Identities of algebraic constraints that the feature's self-describing attributes or parameters participate in
		Rule-based constraints	Identities of rules that the feature or its self-describing attributes, parameters, numerical constraints participate in
	Geometric references		Topological entities
Methods	Geometry construction	*createGeometry()*	Generate the feature geometry
	Interface to geometric model	*getCell()*	Find out the feature's member topological entities
		setCell()	Assign a topological entity as the feature's identity
		insertGeometry()	Notify the geometric model to insert the feature geometry
		deleteGeometry()	Notify the geometric model to delete the feature geometry
	Interface to expert system	*getFact(), setFact()*	Retrieve or create the corresponding facts
		getRule(), setRule()	Retrieve or assign the corresponding rules
		checkRule()	Check whether the related rules are satisfied or not
	Interface to relation manager	*addToJTMS()*	Add the feature or its self-describing attributes, parameters to the relation manager as nodes
		validityChecking()	Call the relation manager for feature validation
	Interface to database	*saveFeature(), retrieveFeature()*	Store a feature in or retrieve a feature from the database

the *UnifiedFeature* class. Dashed and directed lines represent dependency relations. The lines are directed from the depending class to the class it depends on. Solid and directed lines with triangular open arrowheads represent generalisation relationships, pointing to the more general class that defines basic properties. Solid and directed lines with open diamonds represent aggregation relationships, pointing from the parts to the whole, aggregated object. Composition (indicated by a filled diamond) is a variation of simple aggregation relationship. It describes strong ownership and coincident lifetime between the parts and the whole. The ranges aside the origin and target of an aggregation (or composition) arrow indicate how many parts can or must be in a whole. For example, a unified feature may include none or many other unified features. A circle attached to a class represents an interface (such as the *IAttribute*) realised by (undirected lines) the class. Other classes can use this interface, *e.g.* the *UnifiedFeature* class uses the *IAttribute* interface.

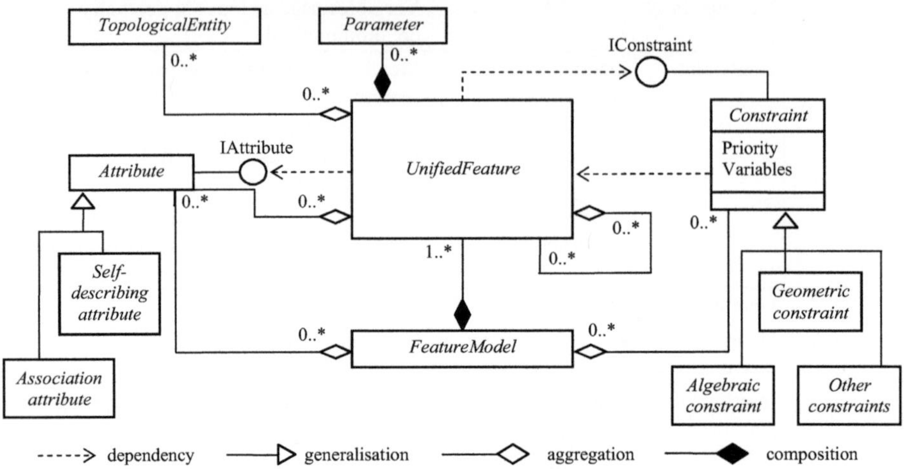

Figure 3.1. Unified feature

3.3.1 Fields

The unified feature class has four main kinds of fields.

(1) *Non-geometric attributes* represent feature properties that are attached to the feature or to the feature's geometric entities. They do not directly describe a feature's shape. Attributes are further classified into self-describing attributes and association attributes. Self-describing attributes represent properties that are special to a particular feature class. Examples of self-describing attributes are material type, surface finish, and feature nature (adding or removing material). Association attributes are references to the entities associated to this feature, such as other features, corresponding facts in the expert system, *etc.* In addition, association attributes are used to refer to non-geometric entities. For example, they refer to functions and behaviours in the conceptual design stage, or machine tools and machining operations in the process planning stage.

(2) *Geometric parameters* describe a feature's geometric shape, dimension, position, and orientation, such as the origin position and length, width, height of a block feature. Geometric parameters are used as input to the geometry creation methods provided by the geometric modelling kernel.

(3) *Constraints* can be classified according to the elements they constrain: (a) *intra-feature constraints* restrict the field values in a feature. For example, a

pocket's width equals to its length or a blind-hole's bottom face must be on the part boundary; (b) *inter-feature constraints* specify relations between two or more features; and (c) *semantic constraints* can also be specified between a feature and other entities. For example, a process planning rule is used as the constraint to specify whether a cutter can be used to create a feature with the specified shape, dimension, tolerance and surface finish. Constraints can also be classified according to their types, *i.e.* (a) *algebraic constraints*; (b) *geometric constraints*; and (c) *rule-based constraints*, which are used to restrict a feature's presence or the values of feature properties directly based on engineering rules. Constraints are prioritised.

(4) *Geometric references* are pointers to topological entities in a geometric model. Since features are used to describe specific relations between topological entities, a feature's geometry is not necessarily volumetric, connected, or two-manifold.

3.3.2 Methods

Interfacing functions, which deal with geometric modeller, knowledge engineering module, relation manager and database, are defined in the unified feature class.

(1) *Creating and editing feature geometry*. In the proposed scheme, conceptual design and detailed design features are created from predefined and parameterised geometric templates. The values of these parameters are specified to generate feature geometry. In the process planning stage with a design feature model as input, a process planning application analyses all machining faces for suitable process planning features. The properties of these faces are then used to determine the parameters of process planning features. Feature parameters are used to create product geometry with the help of functions provided by a geometric modeller. Because the definition of geometry is application specific, the way geometry is created is delegated to the specific application features. Feature geometries can be 2D faces or 3D solids in the developed scheme. The geometries of different dimensional features are represented uniformly in a non-manifold geometric model (Chapter 3.5). When an application feature is created, its geometry is inserted into the geometric model. When a feature is changed, it notifies the geometric model of modifications. In both cases, the geometric model will update itself accordingly.

(2) *Supporting knowledge embedment* [3.33]. A fact table corresponding to a set of associated features is created as a subset supporting a knowledge base. When an application feature is created, a corresponding fact is generated and inserted into the corresponding fact table and then accessible from the knowledge base. The fact of a feature describes the feature's identity, its parameters and self-describing attributes. The fact generation and insertion methods are defined in the unified feature class. When a feature is altered, it notifies the knowledge base. Matching rules (if any) are then fired.

(3) *Supporting data associations and validity maintenance*. In a single stage, when an application feature is created, a corresponding node is generated and inserted into a relation manager. The relation manager is responsible for managing the dependency relations among entities. The constraints, which are responsible for the feature's presence or controlling the values of feature parameters or self-describing attributes, are also inserted into the relation manager and are associated to

the corresponding feature node. The node generation, insertion and association methods are commonly defined for different application features. When a feature is modified, it calls the relation manager for change propagation. Related constraints are validated. To support inter-stage data sharing, associations and change propagation, application features as well as their inter-relations are stored in a common database. The methods of storing features into the database are defined in the unified feature class.

Two points about the above proposed unified feature definitions are worth noting. First, traditionally, numerical constraints are used to represent engineering intent. As an extension, the unified feature definition also defines associations to knowledge base, geometric model and other non-geometric entities in order to represent and maintain engineering intent. Second, from the viewpoint of software engineering, data sharing is difficult because one application does not know the data structures of other applications.

Hence, applications cannot manipulate the data created by other applications. With the unified feature definition, the issue of sharing feature data among applications is considerably improved. An application feature may have its specific properties, which are not included in the unified feature definition. However, with both application features defined as sub-classes of the unified feature class, an application understands the generic part of feature objects of other applications. These generic data is then used to reconstruct unified feature objects (Figure 3.2). In the proposed scheme, each application stores the data in a central relational database. An application can access the database to retrieve the data that is authorised.

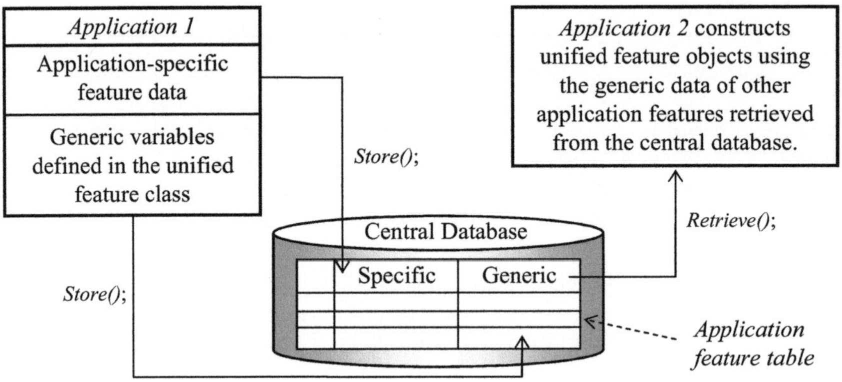

Figure 3.2. Data access methods via generic fields of application features

3.4 Entity Associations

The prime purpose of the unified feature-based product modelling scheme is to maintain the validity, consistency, and integrity of product models. Traditional CAx systems have limitations in serving this purpose. Two major problems are (1) engineering intent is not well represented and managed, and (2) inter-stage, non-geometric relations are not well maintained. The unified feature-based product modelling scheme tackles these two problems via establishing and maintaining geometric and non-geometric data associations, within a single or across different stages. For example, in the conceptual design stage, the geometry of a feature is usually not fully defined. The resulted entities could be, for instance, only surface shapes, abstract mechanism concepts, or parameterised volumes without assigning detailed properties. An abstract conceptual design feature has its concrete counterparts in the detailed design feature model. Because a conceptual design feature represents a primitive design function that is usually realised through the interactions between a few components, it is likely that an individual conceptual design feature is transformed into several features belonging to different components in the detailed design stage. On the other hand, one detailed design feature may also participate in the realisations of several conceptual design features. Such *feature object dependency* associations are one kind of non-geometric associations between features as discussed in [3.34]. Feature attributes, parameters, or constraints specified in the conceptual design feature model are transformed into attributes, parameters, or constraints for corresponding detailed design features. For example, a parameter of a conceptual design feature may be transformed into a constraint between two detailed design features of different components. A conceptual design constraint could be related to several constraints in the detailed design feature model. Such *feature property dependency* associations are another kind of non-geometric associations across features of different stages [3.34]. These associations are generalised as constraint-based associations and sharing associations (Figure 3.3). Constraint-based associations are established on the basis of intra- or inter-stage, numerical or rule-based constraints. Sharing associations are established based on the unified cellular model.

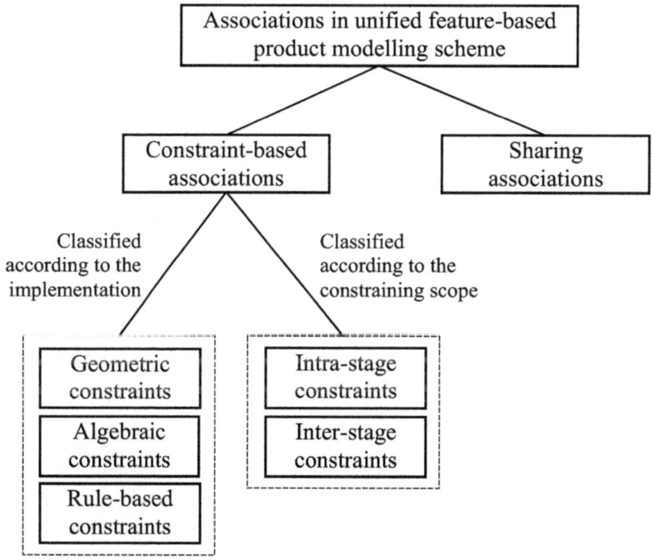

Figure 3.3. Associations in the unified feature-based product modelling scheme

3.4.1 Implementing the Constraint-based Associations

Together with a rule-based expert system and a numerical constraint solver, a justification-based truth maintenance system (JTMS) is used to implement the constraint-based associations as introduced in [3.34]. A JTMS dependency network consists of a series of related nodes that represent the belief status of entities. *Assumption* nodes are believed without any supporting justifications. *Simple nodes* are only believed if they have valid justifications. An assumption node can be converted into a simple node, which then needs to be supported by justifications. A justification consists of antecedent nodes and consequent nodes. A node is said to be justified by a supporting justification if all antecedents of the justification are justified.

Whenever a constraint-based association is generated, the corresponding JTMS nodes and justifications are inserted into a JTMS dependency network. After the insertion process, each node records three items: (1) a reference to its direct supporting justification; (2) references to the justifications that use this node as antecedent (for later change propagation); and (3) its current belief status. Whenever a modification to the JTMS dependency network occurs, such as adding or retracting assumptions, modifying nodes or adding justifications, the JTMS dependency network is searched for the affected nodes as well as the related justifications. If it is a rule-based constraint to provide the justification, the system refers to the knowledge base to validate the modification. If it is a numerical constraint to provide the justification, the system refers to the numerical constraint solver to validate the modification. These checking and change propagating processes are automated. The result is a new status of each affected JTMS node or a rejection of the modification on the basis of contradicting beliefs. The data structures and algorithms of JTMS are generic. Therefore, it handles geometric and non-geometric constraints uniformly.

A relational database is used for all applications to store and publish their data. An application can access and enquire the database for data published by other applications. When an inter-stage constraint-based association is established, this association and the involved data are stored in the database. When an application modifies its model, it must check the database for relevant inter-stage associations. If such associations exist, a validity checking process is triggered. The applications involved are responsible for maintaining the consistency (between associated stages) while the database is a medium for storing the repository data, inter-stage associations, and propagating changes. Figure 3.4 illustrates the constraint-based associations between the conceptual and the detailed design feature models. The constraint-based associations between the detailed design and the process planning feature models are established in a similar way.

3.4.2 Implementing the Sharing Associations

Two methods are developed for sharing associations using a unified cellular model implemented in a database.

(1) *Generating a new application feature.* Each application feature class has its geometry creation and manipulation functions. When a creation function is invoked, the feature geometry is created and inserted into the application's runtime cellular

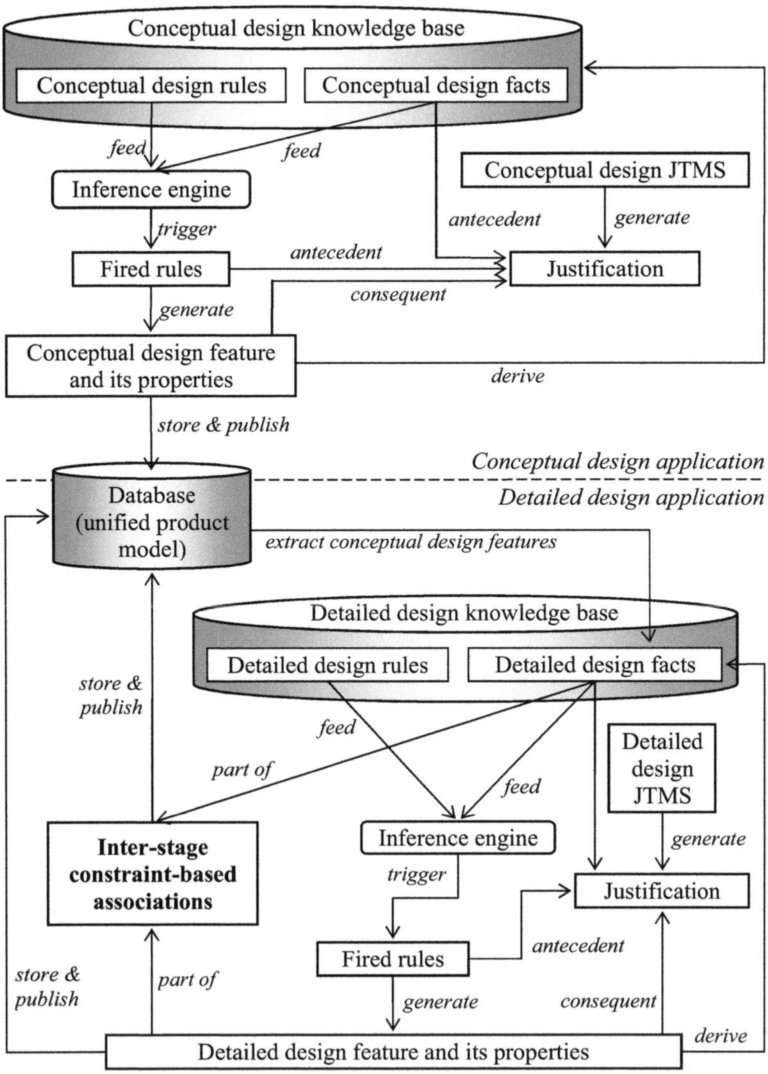

Figure 3.4. Constraint-based associations for conceptual and detailed design stages

model. The topological entities created are associated with the feature through the owning feature attributes and the feature's runtime geometric references. The feature geometry is also inserted into the unified cellular model. If any cell in the unified cellular model is affected by this new feature, *e.g.* overlapping, the owning features of the affected cells are marked for validity checking.

(2) *Modifying an application feature*. When an application feature is modified, in addition to updating the application's runtime cellular model, the application also notifies the unified cellular model about the modifications. The unified cellular model is updated and the affected cells are marked as been modified. The owning

features of the affected cells are then validated by the corresponding applications. The sharing association mechanism enables different application features to be associated with the same geometric or topological entities and hence supports achieving inter-stage geometric consistency.

3.4.3 Evaluation of Validity and Integrity of Unified Feature Model

This subsection introduces a set of criteria, which is used to evaluate the validity and integrity of a unified feature-based product model. The general requirement for a valid product information model is that each application model (corresponding to a particular stage) must be valid and also consistent with other associated application models. The detailed evaluation criteria are classified into feature, intra-stage, and inter-stage levels.

A feature is valid if (i) the feature geometry refers to valid topological entities; (ii) the values of feature parameters are consistent with the product's geometric model; (iii) all constraints on the feature are satisfied; and (iv) any feature property, if included in the JTMS dependency network, has a "believed" status, *i.e.* its supporting justifications are valid.

A product model is valid if (i) all features in the model are valid; (ii) in its knowledge base, the antecedent conditions of all fired rules, which are the justifications for the generated features (or feature properties), are satisfied; (iii) all constraint-based associations between consequent facts and respective features (or feature properties) hold; and (iv) cellular entities, which are referenced by the geometric references of all the existing features, exist and have the correct status (material or void, on the boundary or not on the boundary) according to the feature sequences in their owning feature lists.

Two product models (corresponding to different lifecycle stages) are consistent if (i) sharing associations between their corresponding application features hold; and (ii) constraint-based associations between their corresponding application features or feature properties hold. In particular: (a) each critical feature in the conceptual design is linked to features in the detailed design via valid constraint-based associations; (b) each feature property or inter-feature constraint in the conceptual design has its valid counterparts (may not be one to one relations) in the detailed design; (c) each detailed design feature to be machined is linked to process planning features via valid constraint-based associations; and (d) all the design specifications (such as tolerances and surface finishes) are satisfied by the finish process planning features.

3.4.4 Algorithms for Change Propagation

If users (designers or process planners) modify the product model, the modifications must be checked to make sure that the consistency of the whole product information model is maintained. As indicated in previous sections, a dependency network is established using constraint-based associations and sharing associations. It is implemented through a JTMS and a common database. The purpose of the dependency network is for the propagation of modifications and determining the influence scope of a modification.

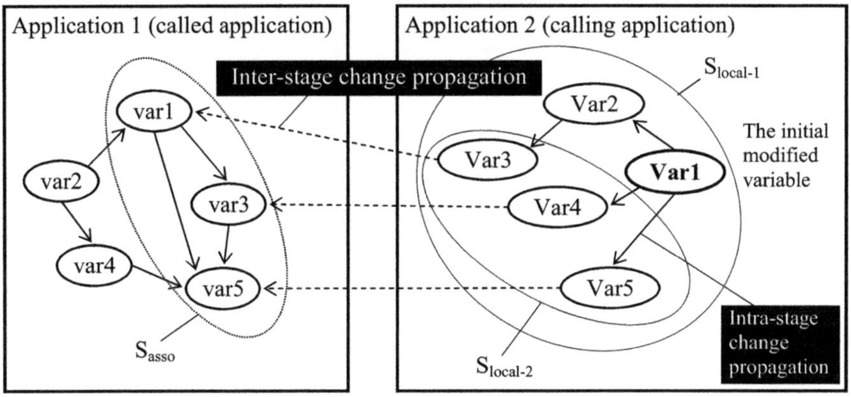

Figure 3.5. Propagation chain of intra- and inter-stage changes

The propagation and checking process is divided into two major generic routines [3.35]: local checking within a specific model and global checking across different models (see Figure 3.5). Assume variable x in an application is changed (as the initial modification). In the figure, arrows are directed from driving to driven variables while "var" represents variables. The change propagation algorithms are developed with reference to constraint-based and sharing associations as well as their corresponding implementations in the unified feature-based product modelling scheme. The algorithm is in iterative manner and starts from a local application domain first; the local change impact is evaluated using a JTMS and a common database to establish inter-stage non-geometric associations. An algorithm for change propagation within a lifecycle stage is presented as follows.

PROCEDURE *Check_Local(x)*
/* checking the intra-stage associations */

(1) Backup the value of the initial modified variable x. Put x into a local set (set_1, which records modified variables v_i that need to be checked for intra-stage associations). For each v_i in set_1, search the JTMS dependency network for variables that associate to v_i using JTMS attributes (antecedent or consequent). The variables, which are antecedents of v_i, are driving variables. The variables, which are consequents of v_i, are driven variables.

(2) Check the constraints between each v_i and its driving or driven variables one by one:

– If the new value of v_i violates the constraints between v_i and any of its related variables:

→ If the related variable is a driven variable

- If the value of the driven variable is fixed by the constraint, *i.e.* without alternative values, then the modification is rejected and run *Abandon()*.
- If the driven variable has alternative values, search one for which the constraint is satisfied:

- If the constraint can be satisfied (and the value of the driven variable is changed), make a backup of the old value and put the driven variable into set_1.
- If no alternative value satisfies the constraint, then *Abandon()*.

→ If the related variable is a driving variable, then *Abandon()*.

– If no constraint has been violated or if some constraints has been violated but can be re-satisfied, the modification is locally accepted.

The *Abandon()* used in the above algorithm is given here:

PROCEDURE *Abandon()*
/* retracting all changes temporarily made */

(1) All modifications made in the calling and called applications are revoked using backup values.
(2) In the database, the data of the called application, whose values are temporarily changed, are set back to their original values.

Next, further check is carried out for v_i in the database. If v_i appears in any inter-stage associations in the database, move v_i from set_1 to set_2 (which records variables that need to be checked for inter-stage associations). Run *Check_Global()*.

PROCEDURE *Check_Global()*
/* checking the inter-stage associations */

(1) For each member of set_2, add all associated features or feature properties in the database to set_3 (which records associated variables in other applications). An initial modification in an application may invoke many modifications in other applications. The members of set_3 are checked (in the next two steps) one by one until set_3 becomes empty.
(2) The values of members of set_3 are temporarily changed in the database using the constraints recorded in the calling application.
(3) For each member of set_3, execute *Check_Local()* in the called application until the modification is found to be locally accepted or rejected. The corresponding message (about whether the modification is locally accepted or rejected in the called application) is sent back to the calling application that initiates the initial modification.
    – If all modifications are accepted, the initial modification in the calling application is globally accepted and committed.
    – If any of the modifications are rejected, the initial modification in the calling application is rejected, then *Abandon()*.

After the iteratively looped change propagation procedures, finally, the algorithm for change propagation finishes if there is no new changes triggered. Then further high-level attributes properties of related objects, logical facts, statuses of indicators and controls are updated accordingly. Eventually, the consistency of product models has been evaluated. This algorithm can be triggered again and again whenever there is a major change decision is to be committed.

3.5 Multiple View Consistency

3.5.1 Cellular Model

Traditional geometric modelling systems use boundary representation (B-rep) or constructive solid geometry (CSG) models for geometry representation [3.36]. They have limitations with respect to the requirements of the unified feature-based product modelling scheme.

Only the final product geometry is stored and managed. Intermediate geometries, which do not belong to the final boundary, are usually not stored. This limitation makes feature modifications difficult. It also results in a persistent naming problem.

CAx systems have different requirements on representing product geometry. A hybrid geometric modelling environment that can accommodate the associative wireframe, surface, and solid models coherently is a natural outcome of the unified feature-based product modelling scheme. CAx systems need to represent the same product geometry in different ways. On one hand, geometry may be represented in different abstraction levels. For instance, a hole can be represented as a central line (plus a radius), a cylindrical face, or a cylinder in different contexts. On the other hand, product geometry may be represented in different ways. For instance, two adjacent faces in one application may be represented as a single face in another application. In addition, it is important for the unified feature-based product modelling scheme that higher level application features can use lower level topological entities to propagate modifications and control the information consistency. Relationships or constraints in higher levels (*e.g.* feature level) may also be specified using lower level (*e.g.* topological entity level) relations.

Some solutions were proposed to solve these problems. Bidarra *et al.* [3.37, 3.38] proposed to use a cellular model to represent intermediate product geometry as well as to support links between different views. However, their methods are confined to 3D features only. Other researchers [3.39–3.42] proposed to use the multi-dimensional non-manifold topology (MD-NMT) to meet the geometric modelling requirements of different applications. However, they did not fully apply the MD-NMT to the feature-based modelling processes. It can be seen that a multi-dimensional geometric modelling environment, which is capable of propagating geometric modifications across feature models, does not exist.

3.5.2 Using Cellular Topology in Feature-based Solid Modelling

The goal of using a cellular topology is to keep a complete description of all the input geometric entities without removing them after set operations on volumes (unite, intersect, and subtract), regardless of whether they appear in the final boundary or not [3.43]. The cellular model uses three mechanisms to fulfil this goal:

1. *Attribute mechanism.* There are two kinds of attributes used in a cellular model: (a) *cell nature* – a cell is either additive or negative depending on whether it corresponds to materials of the product (or the topological entities on the part boundary) or not; and (b) *owner* – each cell records its owing features because a cell may belong to several features due to feature

interactions. The sequence of the owning features is kept to determine the cell nature.

2. *Decomposition mechanism.* Two 3D cells do not overlap volumetrically. Whenever two cells overlap, new cells for the overlap are generated with the merged owner list.
3. *Topology construction mechanism.* In a cellular topology-based, non-manifold boundary representation, an operation on volumes does not remove any input geometry. The cellular model constructs topology, or generates new faces, edges, and vertexes, before classifying the topological entities as "in", "out" or "on" the boundary. All topological entities are marked and filtered for displaying according to the type of the operation.

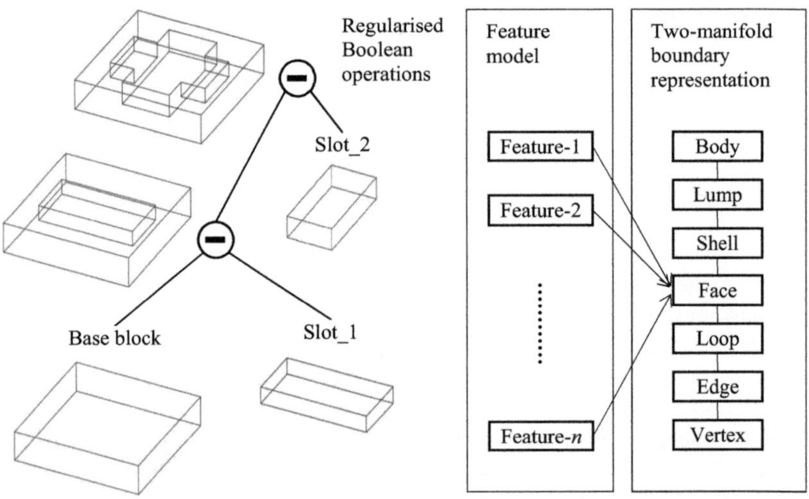

Figure 3.6. Traditional feature model based on a two-manifold boundary representation

Figure 3.6 describes a feature model based on a two-manifold boundary representation. Traditional boundary model does not store intermediate geometries. In other words, according to the types of Boolean operators, "useless" geometries are discarded before the part topology is reconstructed. For example, in Figure 3.6, when inserting the slot_1 feature, its top face is not stored. During the later modelling process, the intersections (due to feature interactions) further split and remove feature geometry from the boundary model. It is hence difficult to relate the feature to its corresponding topological entities in the final boundary model. In the constraint-based, parametric design processes, this limitation makes the feature model history-based. This limitation is also the major reason for the persistent naming problem [3.15].

Alternatively, in cellular topology based non-manifold boundary representations, operations on volumes do not discard any input geometry. Part geometries are represented using cellular topologies. Figure 3.7 shows the decomposed cells of the simple example according to the cellular topology along with the feature modelling process.

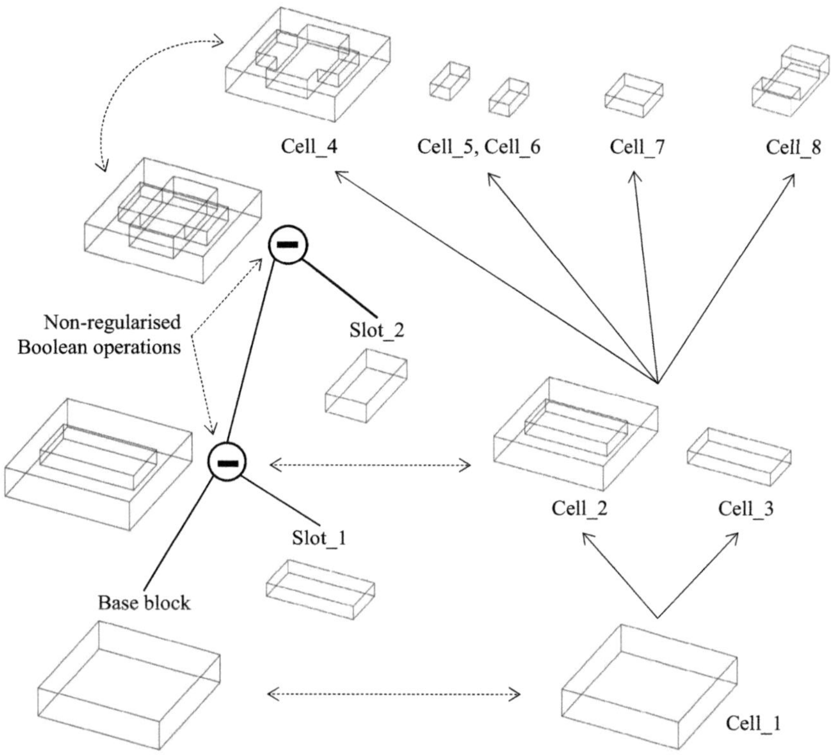

Figure 3.7. Feature model based on the cellular topology

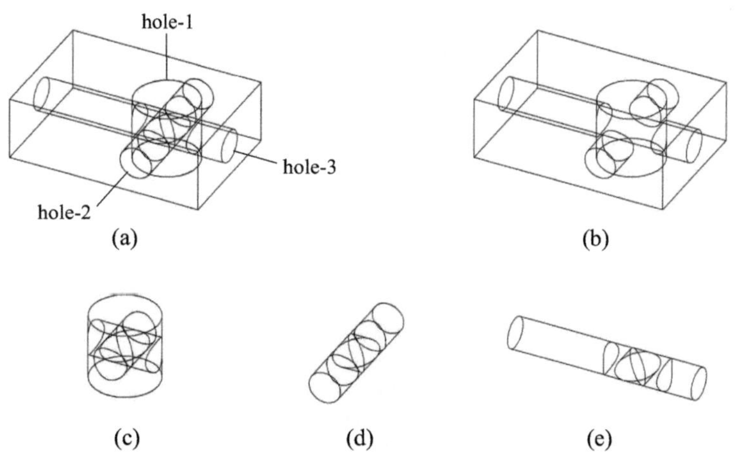

Figure 3.8. Cellular geometry with different cylindrical features: (a) a block with three interacting holes, (b) the cell, which belongs only to the block feature, (c) hole-1 feature, (d) hole-2 feature, and (e) hole-3 feature

The use of cellular topologies simplifies the representation of feature geometry as combinations of cells. When a feature is initialised, it is a single cell that carries only this feature's identifier. Whenever feature interactions occur, new cells that belong to the intersections are generated. The newly generated cells carry a merged owning feature list.

The geometry of a cell can describe any shape. For example, Figure 3.8(a) shows a block workpiece with three intersecting holes. Figure 3.8(b) shows the cell, which carries only the block as its owning feature. Figures from 3.8(c) to (e) show the cellular representations of the three hole features. In this way, all feature geometry, geometric relations between features as well as relations between a feature and its corresponding topological entities are stored in the product model persistently. The persistent naming problem is avoided. It is also possible to modify features based on the dependency relations, not on the construction history because the influence scope of a geometric modification is confined by the inter-cell relations. However, to meet the requirements of the unified feature-based product modelling scheme, this 3D-cell-based multiple-view feature modelling approach needs to be extended. Some case studies are given in [3.43].

3.5.3 Extended Use of Cellular Model

Distinct applications covered by the unified feature-based product modelling scheme have their particular geometry representation requirements. (1) During conceptual design, a designer is concerned about functions and behaviours. Only critical geometries and their relations are specified at that time. These critical geometries may only be represented as abstracted lines, faces, curves, or surfaces. Solid models, detailed topologies and geometries are not specified in this stage. (2) In the detailed design stage, the product geometries or layouts are further materialised. Two-manifold solid model representation is usually preferred. (3) In the process planning stage, features are usually defined as material removal or accessing volumes related to machining operations. Fixtures are also conceptualised in this stage. For these types of features, solid representation with surface manipulation support is more appropriate because, other than the machined volumes, fixture design uses sub-area patches of the part, *e.g.* locating or clamping areas. (4) Similar requirements are applicable to the assembly design stage. In particular, the sub-areas of the part or assembly for interfacing or grasping are concerned.

The geometrical representations discussed above relate to each other. They represent different aspects or abstraction levels of a product. To meet these diverse geometry representation requirements, the current cellular topology-based feature modelling method needs to be extended to support not only 3D solid features, but also non-solid features. A multi-dimensional cellular model, named the *unified cellular model*, is proposed here to integrate all these representations, manage their relations and hence support the multiple-view feature-based modelling processes. The geometric model of each application is a particular aspect (a sub-model) of the unified cellular model. The traditional usage of the cellular topology in multiple-view feature modelling is extended in three aspects: (1) 2D and 3D features are supported uniformly; (2) the unified cellular model is used to share geometric data as well as to propagate geometric modifications (creating new cells, modifying or deleting cells) among views through the cells' owner attributes; and (3) relationships in the cell level are generalised. These relation types can be used as building blocks to establish higher level feature relations.

The unified cellular model ensures the geometric consistency between the application feature models.

1. The geometries of a detailed design and the corresponding process planning model may have different topologies. However, both models correspond to the same final product geometry. In other words, these two application cellular models must correspond to the same B-rep solid model, which represents the final part geometry. This geometric consistency is realised through mapping 2D or 3D application features to the corresponding cells in the shared unified cellular model.
2. Two features may represent the same item at two abstraction levels, *e.g.* a central line or a cylindrical face of a hole. The consistency is maintained through specifying geometric or topological constraints on the related cells in the unified cellular model.

3.5.4 Characteristics of the Unified Cellular Model

A unified cellular model *UCM* includes all geometries from different applications [3.43]. It consists of a set of cells:

$$UC_i\!: UCM = \left(\bigcup_{i=1}^{q} UC_i^0\right) \cup \left(\bigcup_{j=1}^{r} UC_j^1\right) \cup \left(\bigcup_{k=1}^{s} UC_k^2\right) \cup \left(\bigcup_{l=1}^{t} UC_l^3\right) \tag{3.1}$$

In the expression, UC^0, UC^1, UC^2, and UC^3 represent zero-dimensional (0D), one-dimensional (1D), two-dimensional (2D), and three-dimensional (3D) cells, respectively. Similarly, q, r, s, and t are the numbers of 0D, 1D, 2D, and 3D cells, respectively, in the unified cellular model.

Each cell (except 0D cells) is bounded by a set of cells of a dimensionality lowered by one. On the other hand, a cell may exist independently without bounding any higher dimensional cell. The point sets of any two cells (of the same or different dimensionalities) do not overlap: $UC_i^a \cap UC_j^b = \phi$ $(0 \le a < b \le 3$ or $(a = b) \wedge (i \ne j))$. In addition, a cell does not include its boundary, except for 0D cells.

The cellular model obeys the Euler–Poincare formula for non-manifold geometric models [3.41, 3.43]. Each application feature model uses the unified cellular model. The relations among these models are described in [3.43]. An application feature model *AFM* consists of a set of application features AF_i and other non-geometric entities NGE_j:

$$AFM = \bigcup_{i=1}^{m} AF_i \cup \bigcup_{j=1}^{n} NGE_j \tag{3.2}$$

where m and n are the numbers of application features and non-geometric entities in this application feature model.

An application cellular model *ACM* is created at runtime, which consists of a set of application cells AC_i: $ACM = \bigcup_{i=1}^{u} AC_i$, where u is the number of application cells in this application cellular model. Each application feature refers to a set of application cells. An application cell may belong to several application features, *i.e.* it records several features in its owning feature list. The geometries of an application feature correspond to 1D, 2D, or 3D cells.

An application cell can be mapped to one or more cells in the unified cellular model. On one hand, for a particular application, one cell in the unified cellular model is mapped to at most one application cell. On the other hand, each cell in the unified cellular model is mapped to at least one application cell (and therefore at least one application feature). This mapping is realised through the owner attribute mechanism.

The rule for determining cell nature applies to the unified cellular model, *i.e.* the nature of the latest feature in the owner list determines the nature of the cell. Please refer to [3.43] for more details. All applications use this unified multi-dimensional non-manifold cellular model. The geometry of each application feature model is a particular aspect of the unified cellular model.

In Figure 3.9, the design of a cooling system of a plastic injection mould is used as an example to illustrate the idea. In the conceptual design stage, the cooling system is represented as cooling circuits for cooling effect analysis while in the detailed design or process planning stage, the cooling system is in 3D for manufacturability analysis and process planning. The cooling circuits and the cooling channels are representations of the same cooling system in different abstraction levels. The geometries of these two feature models are kept consistent through the unified cellular model.

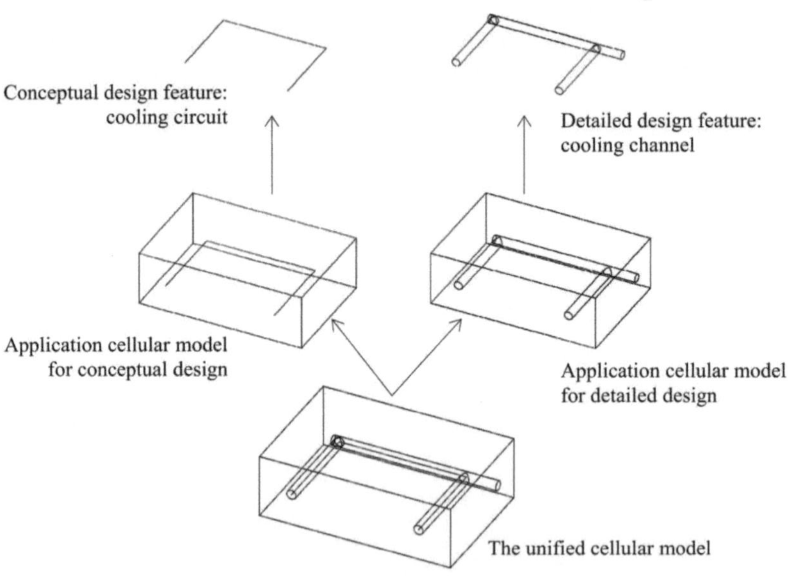

Figure 3.9. Link conceptual and detailed designs using unified cellular model

3.5.5 Two-dimensional Features and Their Characteristics

The idea in this sub-section is that the unified cellular modelling scheme represents 1D, 2D, and 3D cells uniformly (they are referred to as edge, face, and solid cells, respectively hereafter). Currently, the prototype system only handles face and solid cells (corresponding to 2D and 3D features, respectively). 1D features are mentioned here but more research will be done in the future. Examples of 2D application features include: (1) conceptual design features, which represent functional areas in the product; the geometries of this kind of features are usually abstracted as pairs of interacting faces; these faces correspond to partial faces in the detailed design; (2) assembly features, which represent grasping or mating areas of parts in assembly processes; and (3) locating or clamping features, which represent locating or clamping faces during a machining operation.

Similar to solid cells, the major advantage of using face cells instead of geometric faces is that operations on face cells do not remove faces (or parts of faces) even if they do not belong to the final boundary. For example, the non-regularised operations on faces and decomposition mechanism upon overlapping detection are available to face cells. A 2D feature is represented as a group of associated face cells with engineering semantics. For example, a locating feature is defined as a pair of faces associated with the constraints on accessibility, machining accuracy, non-interference, and minimising setup changes. The geometry of a 2D feature is one or more surfaces. Two characteristics of 2D features exist. (1) A 2D feature has a nature attribute (additive or negative) that can be changed by feature interactions. A change of cell nature (from additive to negative or *vice versa*) requires the corresponding features to be validated. For example, a clamping feature represents a local area on a part that is used for clamping. When a clamping feature is altered, its face cells may be split with the natures of some of the resulting face cells inverted. This may jeopardise the clamping feature's stability (sufficient area for clamping). Similar situations are encountered for functional, assembly, and locating features. (2) Face cells corresponding to functional, assembly, locating, and clamping features have the same surface definitions as existing face cell(s). Hence, to simplify the implementation, it is assumed that newly inserted face cells and existing solid cells do not intersect. However, this is not valid for some CAE analysis applications, in which middle faces are commonly used.

When a 2D feature is generated, the corresponding face cell is also generated and inserted into the application and the unified cellular models. Figure 3.10 illustrates a simple example: the integration of a detailed design and a process planning model on the basis of a unified cellular model.

The designed part is a block with a blind hole. The hole has a distance specification with face F2 as datum and a perpendicularity specification with face F1 as datum. The corresponding process planning model begins with a blank feature (larger than the part to allow machining). Other process planning features are (1) a surface-milling feature (due to the perpendicularity specification); (2) a clamping feature (for surface-milling); (3) a drilling feature and a boring feature (to meet the surface finish requirement of the hole); and (4) a locating feature and a supporting feature for the drilling and boring operations. These 2D or 3D features are associated in the unified cellular model.

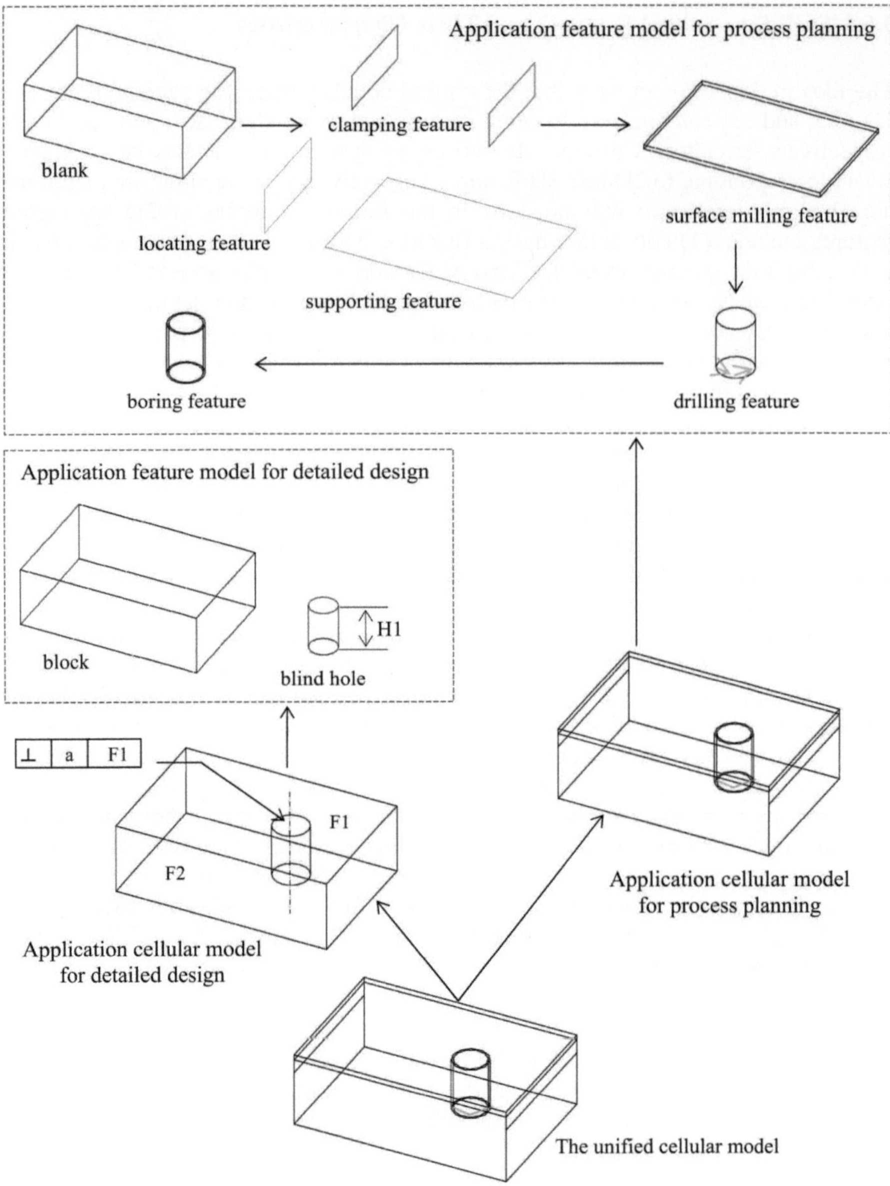

Figure 3.10. Integrating detailed design and process planning feature models

3.5.6 Relation Hierarchy in the Unified Cellular Model

Relations can be established on the cell level, the feature geometry level, and the feature semantic level, respectively. Higher-level relations are established on the basis of two lower-level relations.

The lowest level of relations is between two cells, which cover four cases:

1. The bounding relations among cells
2. The bounding cells that inherit the owner attributes of the bounded cell
3. Two solid cells that are adjacent if they are bounded by one or more common face cells (two face cells are adjacent if they are bounded by one or more common edge cells)
4. Two adjacent edge or face cells that may be part of the same curve or surface

The second level relations are topological relations between the geometries of application features. Note that a feature's dimensionality can be diverse depending on the application nature. Three possible topological relations between two application features are identified here:

1. *Overlap*: After cellular splitting, two n-dimensional features are said to *overlap* each other if they use the same n-dimensional cell(s). An n- and a $(n-1)$-dimensional features are also said to *overlap* each other if they use the same $(n-1)$-dimensional cell(s).
2. *Adjacent*: Two different n-dimensional features are defined as *adjacent* ones if they share $(n-1)$-dimensional cell(s) but do not overlap.
3. In a 3D feature, *adjoining area* refers to one or more faces (represented by the face cells), which are mathematically connected and defined on the same surface.

For two 3D features *A* and *B*, feature *A* is said to be *completely adjacent* to feature *B,* if feature *A*'s adjoining area is fully enclosed by any of feature *B*'s adjoining area. In plastic injection mould design, completely adjacent relations can be used to represent maps from the plastic part to core or cavity inserts as well as electrode geometry. Such maps are commonly encountered in die casting, forging tooling, and fixture design as well. Again, for more details, refer to [3.43]. Other examples are: (i) a single face in the detailed design corresponding to several functional faces in the conceptual design; and (ii) a face in the process planning model corresponding to one or more faces in the detailed design.

Higher level relations are semantic relations between application features. Relation types in this level are application specific. Examples are:

1. *Splitting*. Figures 3.11(a) to (c) show a base block with a *hole* feature; and the *hole* feature is further split by a vertical *through-slot* feature. A similar situation for 2D features is shown in Figures 3.11(d) and (e), in which the original clamping feature is split by a newly inserted *through-slot* feature. The middle face cell of the clamping feature becomes negative. The clamping feature must hence be checked for stability. This kind of relation between two interacting features is defined as a splitting relation [3.38]. Using the above-mentioned two lower levels of relations, the splitting relation can be described as: (i) the nature of the second feature is negative; (ii) the two features overlap; and (iii) the insertion of the second feature splits the original single cell (additive or negative) of the first feature into several (at least three) cells, where the nature of at least one of the middle cell(s) is negative.

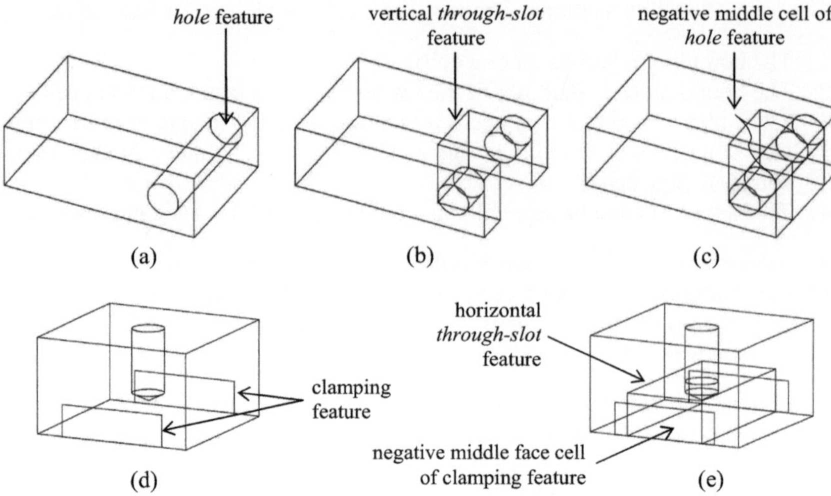

Figure 3.11. Splitting relation: (a) insert the first feature; (b) the second feature splits the first feature; (c) the middle cell of the first *hole* feature becomes negative; (d) two additive face cells of the clamping feature are inserted; and (e) the middle face cell of the *clamping* feature becomes negative

2. *Transmutation.* Figure 3.12 shows a base block with a *blind-hole* feature and a vertical *through-slot* feature. The relation between these two features is defined as a *transmutation* relation in [3.38]. Using the above-mentioned two lower levels of relations, the transmutation relation can be described as: (i) the nature of the two features is negative, but the nature of one of the bounding cells of the first feature, which represents the bottom face of a blind hole, is additive; (ii) the two features overlap; and (iii) the insertion of the second feature splits the original single 3D negative cell into two 3D negative cells. The previous additive bounding 2D cell becomes negative.
3. *Non-interference.* This relation specifies that two features cannot overlap with or are adjacent to each other. This constraint is satisfied if no cell in the unified cellular model has both of these two features in its owner list. This constraint is commonly used in product design or manufacturing activities. For example, a process planning feature cannot interfere with the corresponding clamping features.

3.6 Conclusions

Unified feature theory is a significant contribution to feature level collaboration in future virtual enterprises. In the proposed scheme, unified features provide an intermediate information layer to bridge the gap between engineering knowledge and product geometry. Unified features are also used to maintain geometric and non-geometric relations across product models. The feasibility of the proposed unified feature modelling scheme is demonstrated with a prototype system and case studies.

With the unified feature definition, application feature definitions, the unified cellular model, dependency network, and the change propagation algorithm, the proposed unified feature-based product modelling scheme is able to integrate the conceptual design, detailed design, and process planning applications. For detailed case studies, please refer to [3.35].

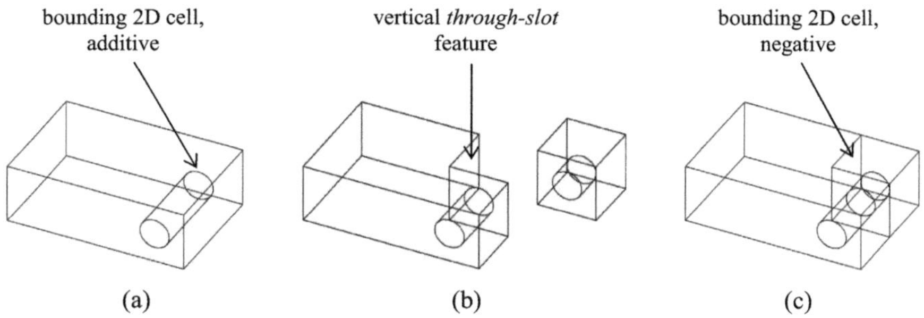

Figure 3.12. Transmutation relation: (a) insert the first feature, a *blind-hole*; (b) the second feature changes the *blind-hole* into a *through-hole*; (c) the bounding 2D cell is changed to negative due to feature interaction

References

[3.1] Ma, Y.-S. and Fuh, J., 2008, "Editorial: product lifecycle modelling, analysis and management," *Computers in Industry*, **59**(2–3), pp. 107–109.

[3.2] Varady, T., Martin, R.R. and Cox, J., 1997, "Reverse engineering of geometric models – an introduction," *Computer-Aided Design*, **29**(4), pp. 255–268.

[3.3] Benko, P., Martin, R.R. and Varady, T., 2001, "Algorithms for reverse engineering boundary representation models," *Computer-Aided Design*, **33**(11), pp. 839–851.

[3.4] Starly, B., Lau, A., Sun, W., Lau, W. and Bradbury, T., 2005, "Direct slicing of STEP based NURBS models for layered manufacturing," *Computer-Aided Design*, **37**(4), pp. 387–397.

[3.5] Kramer, T.R., Huang, H., Messina, E., Proctor, F.M. and Scott, H., 2001, "A feature-based inspection and machining system," *Computer-Aided Design*, **33**(9), pp. 653–669.

[3.6] Rezayat, M., 1996, "Midsurface abstraction from 3D solid models: general theory and applications," *Computer-Aided Design*, **28**(11), pp. 905–915.

[3.7] Ma, Y.-S., Britton, G., Tor, S.B., Jin, L.-Y., Chen, G. and Tang, S.-H., 2004, "Design of a feature-object-based mechanical assembly library," *Computer-Aided Design & Applications*, **1**(1–4), pp. 397–404.

[3.8] International Organisation for Standardization (ISO), 2000, *Industrial Automation Systems and Integration: Product Data Representation and Exchange: Integrated Generic Resource: Part 42 – Geometric and Topological Representation*, Geneva, Switzerland.

[3.9] Shah, J.J. and Rogers, M.T., 1993, "Assembly modelling as an extension of feature-based design," *Research in Engineering Design*, **5**(3–4), pp. 218–237.

[3.10] van Holland, W. and Bronsvoort, W.F., 2000, "Assembly features in modelling and planning," *Robotics and Computer Integrated Manufacturing*, **16**(4), pp. 277–294.

[3.11] Karinthi, R.R. and Nau, D., 1992, "An algebraic approach to feature interactions," *IEEE Transactions on Pattern Analysis and Machine Intelligence*, **14**(4), pp. 469–484.

[3.12] Hounsell, M.S. and Case, K., 1999, "Feature-based interaction: an identification and classification methodology," *Proceedings of the Institution of Mechanical Engineers, Part B – Journal of Engineering Manufacture*, **213**(4), pp. 369–380.

[3.13] Vandenbrande, J.H. and Requicha, A.A.G., 1993, "Spatial reasoning for the automatic recognition of machinable features in solid models," *IEEE Transactions on Pattern Analysis and Machine Intelligence*, **15**(12), pp. 1269–1285.

[3.14] Laakko, T. and Mantyla, M., 1993, "Feature modelling by incremental feature recognition," *Computer-Aided Design*, **25**(8), pp. 479–492.

[3.15] Bidarra, R. and Bronsvoort, W.F., 2000, "Semantic feature modelling," *Computer-Aided Design*, **32**(3), pp. 201–225.

[3.16] Schulte, M., Weber, C. and Stark, R., 1993, "Functional features for design in mechanical engineering," *Computers in Industry*, **23**(1–2), pp. 15–24.

[3.17] Feng, C.-X., Huang, C.-C., Kusiak, A. and Li, P.-G., 1996, "Representation of functions and features in detail design," *Computer-Aided Design*, **28**(12), pp. 961–971.

[3.18] Anderl, R. and Mendgen, R., 1996, "Modelling with constraints: theoretical foundation and application," *Computer-Aided Design*, **28**(3), pp. 301–313.

[3.19] Kim, C.-S. and O'Grady, P.J., 1996, "A representation formalism for feature-based design," *Computer-Aided Design*, **28**(6–7), pp. 451–460.

[3.20] Mukherjee, A. and Liu, C.R., 1997, "Conceptual design, manufacturability evaluation and preliminary process planning using function-form relationships in stamped metal parts," *Robotics & Computer-Integrated Manufacturing*, **13**(3), pp. 253–270.

[3.21] Whitney, D.E., Mantripragada, R., Adams, J.D. and Rhee, S.J., 1999, "Designing assemblies," *Research in Engineering Design*, **11**(4), pp. 229–253.

[3.22] Khoshnevis, B., Sormaz, D.N. and Park, J.Y., 1999, "An integrated process planning system using feature reasoning and space search-based optimisation," *IIE Transactions*, **31**(7), pp. 597–616.

[3.23] Stage, R., Roberts, C. and Henderson, M., 1999, "Generating resource based flexible form manufacturing features through objective driven clustering," *Computer-Aided Design*, **31**(2), pp. 119–130.

[3.24] Brunetti, G. and Golob, B., 2000, "A feature-based approach towards an integrated product model including conceptual design information," *Computer-Aided Design*, **32**(14), pp. 877–887.

[3.25] Park, S.C., 2003, "Knowledge capturing methodology in process planning," *Computer-Aided Design*, **35**(12), pp. 1109–1117.

[3.26] Brunetti, G. and Grimm, S., 2005, "Feature ontologies for the explicit representation of shape semantics," *International Journal of Computer Applications in Technology*, **23**(2/3/4), pp. 192–202.

[3.27] Rossignac, J.R., 1990, "Issues on feature-based editing and interrogation of solid models," *Computers & Graphics*, **14**(2), pp. 149–172.

[3.28] Ma, Y.-S. and Tong, T., 2003, "Associative feature modelling for concurrent engineering integration," *Computers in Industry*, **51**(1), pp. 51–71.

[3.29] Martino, T.D., Falcidieno, B., Giannini, F., Hassinger, S. and Ovtcharova, J., 1994, "Feature-based modelling by integrating design and recognition approaches," *Computer-Aided Design*, **26**(8), pp. 646–653.

[3.30] Ma, Y.-S., Chen, G. and Thimm, G., 2008, "Paradigm shift: unified and associative feature-based concurrent and collaborative engineering," *Journal of Intelligent Manufacturing*, **19**(6), pp. 625–641.

[3.31] Chen, G., Ma, Y.-S., Thimm, G. and Tang, S.-H., 2004, "Unified feature modelling scheme for the integration of CAD and CAx," *Computer-Aided Design & Applications*, **1**(1–4), pp. 595–601.
[3.32] Booch, G., Rumbaugh, J. and Jacobson, I., 1999, *The Unified Modelling Language User Guide*, Addison Wesley.
[3.33] Chen, G., Ma, Y.-S., Thimm, G. and Tang, S.-H., 2005, "Knowledge-based reasoning in a unified feature modelling scheme," *Computer-Aided Design & Applications*, **2**(1–4), pp. 173–182.
[3.34] Chen, G., Ma, Y.-S., Thimm, G. and Tang, S.-H., 2006, "Associations in a unified feature modelling scheme," *Transactions of the ASME, Journal of Computing and Information Science in Engineering*, **6**(6), pp. 114–126.
[3.35] Chen, G., Ma, Y.-S. and Thimm, G., 2008, "Change propagation algorithm in a unified feature modelling scheme," *Computers in Industry*, **59**(2–3), pp. 110–118.
[3.36] Hoffman, C.M., 1989, *Geometric and Solid Modelling: An Introduction*, Morgan Kaufmann, San Francisco.
[3.37] Bidarra, R., Madeira, J., Neels, W.J. and Bronsvoort, W.F., 2005, "Efficiency of boundary evaluation for a cellular model," *Computer-Aided Design*, **37**(12), pp. 1266–1284.
[3.38] Bidarra, R., de Kraker, K.J. and Bronsvoort, W.F., 1998, "Representation and management of feature information in a cellular model," *Computer-Aided Design*, **30**(4), pp. 301–313.
[3.39] Lee, S.H., 2005, "A CAD-CAE integration approach using feature-based multi-resolution and multi-abstraction modeller techniques," *Computer-Aided Design*, **37**(9), pp. 941–955.
[3.40] Sriram, R.D., Wong, A. and He, L.-X., 1995, "GNOMES: an object-oriented non-manifold geometric engine," *Computer-Aided Design*, **27**(11), pp. 853–868.
[3.41] Masuda, H., 1993, "Topological operators and Boolean operations for complex-based non-manifold geometric models," *Computer-Aided Design*, **25**(2), pp. 119–129.
[3.42] Crocker, G.A. and Reinke, W.F., 1991, "An editable non-manifold boundary representation," *IEEE Computer Graphics & Applications*, **11**(2), pp. 39–51.
[3.43] Chen, G., Ma, Y.-S., Thimm, G. and Tang, S.-H., 2006, "Using cellular topology in a unified feature modelling scheme," *Computer-Aided Design & Applications*, **3**(1–4), pp. 89–98.

4

Collaborative Supplier Integration for Product Design and Development

Dunbing Tang[1] and Kwai-Sang Chin[2]

[1] College of Mechanical and Electrical Engineering
Nanjing University of Aeronautics and Astronautics, Nanjing 210016, China
Email: d.tang@nuaa.edu.cn

[2] Department of Manufacturing Engineering and Engineering Management
City University of Hong Kong, Hong Kong, China
Email: mekschin@cityu.edu.hk

Abstract

It is widely acknowledged that current industry is more than ever obliged to improve its product design and development strategy according to the increasing pressure of product innovation and complexity, the changing market demands and increasing level of customer awareness. Due to the complex development cycle, the OEM (original equipment manufacturer) has begun to adopt supplier integration into its product development process. To respond to this trend, the collaboration and partnership management between the OEM and suppliers need to be investigated. Regarding the depth of collaboration, the integration of suppliers into the OEM process chain has been defined in two ways, quasi-supplier integration and full supplier integration. To enable the success of supplier integration, this chapter has investigated how to manage the collaboration between the OEM and its suppliers, through determining an appropriate supplier integration method. The collaboration tools enabling supplier integration for product development have been proposed. Taking the tool supplier as a case study, a Web-based system called "CyberStamping" has been developed to realise collaborative supplier integration for automotive product design and development.

4.1 Introduction

The industrial norm in the Western world is that 70% of the added value of the product comes from suppliers [4.1]. The opportunities for using the specialist skills and knowledge of suppliers in enhancing the design of new products are immense. To cope with the changing market conditions, a strong partner relationship between OEM and supplier has been consistently cited as critical to a win-win situation for both sides [4.2].

It has been found from contemporary research in the fields of concurrent engineering and supply chain management that significant benefits can be achieved if suppliers are integrated or involved in product development processes as early as possible, which is called ESI (early supplier involvement) [4.3–4.8]. The rationale is that suppliers frequently possess a greater depth of domain expertise, which can lead to improvements in product design and development. The traditional OEM-supplier relationship is characterised by a two-step sequential interaction. In the first step, the OEM gives clear product and production requirements to the supplier. In the second step, the supplier delivers the product or service to the OEM. Both parties tend to optimise their own position instead of looking at the cooperative gain, and this behaviour is not based on complementary strengths. Supplier integration/involvement is a new method for incorporating a supplier's innovativeness in the product development process. Supplier integration/involvement strives to create synergy through mutually interacting deliverables and decisions between OEM and suppliers. Both sides take advantage of each other's capability to develop the product as well as to obtain feedback from the other party to improve the product development. Supplier involvement is touted as enhancing a firm's competitive edge, and various supplier involvement practices have been conducted in industry to decipher the impact of supplier involvement [4.7, 4.9–4.13].

The changing market conditions and international competitiveness are forcing the time-to-market to reduce rapidly. For example, an automobile today with complexity several times higher than before needs to be brought to manufacture in less time [4.14]. In this context, tool suppliers are under increasing pressure to respond rapidly to automotive OEM requirements in order to gain their places in the tooling market. Meanwhile, to minimise product cost, the automotive manufacturer (OEM) always tries to reduce the number and cost of tools and dies. For instance, five years ago a car body side required 20 to 30 dies, while today a body side requires only 4 to 10 dies [4.15]. Due to the increasing innovation pressure to meet different needs of customers, the automotive industry is seen to adopt supplier involvement into the development process or to outsource a higher percentage of the product development to suppliers, such as Magna's involvement in Citreon, BMW and DCX, Valeo and ArvinMeteritor in BMW [4.16, 4.17]. To decrease the development cycle as much as possible, the automotive industry tries to focus its time and cost on core competency areas such as styling, body in white (BIW), engine, and transmission, while shifting other portions of auxiliary system development to suppliers, which can lead to a win-win situation for both the automotive manufacturers as OEM, and suppliers. Meanwhile, suppliers are seeking new ways to strictly contain costs without sacrificing innovative, feature-rich products and platforms. With the demands for faster innovation, higher quality and increased regulation, it becomes apparent that the winning suppliers will be those that leverage product innovations to rapidly develop new platforms and win new programs.

In this current climate of continual change, most suppliers consider that it is difficult to guarantee their competitive behaviour. On closer examination, it is evident that their development strategies are far from ideal and there are inherent weaknesses, such as lack of comprehensive awareness of the changing collaboration/co-operation relationship with OEM; non-existence of a flexible organisation conducive to the changing co-operation relationship with OEM; and lack of consistent systematic approaches for continuous improvement.

The partner relationship between OEM and suppliers may change with different types of products and requirement levels, depending on the skill and cost mix. This requirement for dynamic co-operation/collaboration between OEM and suppliers places a heavy demand on adaptive partnership. Therefore, decision models for adaptive supplier integration/involvement need to be drawn up; and new methodologies and concepts are required to support better co-operation/ collaboration between OEM and suppliers. To cope with these issues, the main aim of this chapter is to develop comprehensive initiatives for adaptive partnership development between OEM and suppliers.

4.2 Different Ways of Supplier Integration

An OEM is responsible for providing a product to its customers, who proceed to modify or bundle it before distributing it to their customers. To speed up the product development efficiency and reduce OEM development costs, the OEM development process always incorporates the supplier's activities. It has to be considered, that the more active the involvement of the supplier in the OEM development process chain is supposed to happen, the more complex the co-ordination process will be. The early integration of suppliers into the OEM product development process chain does not only lead to an earlier start of the supplier's usual activities but also to a shift in the focus on activities to be processed. This will cause new challenges for the collaboration between OEM and suppliers. In the current global manufacturing context, the OEM and associated suppliers may be geographically separated. Each geographical location is focusing on certain area of the product lifecycle based on resource strengths and cost-effectiveness [4.17]. For example, as the auto market is currently expanding very rapidly in China, some big automotive companies (such as VW, Ford and GM) put the final assembly in China where manpower is cost-effective, while keeping the design and research with the automotive OEMs. To facilitate supplier integration/involvement in OEM product development, not only technology integration but also process and organisation integration have to be considered. The OEM needs to make the evolving product definition and development process available to their suppliers, while protecting everyone's private data and private process and managing everyone's role. The collaboration between the OEM and the integrated supplier can be defined at different levels according to the depth of collaboration and types of partnership.

Regarding the depth of collaboration, the supplier integration/involvement is defined in different ways. In this research, the integration of the supplier in the OEM process chain can be defined in two ways (see Figure 4.1 for details): quasi-supplier integration and full supplier integration. Quasi-supplier integration means joint development efforts with the supplier taking part only at certain times. The development processes of both OEM and supplier remain semi-connected and essential know-how and information stays with each party's operation, either side only takes advantage of the other side's input and feedback. In full supplier integration, OEM and supplier contribute and share resources to a much larger extent. During the whole product development lifecycle, know-how and information are exchanged freely. The boundaries between their development processes begin to diminish.

Figure 4.1. (a) Quasi-supplier integration, and (b) full supplier integration

To enable supplier integration success, one project task is to control collaboration between the OEM and associated suppliers, through deciding an appropriate way of supplier integration at the beginning of the product development project. The decision needs to be refined such that the degree of collaboration effort by both OEM and supplier is effectively and efficiently managed, and a clear designation and agreement of the responsibilities for collaborative development between both sides is needed. In this research, the preferred method of supplier integration is determined by two dimensions: the development capability comparison between OEM and supplier, and the maturity degree of the product (from very old product to very new product). Based on both dimensions, how to specify the method of supplier integration is explained as follows.

Regarding the comparison of the development capability between the supplier and OEM, the required development capabilities for a product development may be distributed either *one-sided* or *split* between the supplier and OEM. *One-sided* means that the supplier has sufficient capabilities to develop a special type of product, namely, the supplier's capability is higher than the OEM's. For example, seat producers, as suppliers providing automotive seats, have a greater depth of knowledge and expertise within this given product domain whereas the automotive OEM is a seat system integrator. Thus, seat development could be shifted to the supplier. In this context, quasi-supplier integration is preferable. *Split* means that both the OEM and associated suppliers should team up their development capabilities to meet the needs of the product development, and full supplier integration is more likely to be selected.

The other factor affecting the method of supplier integration is the maturity degree of the product: from very old product to very new product. An old product means that the supplier or OEM already has enough experience on the current product development, and quasi-supplier integration is more likely to occur in this case. In contrast, the newer the product development, the more co-operations between the OEM and suppliers are needed.

It is noted that both factors mentioned above should be considered together when deciding the method of supplier integration. Comparison of the development capability of the supplier and OEM can be described by Equation (4.1):

$$P \equiv \{p_1, p_2, \ldots, p_n\} \tag{4.1}$$

where p_1 denotes that the development capability of a selected supplier is relatively moderate compared with the OEM; P_n means that the development capability of a selected supplier is relatively much higher than the OEM.

The maturity degree of the product can be described by Equation (4.2):

$$Q \equiv \{q_1, q_2, \ldots, q_m\} \tag{4.2}$$

where q_1 denotes the very old product; and q_m denotes the very new product.

If we use w to represent the way of supplier integration, the decision mechanism of supplier integration can be described by Equation (4.3):

$$w = F(p_i, q_j) \mid p_i \in P, q_j \in Q \tag{4.3}$$

Combining both factors, Figure 4.2 illustrates which type of supplier integration is preferred in different contexts. In Figure 4.2, the space above the dashed line means quasi-supplier integration, while the space below the dashed line refers to full supplier integration. For example, for the case A, as the developed product is very old, quasi-supplier integration is selected. For the case B, although the product to be developed is moderately new, full supplier integration is selected because the capability of the associated supplier is not very strong. For the case C, quasi-supplier integration is selected on account of the higher capability of the supplier compared with the OEM. For the case D, full supplier integration is chosen because the product to be developed is very new, and tight co-operation between the OEM and associated suppliers is necessary.

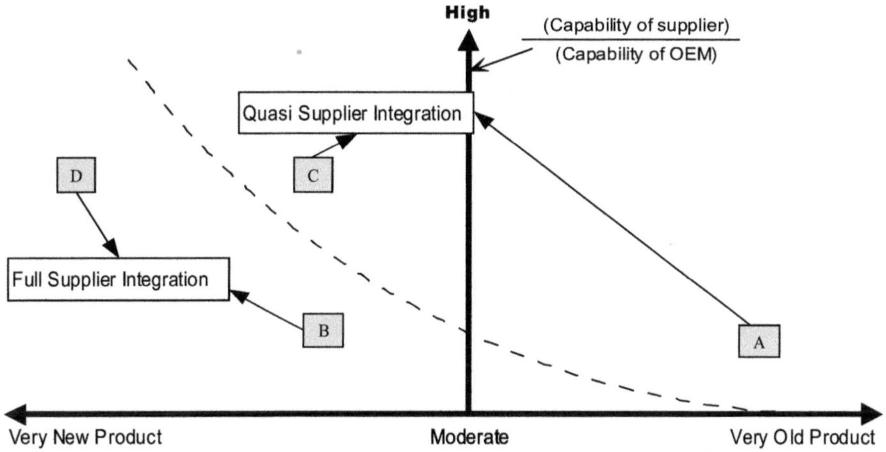

Figure 4.2. Sketch of supplier integration

The most important issue of partnership development is the interaction relationship between OEM and supplier. Based on the rational ESI methodology, an adaptive partnership requires that the supplier should justify its role acting as an active partner along the entire OEM product development chain. The ESI model can be further classified into two types: decoupled ESI model and integrated ESI model, which correspond to the quasi-supplier integration and full supplier integration, respectively.

4.3 Know-how Sharing for Supplier Integration

According to ESI partnership, the supplier can become an arranging partner in the entire OEM process chain of product development and attain co-operation or collaboration with the OEM product designer. On one hand, co-operation is realised by information sharing and exchange. For example, product modelling by means of CAD systems will be supported by suppliers, so that transfer of data for collaborative engineering becomes possible without complex editing. On the other hand, co-operation is guaranteed by know-how sharing between OEM and suppliers, which is the most important factor in the context of ESI. For example, through know-how sharing, feasibility studies can be conducted on the basis of product sketches, and possible problems of production can be identified by suppliers and consequently avoided. Know-how sharing during product development can be considered as an enabler for active partnership implementation.

Product development can be seen from different viewpoints. For example, an OEM product designer may look at a product with a view to achieving functionality by satisfying the product specifications, while a tool and die supplier may consider product manufacturability, tool and die design, as well as tool and die cost [4.18]. Each viewpoint requires different constraints on product design. In this case, some problems may arise because the OEM product designer and suppliers may think in different ways. Know-how sharing, thus, extends the scope of both parties to share not only design information but also design knowledge and to achieve consistency in product development.

The product design cycle can be divided into different design stages, such as conceptual design, embodied design, and detailed design. At different design stages, supplier involvement and know-how sharing focus on different levels. Taking a metal stamping product as an example, the tool supplier's know-how about the product design mainly refers to near-net shape manufacturability analysis, near-net shaping process selection and planning, tool and die cost estimation, tool and die manufacturability analysis, *etc.* (Figure 4.3). Know-how is provided in knowledge objects, which can be case studies, explanations, references, articles, abstracts, manuals, *etc.* Know-how mining, searching, and navigating are performed based on the knowledge meta-data, which is high-level information about a knowledge object. By adding keywords to the knowledge meta-data, a search of the specific knowledge area can be performed. The meta-data enables a user to move quickly in the knowledge object database and to browse through keywords. The know-how includes not only knowledge about product development, but also knowledge of tool application and usage [4.6]. For example, the conclusions drawn from tool maintenance and repair results can be integrated into the know-how database. They will be shared by the OEM product designer and considered during the optimisation of process parameters or during the production of back-up tools.

Furthermore, according to different partnerships there are different ways of know-how sharing between OEM and associated suppliers (Figure 4.4), which are explained as follows:

- The traditional partnership is confined to a buyer-seller interaction, thus know-how sharing between OEM and suppliers is at the lowest level.

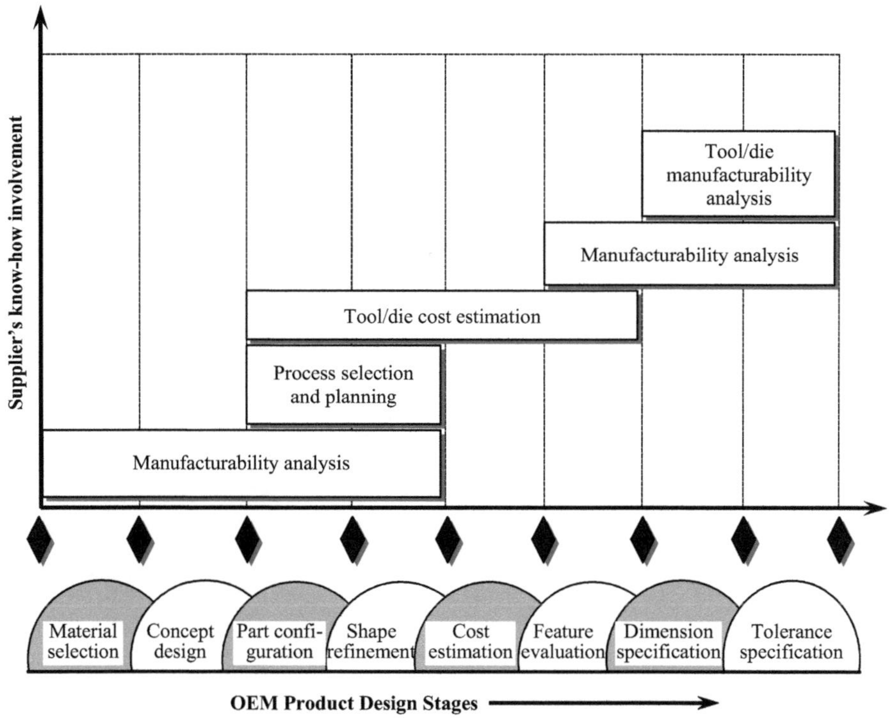

Figure 4.3. Tool supplier's know-how shared in different product design stages

- The decoupled ESI partnership means a little closer co-operation between both sides. In this context, the know-how sharing commitment means that the suppliers will be involved into several important design stages of OEM product development.
- The integrated ESI partnership requires the closest collaboration between OEM designer and suppliers, where company boundaries become blurred and the know-how of suppliers can be integrated into the whole product development process.

4.4 Collaboration Tools for Supplier Integration

As illustrated above, to realise supplier integration/involvement, it is important to provide supporting tools to enable appropriate collaboration between the OEM and its involved suppliers. Using the collaboration tools, the suppliers can conduct product design and development for OEM in an appropriate role (Figure 4.5).

The detailed enabling tools are shown in Figure 4.6, mainly including classical project tools and supplier integration tools. The classical project tools include tools for engineering information management, process management, configuration management, and project management, *etc.* They can be used for (1) storage and

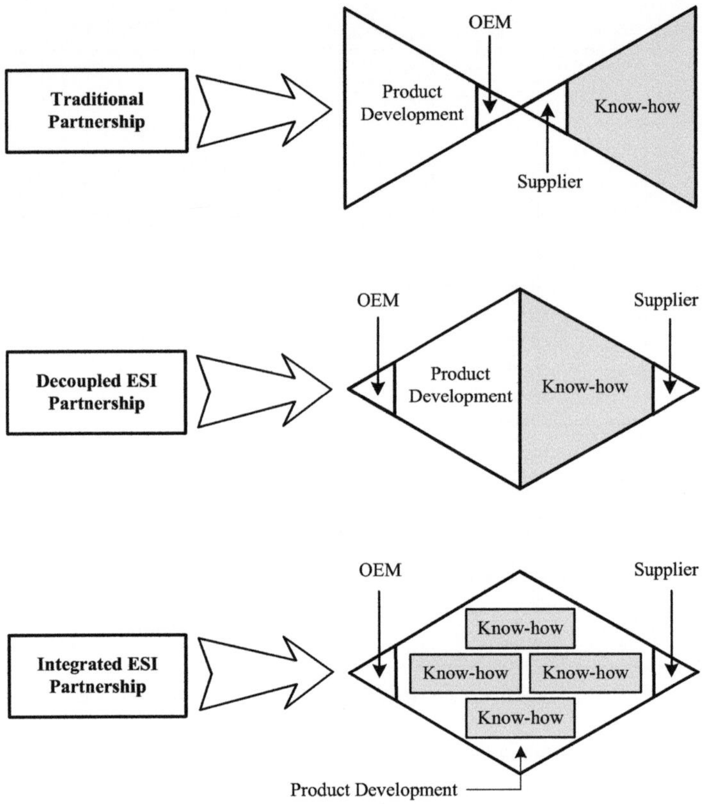

Figure 4.4. Different forms of know-how sharing according to different partnerships

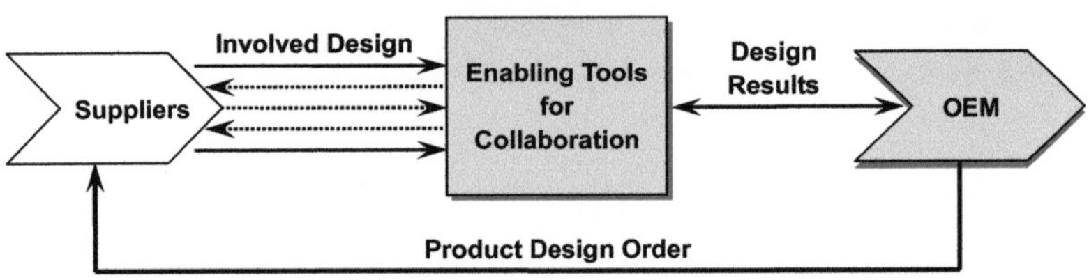

Figure 4.5. Enabling tools for collaborative supplier integration

management of technical objects, configuration data, product model; (2) definition and management of development process; and (3) management of technical objects using check-in/check-out mechanisms and maintaining data revision and status.

Besides common and classical project tools, the global aim is to develop and validate supporting tools to enable supplier integration with appropriate partnership management all along the OEM product development lifecycle. As shown in Figure 4.6, the supplier integration tools mainly include supplier selection, partnership management between the OEM and suppliers, information sharing (mainly referring

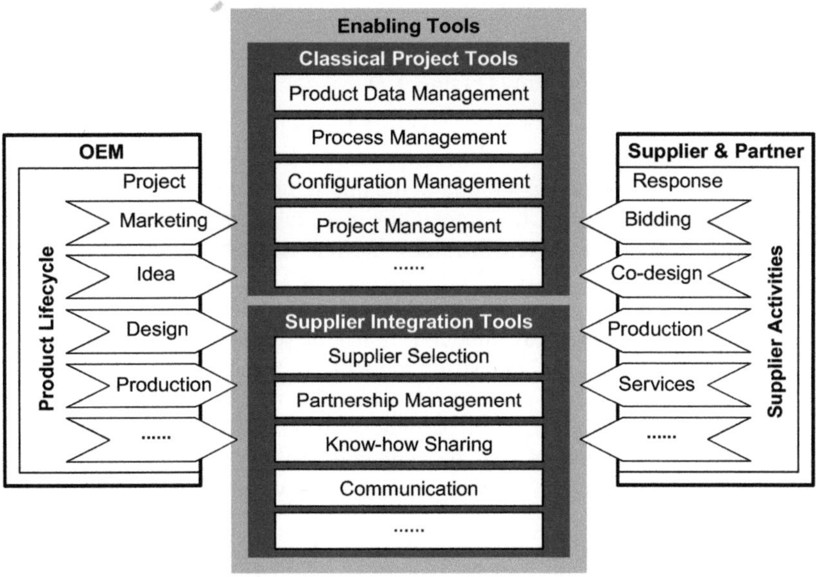

Figure 4.6. Framework focusing on supplier integration for OEM product development

to know-how sharing) between the OEM and suppliers, and communication utility to enable interaction between the OEM and suppliers, *etc*. These tools are easy to understand except know-how sharing, which is explained as follows.

Know-how sharing is aimed at supporting suitable supplier integration into the OEM product development process, while certain know-how issues such as the physical distribution of information, access rights to shared know-how, know-how visibility levels, as well as partner know-how interoperability bring new challenges to know-how management. Know-how management is based on such facts: (1) the partners (OEMs or suppliers) are autonomous; and (2) not all partners play the same role and not all of them have the same access level to the know-how stored in other partners.

In order to facilitate appropriate know-how sharing, the first step is to analyse and classify the know-how depending on the application. The know-how, hereby, is categorised as follows:

- *Private know-how.* This type of know-how is not shared with other partners; it is intended to be accessed only for local processing. For example, the know-how related to the core competence of the OEM is of this type.
- *Public know-how.* This type of know-how is accessible by both the OEM and the associated suppliers.
- *Exchanging know-how.* This refers to the know-how between the OEM and suppliers, such as sending and receiving messages.
- *Interoperable know-how.* The know-how not only can be remotely accessed, but also can be interoperated and changed remotely by other partners. For example, through full supplier integration, the product model designed by the OEM could be improved on-line by suppliers in a co-design way.

Due to the know-how classification, four types of know-how interactions are defined between the OEM and associated suppliers: browsing, exchanging, quasi-interoperating, and interoperating (Figure 4.7):

- *Browsing.* This is the lowest level of know-how interaction. For instance, through the Internet, general description of a supplier in a way that advertises the company is made accessible to the public including the OEM.
- *Exchanging.* Through the interaction of exchanging, one side can obtain acquaintance know-how from the other side to serve internal purposes. For example, an OEM owns the end product; he/she can download the standard part (such as bolt, nut) model from other outsourcing supplier enterprises and uses it as his/her own part model to finish the product development process.
- *Quasi-interoperating.* Quasi-interoperating interaction means that the supplier can get some product related issues from the OEM. After changing or modifying these issues, the supplier can transmit them back to the OEM. Meanwhile, a message will be sent concurrently as a notification. Quasi-interoperating is a general means of know-how interaction for quasi-supplier integration.
- *Interoperating.* This is the highest level of know-how interaction. Through interoperating interaction, the supplier can directly access the required product model from the OEM and has the full right to operate it on-line. This type of know-how interaction is for full supplier integration.

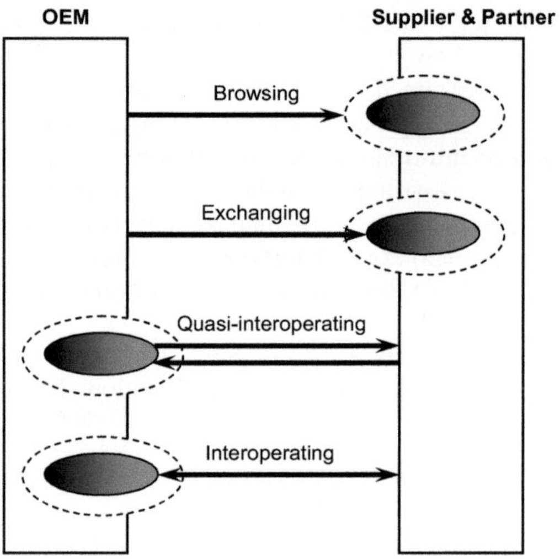

Figure 4.7. Four types of know-how interactions between OEM and supplier

4.5 System Development

Taking a tool supplier as a case study, a web-based system called *CyberStamping* has been developed to realise collaborative supplier integration for automotive product design and development [4.2]. It can act as a supporting environment and an interface between an automotive OEM and associated tool suppliers. Java™ is used as the main programming language for this system. The main interface of CyberStamping is shown in Figure 4.8, and the CyberStamping system can provide the following fundamental functions:

- *Partnership chain definition.* The relationship between automotive OEM and various tool suppliers can be defined through the partnership chain.
- *Tool supplier selection.* The partner selection module can assist the automotive OEM to evaluate and select appropriate tool suppliers.
- *Know-how sharing.* The know-how sharing facilitates know-how and data exchange between automotive OEM and tool suppliers.
- *Utility facilities.* Utility facilities provide supporting functions to the CyberStamping environment, such as user login management, on-line discussion, web site administration, *etc.* The administration function is to manage the user's registration. Both automotive OEM engineers and die suppliers enter into CyberStamping as users. The supporting tools provide some functions such as online discussion, messaging, as well as file downloading and uploading.

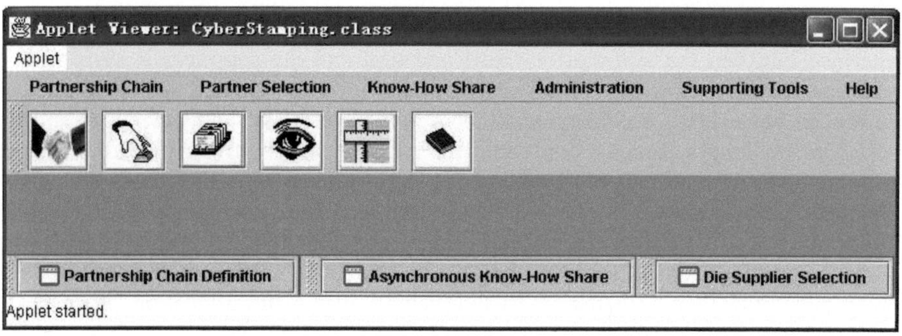

Figure 4.8. The main interface of CyberStamping

Partnership is to define the interface between automotive OEM and the selected tool and die suppliers. Generally, the organisation interface between automotive OEM and tool and die suppliers is mainly modelled based on the product's BOM (bill of materials). The BOM-oriented organisation interface is characterised by a 1-to-*n* relationship (Figure 4.9(a)). This means in detail that, according to the product development, the automotive OEM derives the tool requirements later on, as soon as they are necessary for production. Afterwards, the tool and die suppliers (or tool makers as shown in Figure 4.9) are assigned with appropriate tool orders. Investigations, however, reveal some limitations of the BOM-oriented interface between OEM and tool and die suppliers. For example, in this context, the burden of the automotive OEM increases with the number of tool and die suppliers, because the entire responsibility and execution of the technical and organisational tool project co-ordination lies with the automotive OEM.

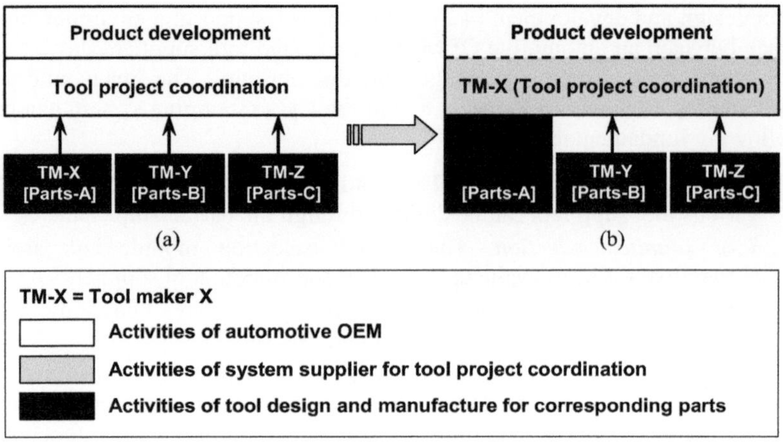

Figure 4.9. Two different interfaces between automotive OEM and tool and die suppliers

As presented in the introduction, the OEM, particularly in the automotive industry, aims at reduction of the number of tool and die suppliers and of their own expenditure on co-ordination of tool procurement. To reduce the complexity of co-ordination and the variety of interfaces, the traditional BOM-based organisation interface between the automotive OEM and tool and die suppliers is shifting to a new one (Figure 4.9(b)). Few capable and effective tool and die suppliers are chosen to have direct connections to the automotive OEM, and other tool and die suppliers no longer directly communicate with the automotive OEM, but instead with an "intermediate" system supplier that works closely with the automotive OEM and deals with the task of tool project co-ordination. The system supplier plays more important roles, is responsible for considering the allocation of orders to other tool and die suppliers, and is faced with the requirement to co-design relevant parts of the product together with the OEM engineers. During the allocation process, the tools for core components/parts are manufactured by the system supplier, and the single or individual tools for other parts are allocated to the other individual tool and die suppliers, called part suppliers.

To conclude, the automotive OEM is aiming at a reduction of the number of tool suppliers and of their own expenditure on co-ordination of tool procurement. In this context, few system tool suppliers have a direct connection to the automotive OEM, and are responsible for considering the allocation of tool orders to part suppliers who are actually second-tier suppliers. This kind of multi-level partnership chain brings two substantial modifications into the relationship of the automotive OEM and tool makers. First, the number of direct tool suppliers can be significantly reduced. Second, the co-ordination of tool development is shifted from the product manufacturer to the system supplier. For example, in the generation of a tool and die system for a car body panel (Figure 4.10), the system supplier always allocates itself to the production of large dies for the body panels that are the large core items. Tools and dies for small interior parts are assigned to other part suppliers.

In CyberStamping, the know-how sharing can be implemented in two ways. One way is to provide the automotive product designer with existing knowledge that is

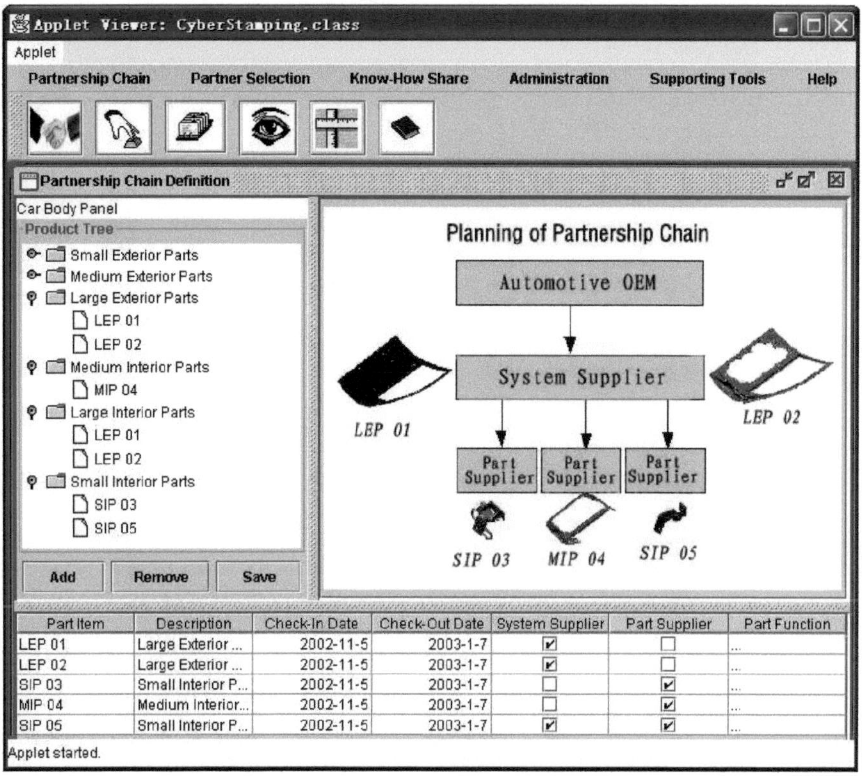

Figure 4.10. An example of partnership chain definition

furnished by tool suppliers, and know-how sharing is implemented through the Web/Internet, asynchronously. In asynchronous know-how sharing, know-how of the die supplier is provided in web-based knowledge bases that can be rules, case studies, explanations, references, *etc*. The authors have developed web-based design services (concept design and evaluation, material selection, cost evaluation, part shape evaluation, process design and evaluation, die sets selection and time evaluation) to carry out necessary analysis at appropriate stages of the stamping product development process [4.19] (Figure 4.11). The knowledge base from the die supplier equips the Web-based services to handle various design problems. On one hand, these Web-based design services can provide the product designer with all opinions and consequences for cost, process and die design as part geometry evolves. On the other hand, based on these design services, the die supplier can integrate its know-how into the product development process. The other way is to provide an on-line communication environment for message and data exchange between the product designer and tool supplier, and know-how sharing is implemented synchronously.

Know-how mining, searching, and navigating are performed based on the knowledge meta-data, that is, high-level information about a knowledge object. By adding keywords to the knowledge meta-data, a search of the specific knowledge

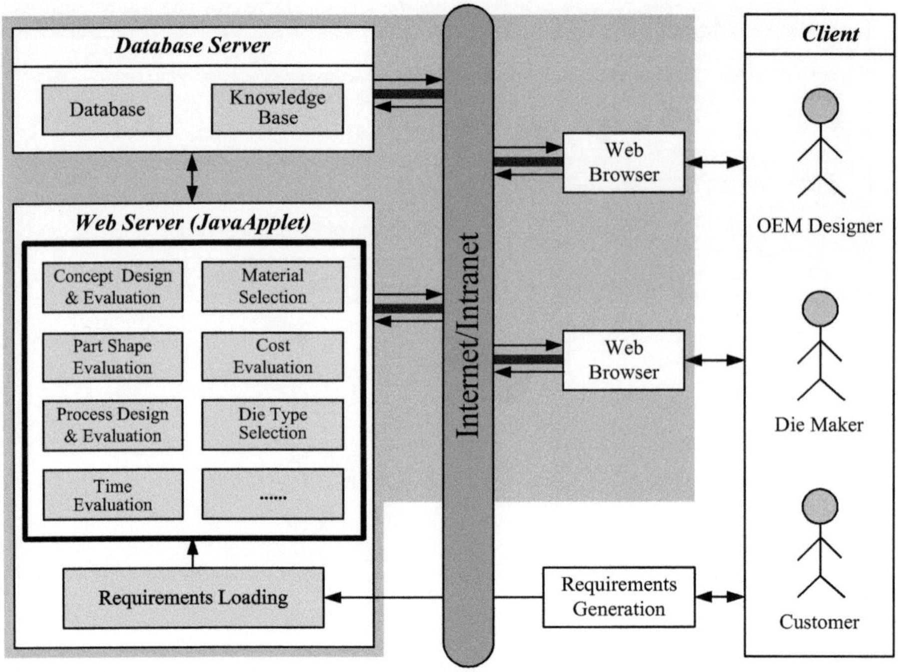

Figure 4.11. Web-based design services for know-how sharing

area can be performed. The meta-data enables the user to move quickly in the knowledge database and to browse through keywords. Figure 4.12 shows an example of the knowledge about part stampability evaluation. The knowledge meta-data is the feature "bending". According to the feature classification and feature characteristics, the part designer can navigate the rules stored in the Web-based knowledge bases for bending part shape analysis, which can help to decide the configuration and some basic specifications of a sheet metal part.

In CyberStamping, synchronous know-how sharing between the product designer and die supplier is performed by an agent communication platform, which is based on Java Agent Template Lite (JATLite [4.20]). JATLite provides a basic infrastructure in which agents register with AMR (agent message route) using a name and password, transfer files, and invokes other programs or actions on the various computers where they are running. Also it is in charge of wrapping existing part and die design programs by providing them a front-end that allows automatic communication with other programs. In this research, JATLite facilitates the constructions of agents that send and receive KQML/XML messages. A KQML (knowledge query and manipulation language) message including XML (eXtensible Markup Language) contents from an agent is passed to the main application of the receiving agent that performs the required operations according to the content of the message. As the XML format is the data exchange standard on the web and KQML is a language for agent communication, the message receiving agent can interpret the XML-based message content and use it as readable data for its application system directly,

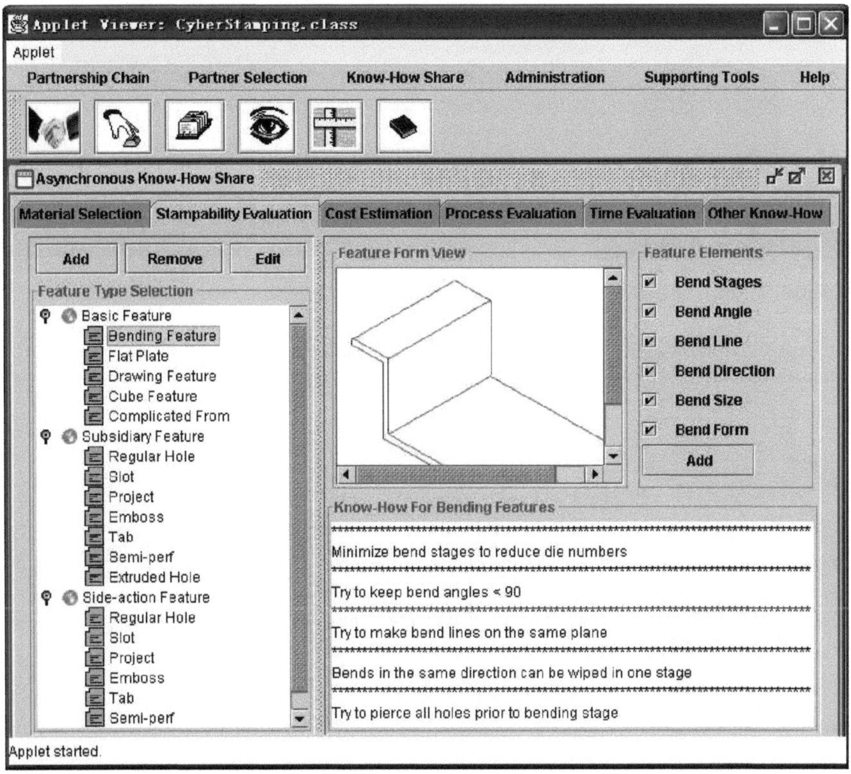

Figure 4.12. An interface of Web-based know-how sharing

minimising data communication errors. Meanwhile, the product model can be viewed in VRML (Virtual Reality Modelling Language) version, and the product data exchange is via STEP (STandard for the Exchange of Product model data).

The OEM product designer, as an agent, uses the part features as design units; and the die supplier, as another agent, uses the features as evaluation units. Through feature templates, the designer can define the part features and generate an XML document to describe the part features. The KQML/XML message is then sent to the die supplier to perform part design evaluation. Through the KQML/XML parser, the die supplier agent can read and parse the part feature information from XML documents. Figure 4.13 illustrates an example showing how this scenario proceeds. Through the JATLite facilitator, the part designer agent sends the XML-based part feature information and the VRML-based part model to a die supplier for the evaluation request at different design stages. At the conceptual design stage, a conceptual model is configured to generate part concept and material selection, and die makers can help to identify the best suited material according to the sheet metal properties (or forming qualities) and cost requirements. At the embodiment design stage, the die cost, part stampability and stock utilisation of a preliminary concept sketch can be evaluated and fed back by die makers. Cost evaluation at the embodiment design stage can be performed on a part sketch that focuses

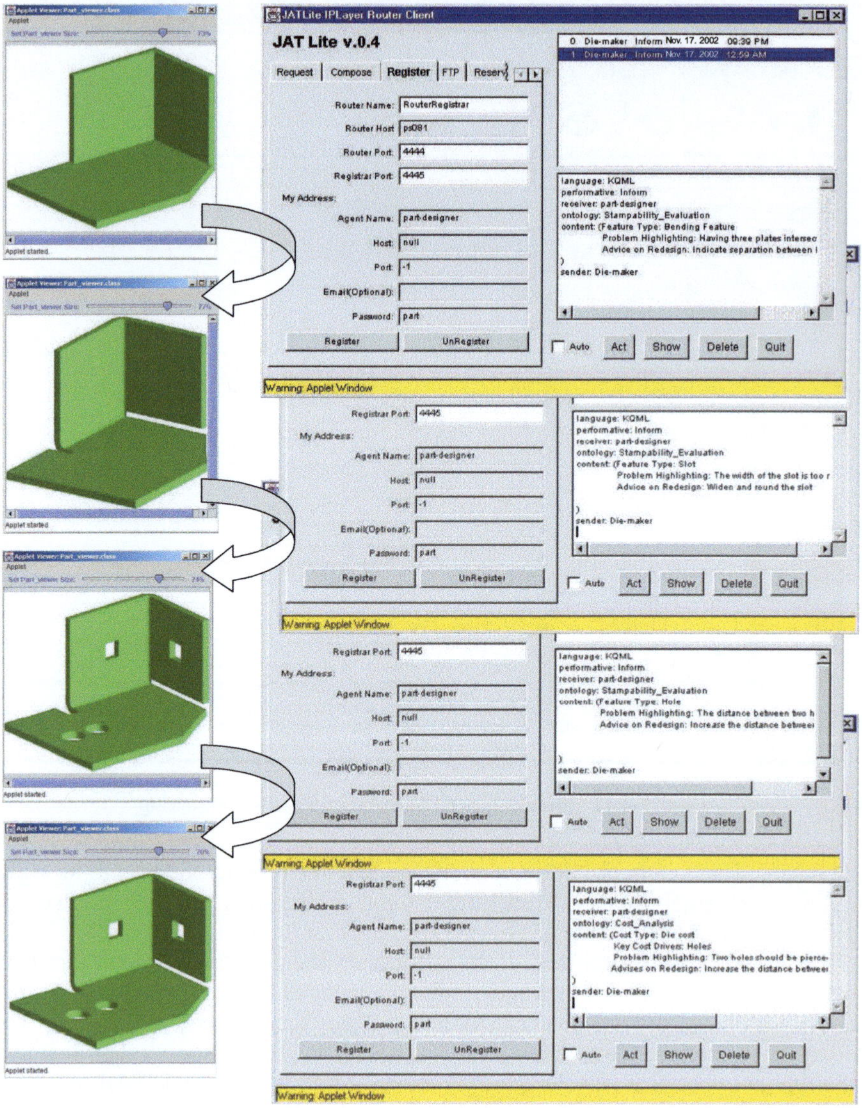

Figure 4.13. A case of synchronous know-how sharing

preliminarily on configuration level, not parametric detail. At the next step, detailed design deals with part shape refinement, dimensions and tolerances specification. From the view of the part designer, the part geometry consists of features that meet the part functional requirements. For example, holes are used for a variety of purposes, such as to assemble with fasteners, to guide or align other components, and to reduce weight. From the view of the die maker, stamping part design should not only be functionally acceptable but also be compatible with the selected stamping process, and can also achieve good stampability, lower cost, shorter lead times and higher quality. To achieve these objectives, all related factors such as die types, number of dies and die manufacturing cost can be considered by the die maker based on feature attributes such as feature form complexity, size, tolerance, *etc*. Any design flaw detected in the design process is notified from the die-maker agent to the part design agent by the facilitator for design modification. The communication windows to the right of Figure 4.13 show the messages from the die-maker to the part designer at different stages.

4.6 Conclusions

It is widely acknowledged that the OEM industry is more than ever obliged to improve its development strategy according to the increasing pressure of product innovation and complexity, the emergence of new technology, the changing market demands and increasing level of customer awareness. As the OEM has begun to adopt supplier integration into the development process, new challenges will arise to support the collaboration and partnership management between the OEM and the supplier. The main contributions of this research are as follows.

- Regarding the depth of collaboration, the integration of a supplier into the OEM process chain has been defined in two ways, quasi-supplier integration and full supplier integration. A strategy to select the appropriate method of supplier integration has been proposed.
- To reduce expenditures for partnership management and co-ordination, it is suggested that suppliers are divided into system suppliers and part suppliers.
- A web-based system *CyberStamping* has been developed to enable the involvement of die suppliers in the automotive OEM development process.

Meanwhile, the following lessons for the automotive industries are learned within the CyberStamping implementation:

- Supplier integration into the OEM's value chain will be the crucial factor of new product development success.
- The collaboration situation cannot be realised by the supplier's effort alone. Instead, a reorientation on the side of OEM is necessary.
- The earlier the integration of the supplier into the OEM process chain is supposed to happen, the more complex the reorientation process will be.

Acknowledgement

This research was supported by NSFC (Natural Science Foundation of China) research grants under projects no. 50505017 and no. 50775111.

References

[4.1] http://www.cranfield.ac.uk/sims/cim/people/people_frames/ip_shing_fan_frame.htm.

[4.2] Tang, D., Eversheim, W., Schuh, G. and Chin, K.-S., 2004, "CyberStamping: a Web-based environment for cooperative and integrated stamping product development," *International Journal of Computer Integrated Manufacturing*, **20**(6), pp. 504–519.

[4.3] Huang, G.Q. and Mak, K.L., 2000, "Modelling the customer-supplier interface over the world-wide web to facilitate early supplier involvement in the new product development," *Proceedings of IMechE, Part B, Journal of Engineering Manufacture*, **214**, pp. 759–769.

[4.4] Huang, G.Q. and Mak, K.L., 2000, "WeBid, a web-based framework to support early supplier involvement in new product development," *Robotics and Computer Integrated Manufacturing*, **16**, pp. 169–179.

[4.5] Peter, M., 1996, *Early Supplier Involvement (ESI) in Product Development*, PhD Dissertation, der University St. Gallen, Switzerland.

[4.6] Tang, D., Eversheim, W. and Schuh, G., 2004, "A new generation of cooperative development paradigm in the tool and die making branch: strategy and technology," *Robotics and Computer-Integrated Manufacturing*, **20**, pp. 301–311.

[4.7] Lyu, J. and Chang, L.Y., 2007, "Early involvement in the design chain – a case study from the computer industry," *Production Planning & Control*, **18**(3), pp. 172–179.

[4.8] Jayaram, J., 2008, "Supplier involvement in new product development projects: dimensionality and contingency effects," *International Journal of Production Research*, **46**(13), pp. 3717–3735.

[4.9] Chang, S.-C., Chen, R.-H., Lin, R.-J., Tien, S.-W. and Sheu, C., 2006, "Supplier involvement and manufacturing flexibility," *Technovation*, **26**(10), pp. 1136–1146.

[4.10] McIvor, R., Humphreys, P. and Cadden, T., 2006, "Supplier involvement in product development in the electronics industry: a case study," *Journal of Engineering and Technology Management*, **23**(4), pp. 374–397.

[4.11] Tang, D.B. and Qian, X.M., 2008, "Product lifecycle management for automotive development focusing on supplier integration," *Computers in Industry*, **59**(2–3), pp. 288–295.

[4.12] Song, M. and Di Benedetto, C.A., 2008, "Supplier's involvement and success of radical new product development in new ventures," *Journal of Operations Management*, **26**(1), pp. 1–22.

[4.13] Vairaktarakis, G.L. and Hosseini, J.C., 2008, "Forming partnerships in a virtual factory," *Annals of Operations Research*, **161**(1), pp. 87–102.

[4.14] Prasad, B., 1997, "Re-engineering life-cycle management of products to achieve global success in the changing marketplace," *Industrial Management & Data Systems*, **97**, pp. 90–98.

[4.15] Gerth, R.J., 2002, "The US and Japanese automotive die industry – some comparative observations," *Aachener Werkzeug – und Formenbau Kolloquium*, 1–2 October, 2002, Aachen, Germany.

[4.16] Sapuan, S.M., Osman, M.R. and Nukman, Y., 2006, "State of the art of the concurrent engineering technique in the automotive industry," *Journal of Engineering Design*, **17**(2), pp. 143–157.

[4.17] Sharma, A., 2005, "Collaborative product innovation: integrating elements of CPI via PLM framework," *Computer-aided Design*, **37**, pp. 1425–1434.

[4.18] Tang, D., Eversheim, W., Schuh, G. and Chin, K.-S., 2003, "Concurrent metal stamping part and die development system," *Proceedings of IMechE, Part B, Journal of Engineering Manufacture*, **217**, pp. 805–825.

[4.19] Chin, K.-S. and Tang, D., 2002, "Web-based concurrent stamping part and die development," *Concurrent Engineering: Research & Applications*, **10**(3), pp. 213–228.

[4.20] http://java.stanford.edu.

5

Reconfigurable Manufacturing Systems Design for a Contract Manufacturer Using a Co-operative Co-evolutionary Multi-agent Approach

Nathan Young and Mervyn Fathianathan

Georgia Institute of Technology
210 Technology Circle, Savannah, GA 31407, USA
Emails: nyoung6@mail.gatech.edu, mervyn.fathianathan@me.gatech.edu

Abstract
This chapter presents a method for designing the structure of reconfigurable manufacturing systems for a contract manufacturer based on the use of co-operative co-evolutionary agents. The aim is to determine the structure of a reconfigurable manufacturing system that can be converted from one configuration to another to manufacture the different products of the customers of the contract manufacturer. The approach involves multiple computational agents where each agent is allocated to each partner for whom the contract manufacturer is to manufacture parts. Each agent synthesises a machine configuration for manufacturing the part for the company it is allocated to achieve minimum machining cost and at the same time co-operates with other agents to co-evolve the structure of the machine to minimise reconfiguration cost between different machine configurations.

5.1 Introduction

Faced with a rapidly changing global environment, product development companies today are reformulating their strategies to be globally competitive. One strategy that companies have adopted is to concentrate on their core competencies and outsource non-core activities to appropriate partners. In the manufacturing industry, this has resulted in the emergence of contract manufacturers whose main role is to manufacture products for their partners. An example of a contract manufacturer is Flextronics International Ltd, which produces Microsoft's Xbox game machine, cell phones for Ericsson, routers for Cisco and printers for HP among other products [5.1]. Contract manufacturers deal with multiple companies concurrently and are faced with the task of designing their facilities and planning appropriate schedules to meet the needs of the different companies. The aim of this chapter is to present a method for designing the structure of a reconfigurable manufacturing system of a contract manufacturer to meet the demands of their customers.

Contract manufacturers face a highly uncertain environment due to changing demands of current customers as well as incoming orders from new customers. To maintain competitiveness in this uncertain environment, contract manufacturers need to possess agility to dynamically and effectively adapt to the changing environment. Agility is necessary at various levels from the business processes to the manufacturing systems. A leap forward in achieving agility at the manufacturing systems level is the concept of reconfigurable manufacturing systems [5.2]. Koren *et al.* [5.2] defined a reconfigurable manufacturing system (RMS) to be one that is "designed at the outset for rapid change in structure… in order to quickly adjust production capacity and functionality." This implies that RMSs should not only possess the necessary flexibility to manufacture a large variety of parts, but also be able to achieve high throughput. The authors distinguish between RMSs and flexible manufacturing systems, stating that flexible manufacturing systems (FMS) are based on general purpose computer numerical control (CNC) machines. The general purpose nature of the CNC machines in FMSs is attributed to the fact that it is not designed around parts to be manufactured, resulting in their high costs. RMSs address this problem by being designed around a group of parts. Therefore, an RMS may be optimised for cost effectiveness for a certain variety of parts.

The question that then arises is how the structure of an RMS should be designed. Different ranges of product varieties would imply different appropriate RMS structures. Further, the range of products that a contract manufacturer manufactures evolves and hence, the structure of the RMS should also evolve accordingly. To meet these needs, a general and flexible method for designing the structure of an RMS is necessary. In this chapter, we present a method for designing the structure of an RMS for a contract manufacturer based on the use of co-operative co-evolutionary agents. The aim is to determine the structure of an RMS that can be converted from one configuration to another to manufacture the different products of the customers of the contract manufacturer. The approach involves multiple computational agents where each agent is allocated to each partner for whom the contract manufacturer is to manufacture parts for. Each agent synthesises a machine configuration for manufacturing the part for the company it is allocated to achieve minimum machining cost and at the same time co-operates with other agents to co-evolve the structure of the machine to minimise reconfiguration cost between different machine configurations.

The rest of this chapter is organised as follows. Section 5.2 reviews the literature in design methods for RMSs. Section 5.3 presents an overview of the co-operative co-evolutionary multi-agent approach to RMS design. Section 5.4 applies the method to the design of reconfigurable milling machines. Section 5.5 discusses a case example of applying the method and Section 5.6 concludes the chapter.

5.2 Related Research

Koren *et al.* [5.2] identified that one of the challenges in RMS is the development of a design methodology for RMS. A number of design methodologies for RMS have since been proposed in the literature.

Koren and Ulsoy [5.3] presented a sequence of steps for the design of RMS based on key characteristics of an RMS that they have defined. These characteristics include modularity, integratability, customisability, convertibility and diagnosability. The design methodology involves the justification of a need for RMS through lifecycle assessment. If the RMS is required, a method is proposed that first addresses system-level design concerns for product family demands followed by machine-level design. Machine-level design involves the identification of acceptable machine modules, configurations of modules, and process planning for that machine configuration. Moon and Kota [5.4] presented a systematic methodology for the design of reconfigurable machine tools that takes in a set of process requirements as inputs and generates a set of kinematically viable reconfigurable machine tools that meet the requirements. The key feature of the methodology is the use of screw theory based representations to transform machining tasks to machine tools for performing the machining tasks.

Zhao *et al.* [5.5–5.8] presented a theoretical, numerical framework for the determination of a configuration for an RMS. The framework was implemented as a stochastic optimisation process to identify an optimal configuration for a product family based upon the average expected profit. Spicer *et al.* [5.9] discussed the design of scalable RMS. Scalable manufacturing systems are systems that are able to satisfy changing capacity requirements efficiently through reconfiguration. They presented a method to determine the optimal number of modules to be included on a modular scalable manufacturing system. The design of scalable RMS was extended to include reconfiguration cost in [5.10]. Abdi and Labib [5.11] proposed the use of analytical hierarchical process (AHP) to configure products into families to facilitate the redesign strategy of manufacturing systems to support management choices. Chen *et al.* [5.12] presented a method for selecting an optimal set of modules necessary to form a reconfigurable machine tool for producing a part family. In their method, machining features are defined as the functional requirements and machine tool modules are defined as design parameters. The functional requirements are mapped to design parameters to define a selection space from which machine tool modules are selected.

In another design method, Youssef and ElMaraghy [5.13] organise a machining system into machines, machining clusters, and operational cluster setups. In this case, machining clusters represent sets of machines that are combined into operation clusters. To account for convertibility, the authors introduce a new metric referred to as reconfiguration smoothness. The smoothness metric is used to evaluate configuration closeness based upon cost, time, and the effort required to convert between operating cluster configurations. With this metric, the authors use a genetic algorithm (GA) to identify the feasible enterprise configuration that yields a minimal capital cost for the configuration at the current time for a given workpiece. A more comprehensive literature review on current design methods for RMS can be found in [5.14].

The related research review presented a number of methods for designing the structure of a manufacturing system such that it can be reconfigured and adapted to changing conditions. These methods include general and detailed approaches for designing reconfigurable structures. Metrics for measuring convertibility between manufacturing machine configurations have also been proposed. Although all of the proposed methods for designing the structure of reconfigurable systems are viable, an area that has not been sufficiently addressed is a method for the design of evolving machine structures. As mentioned earlier, RMSs evolve as the range of products to be manufactured evolves. Therefore, *there is a need for a method for designing the evolving structure of reconfigurable manufacturing machines.* A design method for evolving machine structures based on co-operative co-evolutionary agents is presented in this chapter.

5.3 Co-operative Co-evolutionary Multi-agent Approach to Reconfigurable Manufacturing Systems Design

In this section, an overview of the design method for RMS is presented. For simplicity, the discussion is based on the structure of a single machine, *i.e.* the factory comprises of a single machine that has to be reconfigured to manufacture different parts. The RMS design problem can be defined as follows: given a range of parts to be manufactured, design the structure of a reconfigurable manufacturing machine that will minimise the costs of manufacturing the parts and the costs to reconfigure the machine from one configuration to another. The aim is to identify the structure of a single machine that will be used to manufacture a variety of parts. The machine will be reconfigured appropriately to manufacture the different parts.

The co-operative co-evolutionary approach is implemented based on multiple co-evolving computational agents. In this approach, each agent synthesises the configuration of a machine for a part it is to manufacture and co-operates with other agents that are synthesising machine configurations for other parts to reduce reconfiguration cost in converting from one configuration to another. The co-evolution between multiple agents in the synthesis of the machine structure is shown in Figure 5.1. A single agent synthesises a manufacturing machine configuration from a set of components defined in a component base. The component base contains the necessary components to construct a manufacturing machine. Each agent synthesises a machine based on evolutionary algorithms. The input to the algorithm is a manufacturing order from one of the contract manufacturer's customers. The manufacturing order is characterised by the design of the part in terms of features to be manufactured, material of the part and the batch size. The objective function of the evolutionary algorithm is to minimise costs to manufacture the part and reconfiguration cost. Reconfiguration cost is based on the other agents and will be discussed in the following paragraph. The output of the evolutionary algorithm of each agent is the configuration of the manufacturing machine for the part it is allocated.

All agents synthesise machine configurations in parallel from the same component base. At each iteration cycle of the evolutionary algorithm, each agent calculates the cost of reconfiguring from its current configuration to the current configurations of all the other synthesised machines. This reconfiguration cost is used to update the fitness function of each agent. The inclusion of the reconfiguration cost would then alter the fitness of the current set of synthesised manufacturing machine configurations. This process is depicted in Figure 5.1 as the exchange of information on the best synthesised machine configurations. This can be seen as the vertical arrows depicting information being sent on configuration *A* from agent *A* to agent *B* and information on configuration *B* being sent to agent *A*. This exchange of information is carried out between all agents, *i.e.* agent *A* receives and sends information to all other agents. Accordingly, each agent then updates the evolutionary algorithm and synthesises a new manufacturing machine.

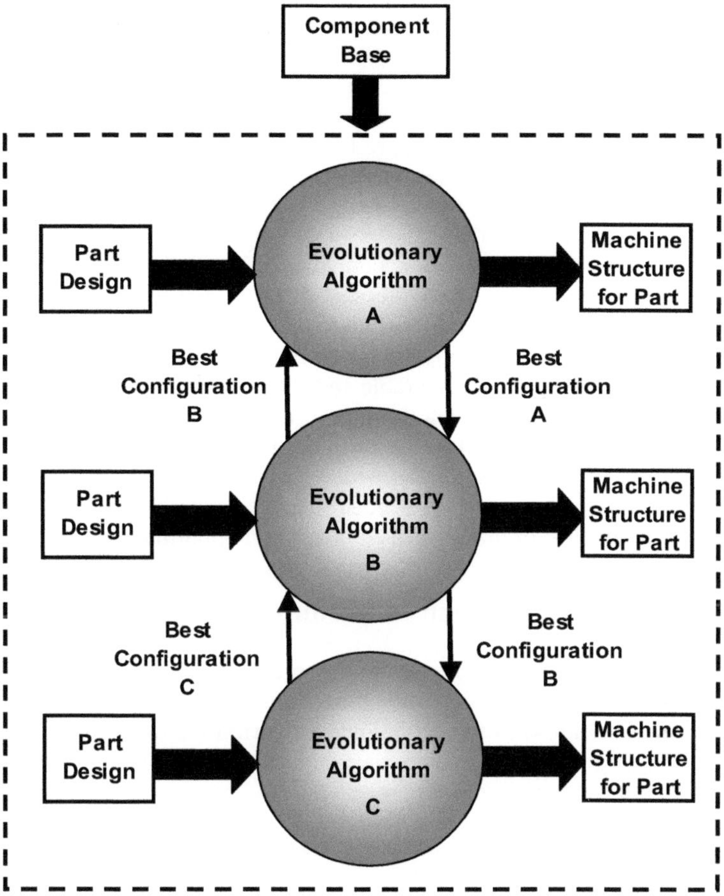

Figure 5.1. Co-evolving agent design synthesis

The co-evolution of agents is continued until a termination criterion is fulfilled. The co-evolutionary algorithm can be terminated when a certain acceptable fitness value is reached for each agent. At the end of the co-evolutionary algorithm, a manufacturing machine configuration would have been identified for each part from the customers. The synthesised machines would have been designed accounting for the tradeoffs between reconfiguring between configurations and the minimisation of a single part cost. The final manufacturing machine would then reveal the components necessary for constructing each machine configuration. Upon removal of duplicated components between the various machine configurations, the necessary structure for the reconfigurable manufacturing machine can be identified. This structure represents a set of components that will be used to construct the manufacturing machine when a change in the part design or quantity occurs.

The co-evolutionary multi-agent design approach to reconfigurable manufacturing machine design has several salient features. First, the approach allows for convertibility by minimising the reconfiguration setup time between the machine configurations. In addition, the agent-based architecture of the algorithm allows reconfigurable manufacturing machines to be synthesised accounting for changes in the range of products that are to be manufactured by a company. For example, agents can be added or deleted depending on the evolving nature of the customers of the contract manufacturer. The structure of the machine can then be altered according to market demands, creating an evolving factory.

From these salient features, the co-evolutionary multi-agent-based approach has one primary characteristic that makes it unique among the other proposed reconfigurable system design methods: co-evolution. Incorporating co-evolution into the approach provides a means to synthesise machine configurations for adaptation to a changing product demand and evolving machine structure. Hence, the machine structure adapts as the product range evolves; thus, fulfilling the need for designing the changing structure of reconfigurable machines within an evolving factory. A contract manufacturer can then plan the appropriate structure necessary to machine a specific product range. Hence, the enterprise has the capability to adapt its factory's architecture to uncertainties associated with a dynamic and volatile market demand. These uncertainties include various product changes such as geometry or demand, economic developments, or unforeseen collaboration opportunities.

5.4 Application of Approach to Reconfigurable Milling Machines

In this section, the automated co-evolutionary multi-agent design method is applied to the design of reconfigurable milling machines (RMMs). In this section, the representation of solutions in the evolutionary algorithm is first presented. Next, the evaluation of solutions in the evolutionary algorithm is discussed. Finally, the co-evolutionary algorithm is presented.

5.4.1 Solution Representation

The solution representation is based on establishing a hierarchy of RMM components as shown in Figure 5.2. An RMM comprises of a single base structure and multiple possible columns to which functional units are attached. Two types of functional units can be connected to a base and column structure: (a) tool holding and movement unit and (b) work holding and movement unit.

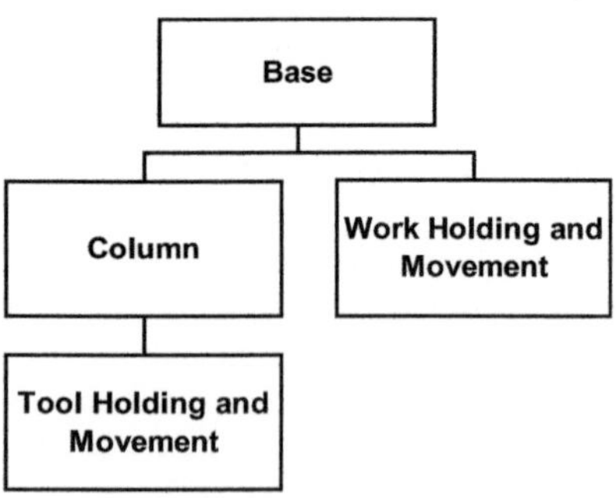

Figure 5.2. Reconfigurable milling machine representation

A tool holding and movement functional unit is an assembly of components that provide translation and rotation to the tool. This assembly of components is comprised of a spindle, drive motor, lead screw, indexing motor, and tool, as discussed in the following:

- Spindle – transmits rotation from the drive motor to the tool.
- Drive motor – provides rotation to the spindle.
- Lead screw and indexing motor – provides z axis translation to the tooling.

A work holding and movement functional unit is an assembly of components that provide support and translation to the workpiece. This assembly is comprised of a table, fixture, two lead screws, and two indexing motors, discussed as follows:

- Table – transmits motion to the workpiece.
- Fixture – locates, restrains and supports the workpiece during cutting.
- Two lead screws and indexing motors – provide x and y axis motions to the workpiece.

Each solution is generated through the creation of a base and random numbers of columns. Random numbers of tool holding and movement and work holding and movement units are attached to the columns and base, respectively. An example solution representation is shown in Figure 5.3. This example depicts a machine that has two columns with one tool holding and movement unit on one column and two tool holding and movement units on another column. The three total tool holding units concurrently machine products located on the two work holding and movement units. Hence, one work holding unit supports a part that is machined by two tool holding units.

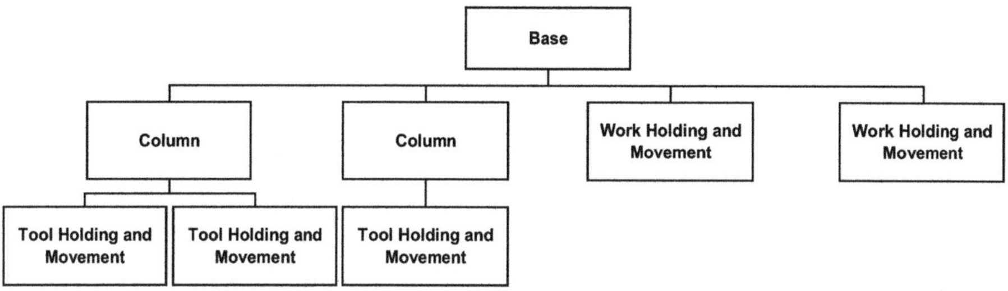

Figure 5.3. Example machine representation

5.4.2 Solution Evaluation

In the developed algorithm, the fitness function includes metrics to evaluate the machining, capital, and reconfiguration cost between machine configurations. The quality of solution is evaluated by simulating the behaviour of each milling machine configuration within the co-evolutionary design algorithm. Overall, the algorithm uses the behaviour of the synthesised machines to simulate the cutting time and a comparison between system configurations to assign a fitness value. The fitness function is discussed below.

Fitness Function

The fitness function is the average cost per part and is formulated as follows:

$$F = \frac{C_B + C_R}{B_S} + C_C \tag{5.1}$$

where C_B, C_R, C_C, and B_S are the machining cost per batch ($), cost to reconfigure ($), capital cost per piece ($) and the batch size, respectively. The details and assumptions associated with calculating these values are as follows.

Machining Cost

Determining the machining cost involves two calculations: (a) batch processing time and (b) the manufacturing cost per batch. Batch processing time calculation involves three sub-routines that include (a) inputting variables, (b) a machinable feature check, and (c) an iterative calculation of cutting time. The three sub-routines are sequentially carried out to determine the cutting time.

To begin the process, the input variables must be supplied. In this implementation, the workpiece features are classified into three surface types: (a) flat, (b) cylindrical (only internal is considered), and (c) irregular. Features are further classified into specific geometries such as flats, holes, and slots for end mill type cutters and t-slots and dove tails for face mill type cutters. The feature dimensions are modelled using a bounding box. For instance, a cylinder with a diameter of one inch and a length of 2 inches would be contained within a bounding box 1 inch by 1 inch by 2 inches. The bounding box is an overestimate, but appropriate for the level of fidelity in this model. After the features are specified, the batch size is supplied.

The second sub-routine is a determination of the machinable features of a workpiece. This sub-routine is accomplished by scanning the geometry and type of every milling tool generated in a solution through the evolutionary algorithm. If all of the features can be machined, the machine is deemed feasible. If all of the features cannot be machined, the machine is deemed infeasible and the number of machinable features is stored. Due to the machine infeasibility, a penalty function is instantiated on the final cutting time.

The final sub-routine in the cut time calculation is an iterative process as discussed in the following.

- First, a counter, initially assigned as zero, is checked for equality to the specified batch size. If this check is satisfied, the cutting time is reported; else the process enters the second step.
- The machinable workpiece features on each fixture are scanned and compared with the current milling tool. If a feature and milling tool are matched, then the algorithm enters step three else the milling tool count is incremented until an appropriate milling tool is located. Once a milling tool is located, the cutting time is estimated that is explained in the next step.
- To obtain the cutting time, another series of calculations are required that include estimating the spindle angular velocity, table feed rate, length of

approach (LOA), length of over travel (LOT), single pass cut time, cut depth, number of required passes, and a cut time incorporating all of the passes required to machine a feature. This series of calculations is performed as follows. The spindle RPM (rotations per minute) is used to determine the table feed rate and is given by [5.15]

$$N_S = \frac{12V}{\pi D} \tag{5.2}$$

where V and D represent the cutting speed and cutter diameter, both of which are defined by the selected cutting tool. In this case, the cutting tool diameter is assumed to be in a range of acceptable diameters from 1/16 to 2 inches in increments of 1/16 inches. All of the data used in the calculations are taken from [5.15]. From the spindle RPM, the table feed rate can be calculated. The feed rate is required to determine the machining operation cut time. The feed rate (inches per minute) of the table is given by [5.15]

$$f_m = f_t N_S n \tag{5.3}$$

where f_t, N_S, and n represent the feed per tooth (inches per tooth), angular velocity of the spindle (RPM), and the number of teeth on the cutter, respectively. After the table feed rate is calculated, the LOA and LOT must be calculated to determine single pass cut time. These geometric features of the cutting tool are different for various classes of milling tools such as vertical or horizontal milling tools. In this application, only vertical milling machines are synthesised; therefore, horizontal milling tools are not considered. For vertical milling tools such as end or face mills, these cutting lengths may be calculated by the following [5.15]:

$$\begin{aligned} L_A = L_O &= \sqrt{W(D-W)} \quad \text{for } W < \frac{D}{2} \\ L_A = L_O &= \frac{D}{2} \quad \text{for } W \geq \frac{D}{2} \end{aligned} \tag{5.4}$$

where L_A, L_O, W, and D represent the LOA (inches), LOT (inches), width of cut (inches), and cutter diameter (inches), respectively. Once the spindle angular velocity, feed rate, LOA, and LOT are known, the single pass time for a feature is calculated. A single pass represents exactly one cut on a feature at the specified length of cut. This valuc may bc calculated in the following manner [5.15]:

$$T_m = \frac{L + L_A + L_O}{f_m} \tag{5.5}$$

where T_m, L, L_A, and L_O represent the cutting time (minutes), length of cut (inches), LOA (inches), and LOT (inches), respectively. After the cutting time for a single pass is calculated, the maximum depth of cut must be calculated to determine the number of passes required to machine a feature. The cut depth is determined as follows [5.15]:

$$C_{D,\max} = \frac{\mu \cdot eff}{\mu_S f_m DOI} \tag{5.6}$$

where μ, μ_s, eff, f_m, and DOI represent the spindle drive motor output (hp), unit power (hp-min/cubic inch), efficiency of the spindle drive motor, table feed rate (ipm), and depth of immersion (inches), respectively. To determine these parameters, several assumptions are required. The motor output and efficiency are assumed to be 5 hp at 80% efficiency. Each milling tool is assumed to have a depth of immersion of 1.25 inches. The unit power represents the required power needed at the spindle to remove a cubic inch of material. From this cut depth calculation, the required amount of passes for a cut can be calculated. The number of passes is determined by the following operation:

$$n_P = \left(\frac{f_d}{C_{D,\max}}\right)\left(\frac{f_w}{C_{W,\max}}\right) \tag{5.7}$$

where n_P, f_d, f_w, $C_{D,\max}$, $C_{W,\max}$ represent the number of passes, feature depth (inches), feature width (inches), max cut depth (inches), and max cut width (inches), respectively. With the number of passes, the total cut time may be calculated by multiplying the single pass cut time by the number of passes to arrive at a final estimation for the time required to machine a feature. Once the total feature cut time is calculated, the value is stored with reference to the cutting milling tool. Then steps two and three are repeated until the machinable features of a workpiece have been cut. To machine all the workpieces, steps one through three must be repeated until the batch size has been met and all workpieces have been machined with reference to their machinable features.

- The final step involves the report of the final cutting time for the milling operation. If the number of machinable features equals the total number of features, the milling tool with the most machining time is reported as the total time required to process the batch. Else, if the number of machinable features is less than the total number of features, a penalty function is used to punish the machine's lack of capability. This operation is described below. After the batch time has been estimated by the milling tool with the highest cutting time on the machine, the following equation is employed to punish infeasible machine configurations as they are undesirable, but necessary for searching the design space

$$B_T = 300B(n_T - n) \tag{5.8}$$

where B_T, B, n_T and n represent the final adjusted batch time (min), calculated batch time (min), number of total features, and number of machinable features, respectively. The multiplication of a constant value of 300 represents a multiplier determined through experiments to dilate the solution in the event of a machine with a single milling tool that can satisfy one feature extremely well. With the characterisation of a penalty function, the algorithm can consider these infeasible configurations for the possible good behaviours or components that they may provide. The next step in the solution evaluation includes estimation of the cost per part, which is discussed below.

After the cutting time for an entire batch is calculated, the total cost per batch may be estimated. This estimation involves the determination of machining and handling cost per piece. In the current model, tool life is neglected; therefore, tooling cost, and tool changing cost are ignored. The equations for the machining and handling cost are shown below

$$C_O = C \cdot n_S \tag{5.9}$$

$$C_1 = (B_T + \frac{B_S}{2})C_O \tag{5.10}$$

$$C_4 = \frac{B_S}{2} \tag{5.11}$$

$$C_B = C_1 + C_4 \tag{5.12}$$

where C_O, C_1, C_4, and C_B represent the total operating cost, machining cost, non-productive handling cost, and overall cost per batch, respectively. To solve the machining cost equations for the overall cost per batch, a few assumptions are made for the operating cost data. From [5.15], it is assumed that the operating cost per spindle (C) is 60 dollars per hour for a machine; therefore, an assumption is made that operating cost would total 1 dollar per minute per spindle. n_s represents the total number of spindles. Also, it is assumed that it requires one half of a minute to load and unload a single part. With this assumption, the machining cost (C_1) is estimated by multiplying the operating cost by the sum of the batch cutting time (B_T) and the time required to transfer the processed parts denoted by the batch size (B_S) divided by two. Furthermore, it is assumed that it takes approximately one half dollar per part for non-productive handling cost (C_4) [5.15]. Hence, it is possible to calculate the total cost per batch (C_B) by adding the machining cost and non-productive handling cost. With a final value for the total cost per batch, an estimation of the reconfiguration cost can be determined. This calculation is explained in the next section.

Reconfiguration Cost

To characterise the reconfiguration setup cost, the configuration of a machine is compared with the next configuration that is required. Through this comparison, a difference in machine components is revealed, which identifies the required structural adaptation of the machine. Once the difference in machine components is determined, assumptions are made to arrive at a final estimation for the cost required to reconfigure the machine. The equation for machine reconfiguration cost is shown below

$$d_T = |n_{AT} - n_{BT}| \tag{5.13}$$

$$d_C = |n_{AC} - n_{BC}| \tag{5.14}$$

$$d_S = |n_{AS} - n_{BS}| \tag{5.15}$$

$$C_R = C_W \left(t_T d_T + t_C d_C + t_S d_S\right) \tag{5.16}$$

In this equation, the reconfiguration cost is modelled as the absolute value of the difference between the number of components (tables, columns, and spindles) of two machine configurations denoted by d_T, d_C, and d_S. This represents the number of machine units that must be added or subtracted to reconfigure to the next machine configuration. Thus, a perfect reconfiguration would be that of zero, which would represent no additional setup.

To estimate the cost associated with setup time, assumptions are made for the time required for installation or disassembly (t_T, t_C, and t_S) and the average worker wage (C_W). The assumed time for installation or disassembly of a table, column, or spindle are two hours, three hours, and one hour, respectively. The assumed average worker wage is approximated as \$15 per hour. Hence, the total cost required to reconfigure a machine is estimated by the labour cost required to reconfigure the machine. After reconfiguration cost is estimated, the final variable in the fitness function can be calculated. This variable is the capital cost and is explained in the following section.

Capital Cost

The capital cost per piece is calculated by accounting for the number of different machine components (n), an assumed cost of each component (mc_n), the machine component types (mt_n), and an assumed number of processed workpieces over the entire lifecycle of the machine (L). The capital cost per piece is expressed as follows

$$C_C = \frac{\sum_{i=1}^{n} mc_n mt_n}{L} \tag{5.17}$$

where the sum of the component costs is divided by the total number of components processed by a machine over its entire lifecycle. The assumed machine component costs are shown in Table 5.1.

The lifecycle number of workpieces was assumed to be one million. This number represents the assumed number of workpieces that can be machined over the entire lifetime of a reconfigurable milling machine. With these assumptions, the total processing cost per piece can be calculated. The cost per batch is added to the

Table 5.1. Assumed machine component costs

Component	Cost ($ × 1000)
Table	5
Lead screw	5
Lead screw motor	10
Fixture	3
Column	3
Spindle	4
Spindle motor	15

reconfiguration cost and divided by the total batch size. The capital cost per part is then added to this value to arrive at the total processing cost per piece.

5.4.3 Synthesising Machine Architecture Using an Evolutionary Algorithm

In this section, the application of the co-evolutionary multi-agent algorithm to the design of the structure of a reconfigurable milling machine is described. The steps of the algorithm are discussed in the following sections.

Input Variables

For each agent, the input variables are a list of features to be machined on a part, list of dimensions associated with each feature, the batch size for the part, and the workpiece material. Due to the nature of machining, the input variables are directly linked to the milling tool and cutting time for the machine configuration. Features, dimensions, and material determine the type of tooling required for the operation. The batch size directly controls the amount of processing time that is required for an entire batch of parts.

Initial Population

An initial population of random machine configurations is generated from a base, columns, work holding, and tooling units. Each solution in the population is generated as follows:

- A machine is synthesised by first instantiating a base structure.
- Next, random numbers of columns and work holding and movement units are attached to the base structure.
- Finally, random numbers of tool holding and movement units are added to each column.

Evaluation and Termination Check

Each solution in the initial population is then evaluated according to the fitness function and assigned a fitness value. A check for termination criterion is then performed. The termination criterion is set to 100 generations. This criterion was determined by testing the algorithm to find the required number of generations to generate acceptable machine configurations. When the selected number of generations is reached, the algorithm terminates and reports the best machine configurations.

Selection and Reproduction

If the termination criterion is not met, a new population of solutions is then formed by means of two selection methods, tournament selection and elitist selection. Tournament selection is a proportionate selection method, *i.e.* solutions with a higher fitness have a greater probability of being selected to be part of the new population. In elitist selection, the parent and child population are combined and sorted. The fittest half of the population of configurations from the combined population is retained. To produce a new population, the selected group of solutions must be processed through evolutionary operators to introduce solution variation.

Evolutionary Operators

From this population pool, another new population of machine configurations is created through the use of evolutionary and topological operators. These operators are instantiated probabilistically. These operators include crossover and mutation.

Crossover is an operation involving an exchange of components between two candidate machines. In this design method, crossover has a 0.7 probability of being selected to generate new solutions. The crossover probability is sub-divided into thirds by the different types of crossover. These crossover types include exchanging work holding units, columns, and tool holding units. These types of crossover allow for the possible trade of structural components of two machine configurations. At the work holding level, crossover involves an exchange of the fixtures, tables, lead screws, and indexing motors between two candidate solutions. By allowing this trade, machines may increase or reduce their number of fixtures; thus, searching the design space with respect to an enhancement in parallel processing capability.

For column crossover, the candidate solutions trade entire assemblies of columns and tool holding units. This type of crossover grants the algorithm the ability to select entire tooling clusters to search for fitter configurations in larger increments through the search space. The larger increments in the search space provide more substantial additions or subtractions in terms of processing capability. This type of crossover is complemented by single tool holding unit crossover, which provides finer tuning of the configuration.

For tool holding unit crossover, the spindle, lead screw, spindle motors, and milling tools are traded between candidate solutions; thus, changing only one milling tool and introducing variation into the processing capability of a machine configuration. This operation is similar to the mutation operator that provides a random variation to the properties of a milling tool.

Mutation is the random selection of a new milling tool from the component base to replace a pre-existing milling tool. The mutation probability is set at 0.2. Hence, when mutation occurs, any milling tool may be selected from the component base to search the design space for a better component.

Another means of incorporating random change into the co-evolutionary algorithm is topological operators. Topological operators represent another form of mutation that changes the solution topology in various ways. In this implementation, a set of machine components may be added, deleted, or duplicated from a machine. Each topology operator has a 0.35 probability of occurrence. If deletion occurs, only work holding units, tool holding units, or columns may be removed. For addition, random work holding units, tool holding units, or columns may be added to expand the topology of a solution. This operation is similar to the duplication procedure that also grows the configuration topology. The duplication operator may clone either a column assembly or an individual tool holding unit. When the operator clones a column of a machine, the algorithm copies a column and its tool holding units from a random machine onto a candidate machine configuration. In a similar fashion, the operator can copy an individual tool holding unit and install it on the configuration. Duplication marks the final type of topological operator that may be instantiated to create a new solution population. To continue in the algorithm, the individuals in the population are evaluated.

Transmitting Co-evolutionary Information and Evaluation

At this point, the co-evolution mechanism is triggered. From each agent within the co-evolutionary network, information on the fittest configuration is sent to other agents within the group of co-evolving agents. Each agent then calculates the reconfiguration cost and updates the fitness values.

Termination Check

Finally, the termination criterion is checked again. If the criterion is not met, the algorithm is reiterated from the *Selection and Reproduction* step to the *Termination Check* step. The algorithm is finally terminated when the termination criterion is met. Upon termination, the algorithm reports each configuration for the products within the specified product range. From these machine configurations, an overall machine structure may be derived, which characterises the necessary level of flexibility required to satisfactorily machine the set of parts. Hence, the minimum number of machine components to meet the needs of the contract manufacturer is identified.

5.5 Case Example

In this section, the co-operative co-evolutionary multi-agent algorithm for reconfigurable milling machines is applied to an industrial case example. In this example, the contract manufacturer has orders for three different kinds of automotive wheels as shown in Figure 5.4. The wheels are each 16 inches in diameter and range from having five spokes to seven spokes resulting in a different number of features. The numbers of features on each wheel are 19, 21, and 23 from left to right as shown in Figure 5.4. To machine these components, the features that would be typically milled are considered. These features include the triangular shapes, slots, surfaces, and holes. An example of the features on a wheel with six spokes is shown in Figure 5.5. The material for all three wheels is aluminium and the required quantity is 5000 for all three wheels. The contract manufacturer would like to determine the structure of a manufacturing machine that could be used to manufacture these three parts. To design the machine structure, an agent is instantiated and allocated to each automotive wheel. Each agent then synthesises the configuration of a machine for each wheel and co-evolves with the other agents to reduce reconfiguration cost. The results of the algorithm are shown in Table 5.2.

(a) (b) (c)

Figure 5.4. Automotive wheel: (a) 5 spokes, (b) 6 spokes, and (c) 7 spokes

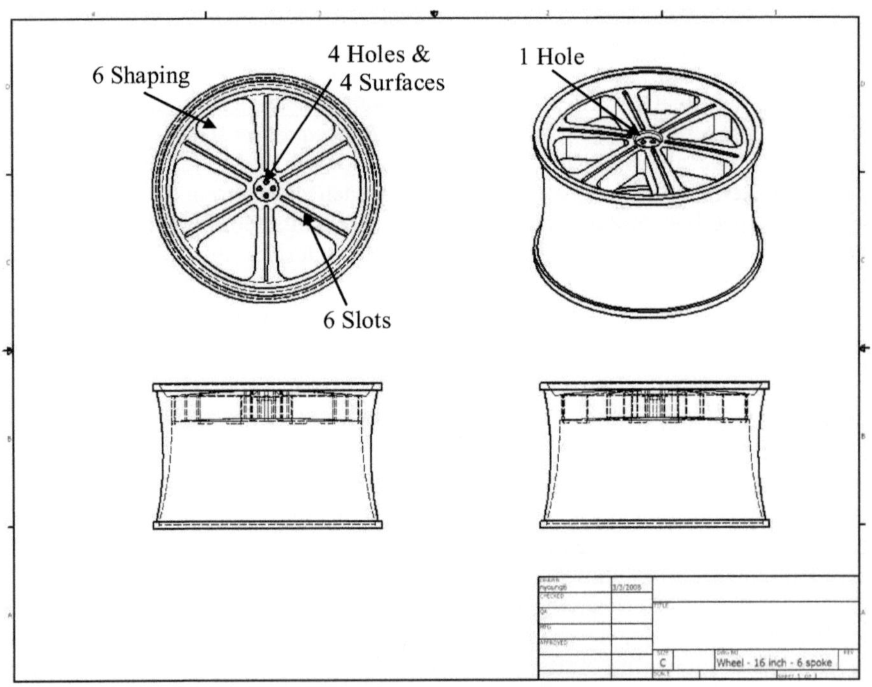

Figure 5.5. Features on the automotive wheel

Table 5.2. Case example results

Machine data			Machine configuration								
Agent	Feature	Cost ($/part)	Base	Table	L.S.	I.M.	Fix.	Col.	Spin.	S.M.	Mills
1	19	10.8	1	10	26	26	10	1	6	6	6
2	21	13.2	1	5	17	17	5	1	7	7	7
3	23	14.3	1	9	32	32	9	2	14	14	14
			Machine architecture								
			1	10	32	32	10	2	14	14	14

Table 5.2 shows the machine configuration for each part synthesised by each agent. The structure of the reconfigurable milling machine is the total number of components needed to construct each machine configuration. Therefore, from this set of components, each of the machine configurations can be constructed. The components shown in Table 5.2 are the base, table, lead screw (L.S.), indexing motor (I.M.), fixtures (Fix.), columns (Col.), spindles (Spin.), spindle motor (S.M.) and milling tools (Mills). The machine configuration for the five spokes wheel has 1 base, 10 tables, 26 lead screws, 26 indexing motors, 10 fixtures, 1 column, 6 spindles, 6 spindle motors and 6 milling tools. This configuration depicts a machine where there are 10 tables with a fixtured workpiece on each table. The machine also has 6 milling tools attached to 6 spindles that in turn are assembled to a single column. The motors and lead screws are present to facilitate motion. The reason for having more fixtured workpieces than milling tools can be explained by the slow cutting on the five triangular features of the workpiece. The largest tooling diameter is 2 inches; therefore, the features require more than one pass to satisfy the required width of cut of 4.75 inches. Furthermore, the length and depth of the features is 4 and 2 inches, respectively. This results in a large required number of passes to fully machine the feature. Hence while the large milling tools are making these machining passes, other milling tools are allowed to continuously carry out machining. This is made possible by the increased number of fixtures.

The machine configuration for the six-spoke wheel has 1 base, 5 tables, 17 lead screws, 17 indexing motors, 5 fixtures, 1 column, 7 spindles, 7 spindle motors and 7 milling tools. This configuration depicts a machine where there are 5 tables with a fixtured workpiece on each table. The machine also has 7 milling tools attached to 7 spindles that in turn are assembled to a single column. This machine configuration has far fewer fixtures than the first machine while increasing the number of spindles by one. This configuration change is credited to the increase in the number of features and required volume removal. The fewer number of work holding units promotes more concurrent machining on a single workpiece.

The machine configuration for the seven-spoke wheel has 1 base, 9 tables, 32 lead screws, 32 indexing motors, 9 fixtures, 2 columns, 14 spindles, 14 spindle motors and 14 milling tools. This configuration depicts a machine where there are 9 tables with a fixtured workpiece on each table. The machine also has 14 milling tools attached to 14 spindles that in turn are assembled to 2 columns. The larger number of features is conducive to concurrent machining operations; hence, there are a large number of work holding and tooling units. The corresponding cost for the five-, six- and seven-spoke wheels are \$10.8, \$13.2, and \$14.3, respectively.

The convergence plot of the three agents is shown in Figure 5.6. The average fitness progressively decreases as the number of features of the product increases. Therefore, the algorithm performs as expected with the increase in volume removal and the number of features.

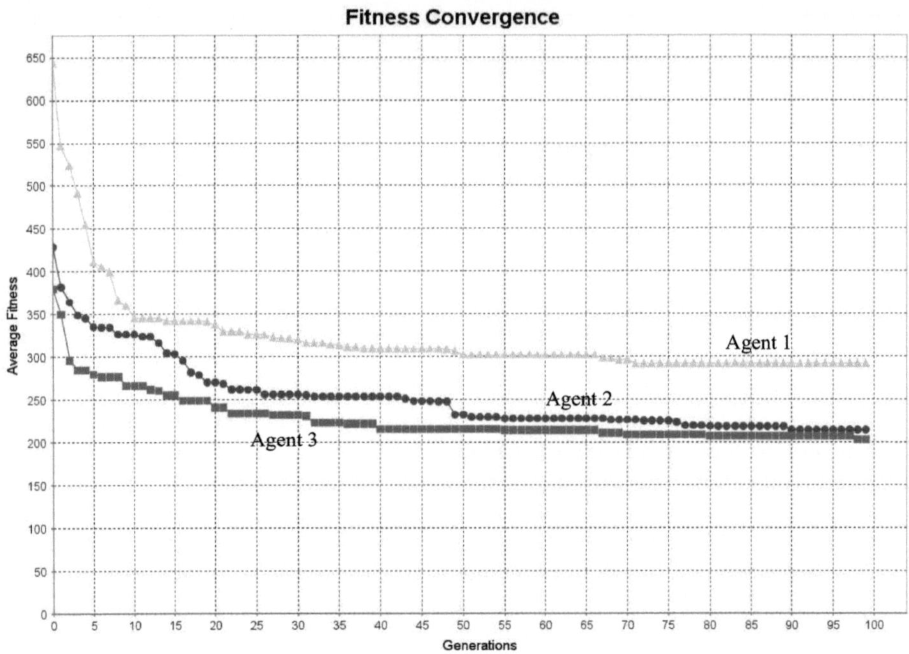

Figure 5.6. Fitness convergence

5.6 Conclusions

In this chapter, we presented a co-operative co-evolutionary multi-agent approach for designing the structure of an RMS. The key features of the approach are summarised as follows:

1. The co-evolutionary multi-agent approach to the design of reconfigurable machines is general and can be applied to different reconfigurable manufacturing machines and systems.
2. The approach allows a reconfigurable machine structure to be synthesised for a defined variety of parts that accounts for tradeoffs between minimising machining cost per part and minimising reconfiguration cost for machine reconfiguration between parts.

3. The modular nature of the algorithm allows it to be applied in dynamic environments with changing product varieties. This can be done by simply changing the input part to an agent or adding and deleting agents.

Future work will look into the geometric details of reconfigurable manufacturing machine synthesis. This will include the synthesis of modular interfaces for reconfiguration of machine structures.

References

[5.1] http://www.flextronics.com

[5.2] Koren, Y., Jovane, F., Moriwaki, T., Pritschow, G., Ulsoy, G. and Brussel, H.V., 1999, "Reconfigurable manufacturing systems," *Annals of the CIRP*, **48**(2), pp. 527–540.

[5.3] Koren, Y. and Ulsoy, A.G., 2002, "Reconfigurable manufacturing system having a production capacity method for designing same and method for changing its production capacity," U.S.P. Office (Ed.), The Regents of the University of Michigan: United States, pp. 1–12.

[5.4] Moon, Y.-M. and Kota, S., 2002, "Design of reconfigurable machine tool," *ASME Journal of Manufacturing Science and Engineering*, **124**(22), pp. 480–483.

[5.5] Zhao, X., Wang, J. and Luo, Z., 2000, "A stochastic model of a reconfigurable manufacturing system, Part 1: a framework," *International Journal of Production Research*, **38**(10), pp. 2273–2285.

[5.6] Zhao, X., Wang, J. and Luo, Z., 2000, "A stochastic model of a reconfigurable manufacturing system, Part 2: optimal configurations," *International Journal of Production Research*, **38**(12), pp. 2829–2842.

[5.7] Zhao, X., Wang, J. and Luo, Z., 2001, "A stochastic model of a reconfigurable manufacturing system, Part 3: optimal selection policy," *International Journal of Production Research*, **39**(4), pp. 747–758.

[5.8] Zhao, X., Wang, J. and Luo, Z., 2001, "A stochastic model of a reconfigurable manufacturing system, Part 4: performance measure," *International Journal of Production Research*, **39**(6), pp. 1113–1126.

[5.9] Spicer, P., Yip-Hoi, D. and Koren, Y., 2005, "Scalable reconfigurable equipment design principles," *International Journal of Production Research*, **43**(22), pp. 4839–4852.

[5.10] Spicer, P. and Carlo, H.J., 2007, "Integrating reconfiguration cost into the design of multi-period scalable reconfigurable manufacturing systems," *ASME Journal of Manufacturing Science and Engineering*, **129**, pp. 202–210.

[5.11] Abdi, M.R. and Labib, A.W., 2003, "A design strategy for reconfigurable manufacturing systems (RMSs) using analytical hierarchical process (AHP): a case study," *International Journal of Production Research*, **41**(10), pp. 2273–2299.

[5.12] Chen, L., Xi, F. and Macwan, A., 2005, "Optimal module selection for preliminary design of reconfigurable machine tools," *ASME Journal of Manufacturing Science and Engineering*, **127**, pp. 104–115.

[5.13] Youssef, A.M.A. and ElMaraghy, H.A., 2006, "Assessment of manufacturing systems reconfiguration smoothness," *International Journal of Advanced Manufacturing Technology*, **30**(1–2), pp. 174–193.

[5.14] Bi, Z.M., Lang, S.Y.T., Shen, W. and Wang, L., 2007, "Reconfigurable manufacturing systems: the state of the art," *International Journal of Production Research*, **46**(4), pp. 967–992.

[5.15] Degarmo, E.P., Black, J.T., Kohser, R.A. and Klamecki, B.E., 2003, *Materials and Processes in Manufacturing*, John Wiley & Sons, Inc., Danvers, MA, USA.

6

A Web and Virtual Reality-based Platform for Collaborative Product Review and Customisation

George Chryssolouris, Dimitris Mavrikios, Menelaos Pappas, Evangelos Xanthakis and Konstantinos Smparounis

Laboratory of Manufacturing System & Automation
Department of Mechanical Engineering & Aeronautics
University of Patras, 26100 Rio-Patras, Greece
Email: gchrys@hol.gr

Abstract

This chapter describes the conceptual framework and development of a web-based platform for supporting collaborative product review, customisation and demonstration within a virtual/augmented reality environment. The industrial needs related to this platform are first outlined and a short overview is given on recent research work. The conceptual framework of the web-based platform is then presented. The design and implementation are discussed, providing insight into the architecture and communication aspects, the building blocks of the platform and its functionality. An indicative "use case" illustrates how this approach and tools have been applied to the collaborative review, demonstration and customisation of products from the textiles industry. Finally, conclusions are drawn with respect to the potential benefits from the use of the web-based platform in addressing the needs of collaborative product development activities.

6.1 Introduction

Manufacturing companies need to innovate, both by designing new products and by enhancing the quality of existing ones [6.1]. Time and cost-efficient product innovation heavily relies today on the swift and effective collaboration of numerous dispersed actors, such as multi-disciplinary engineering teams, suppliers, sub-contractors, retailers, and customers as well. Knowledge sharing is an important issue since these actors typically share a large number of drawing files and assembly models. Quite often, different groups of engineers, being located at geographically different locations, are involved in the design of the various components or sub-assemblies of the product. Moreover, companies are frequently outsourcing engineering activities, in order to accelerate the design and product development process. Nowadays, 50–80% of all the components produced by original equipment manufacturers are outsourced to external suppliers [6.2]. This practice often creates problems due to the lack of tools to support sharing of product design knowledge and collaborative design and manufacturing activities. The respective problems are typically resolved through physical meetings or via e-mails and phone discussions.

Distributed product development lifecycle activities, in a globally integrated environment, are associated with the use of the Internet as well as Web technologies. Focusing on the collaboration aspect of engineering activities, several platforms for collaborative product and process design evaluation have also been presented in the scientific literature. The Distributed Collaborative Design Evaluation (DiCoDEv) platform enables the real-time collaboration of multiple dispersed users, from the early stages of the conceptual design, for the real-time validation of a product or process, based on navigation, immersion and interaction capabilities [6.3]. In order to support collaborative work on shape modelling, a Detailed Virtual Design System (DVDS) has been developed, providing the user with a multi-modal, multi-sensory virtual environment [6.4]. An asynchronous collaborative system, called Immersive Discussion Tool (IDT), which emphasises on the elaboration and transformations of a problem space and underlines the role that unstructured verbal communication and graphic communication can play in design processes [6.5], has also been presented. Another system for dynamic data sharing in collaborative design has been developed, ensuring that experts should use it as a common space to define and share design entities [6.6]. Various collaborative design activities are facilitated by a web-enabled PDM system, which has been developed and provides 3D visualisation capabilities as well [6.7]. Moreover, an Internet-based virtual reality collaborative environment, called Virtual Reality-based Collaborative Environment (VRCE) developed with the use of Vnet, Java and VRML, demonstrates the feasibility of collaborative design for small to medium sized companies that focus on a narrow range of low-cost products [6.8]. A web-based platform for dispersed networked manufacturing has also been proposed, enabling authorised users in geographically different locations to have access to the company's product data and carry out product design work simultaneously and collaboratively on any operating system [6.9]. A cPAD prototype system has been developed to enable designers to visualise product assembly models and perform real-time geometric modifications, based on polygonised representations of assembly models [6.10]. Another system, called IDVS (Interactive Design Visualisation System), has been developed, based on VRML techniques, in order to help depict 3D models [6.11]. An agent-based collaborative e-engineering environment for product design has been developed, based on the facilities provided by the AADE — a FIPA-compliant agent platform validated through a real-life industrial design case study [6.12]. Finally, addressing the needs for IT systems to support collaborative manufacturing, a new approach to collaborative assembly planning, in a distributed environment, has been developed [6.13]. Comprehensive reviews on systems, infrastructures and applications for collaborative design and manufacturing have also been presented in the scientific literature [6.14, 6.15].

Further to the research work on web-based collaborative product design, a few commercial tools are available to support such functionalities. OneSpace.net [6.16] is a lightweight web collaboration tool that supports online team collaboration for project development. It combines architecture for web services with familiar concepts, such as organised projects, secure messaging, presence awareness and real-time online meetings. IBM's Product Lifecycle Management Express Portfolio is designed specifically for medium-sized companies that design or manufacture products. This system mainly focuses on business processes but also allows design engineers to share 3D data, created with diverse authoring tools and consequently, product development can be managed. It includes CATIA V5 Instant Collaborative Design software and ENOVIA SmarTeam [6.17] for product data and release management. ENOVIA MatrixOne [6.18] is designed to support the deployment of all sizes. It includes PLM business process applications that cover a wide range of processes, including product planning, development and sourcing, and program management. Furthermore, it allows diverse design disciplines to be synchronised around design activities and changes, by reducing the critical errors and cost associated with poor collaboration. SolidWorks eDrawings [6.19] is an email-enabled communication tool that eases the review of 2D/3D product design data across extended product development teams. UGS [6.20] TeamCenter's portfolio of digital lifecycle management solutions is built on an open PLM foundation.

In the past, many research approaches and applications were focused on the use of virtual reality (VR) in order for the complexity of product design and manufacture to be overcome [6.21]. They may include human simulations for performing an ergonomic analysis of virtual products or assembly processes [6.22]. In recent years, research has also presented several applications of augmented reality (AR), ranging from games and education to military, medical and industrial applications [6.23, 6.24]. Dempski presented the idea of context-sensitive e-commerce using AR. She used furniture shopping as an example to show 3D and full sized representations of virtual objects in a physical living room [6.25]. Around the same time, Zhang *et al.* [6.26] developed a direct marketing system based on AR technology. A marker plate is shown in a live video stream. The system recognises this plate and superimposes animated 3D models of real products.

Despite the investment made in the last years, both in research and in industrial applications, the global market still lacks collaboration tools capable of providing AR and VR techniques with the possibility of product design evaluation. Most collaborative tools are more related to a PLM environment and less to shared virtual environments (VEs). Thus, the developments in the context of this work are focused on the conceptualisation and pilot implementation of a web-based platform for product collaborative design and evaluation, called Collaborative Product Reviewer (CORE). The platform includes links to VR and AR viewers in order to support the advanced visualisation of a product, as well as to provide multi-user navigation and interaction capabilities. CORE is part of an integrated collaborative manufacturing environment (CME) framework.

6.2 Collaborative Manufacturing Environment Framework

In the context of a European Commission funded research project, called DIFAC, a VR-AR based CME has been conceptualised for the next generation of digital manufacturing. Its aim is to support real-time collaboration within a virtual environment for product design, prototyping, manufacturing and worker training. The CME is expected to provide users with advanced capabilities, including visualisation, group presence, interaction, information sharing, knowledge management and decision making.

The CME addresses the major activities in digital factories, namely product development, factory design and evaluation, as well as workforce training. Three pillars, *Presence*, *Collaboration* and *Ergonomics*, underpin the methodological and technical realisation. Since the digital factory of the future is a human-centred CME, it will be the human factors that will play the critical role for the foundation of the three pillars. Six basic modules will be integrated into the proposed CME (three of them at the system level and another three at the application level) in order for the respective activities to be supported (Figure 6.1).

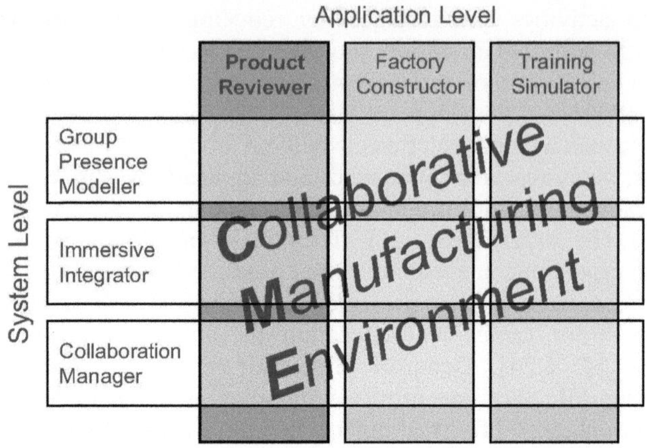

Figure 6.1. Six basic modules of CME and their interrelations

The system level consists of

- *Group Presence Modeller*, which aims at making groups of workers interact in a real collaborative working environment. It deals with the software aspect related to the visualisation, perception and interaction.
- *Immersive Integrator*, which enables all the work groups and group workers to collaborate with each other, realising a hardware integration layer and a VR interaction metaphors toolset. This is important for the developers of the application components.
- *Collaboration Manager*, which supports both group decision making and collaborative knowledge management.

The application level consists of

- *Product Reviewer*, which provides the functionality for a web-based collaborative product and process design evaluation.
- *Factory Constructor*, which provides the simulation of a complete virtual plant that will have the capability of completely emulating the real factory operations.
- *Training Simulator*, which is used for training new workers and re-training experienced workers who will be working in a new or reconfigured group or with new facilities.

6.3 Collaborative Product Reviewer

As part of this CME integrated framework, the CORE platform aims to enable dispersed actors to demonstrate, review and customise a product design in a collaborative way (Figure 6.2). A user requirements analysis, especially addressing the needs of small and medium sized manufacturing companies, showed that such a tool should help the appropriate actors to share and manage design data, to review the data by visualising and interacting with the virtual product, to communicate their views and make remarks online, and finally to keep track and manage the outputs of the review session in order to provide feedback to the design process.

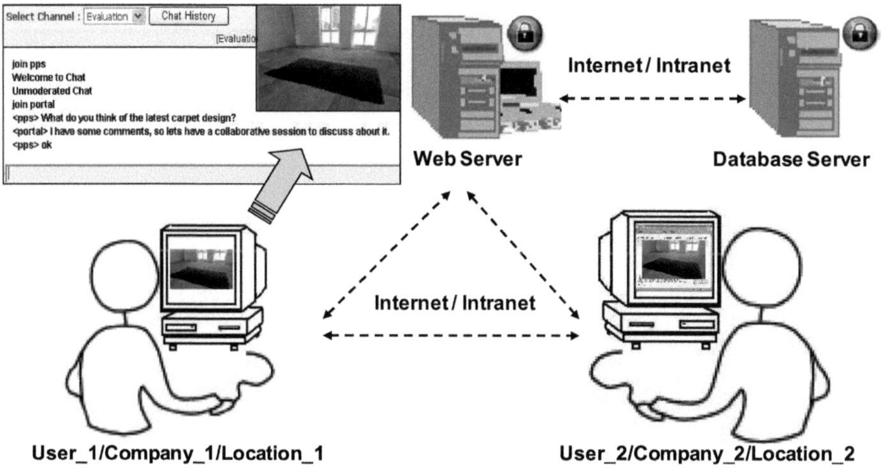

Figure 6.2. The CORE platform concept

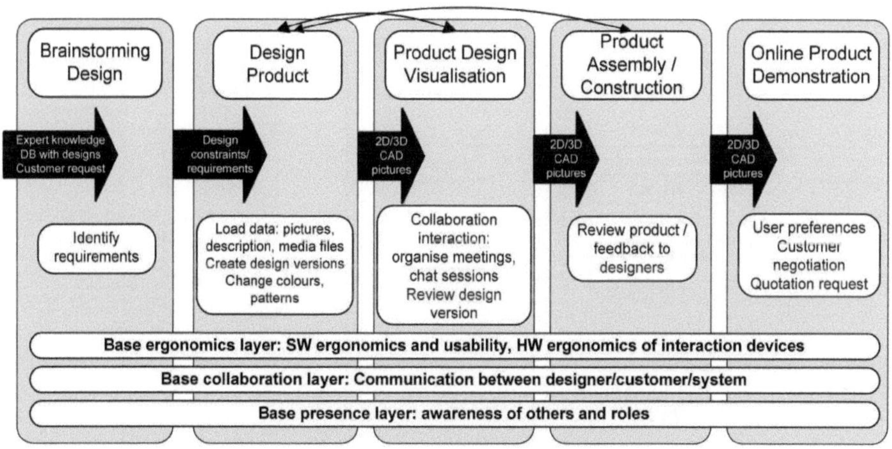

Figure 6.3. Workflow of CORE

The CORE workflow is a broad description of the key phases of any product design application, inside the digital factory, and specifies the way this platform answers to design needs (Figure 6.3).

From a technological point of view, the CORE concept lies in the integrated use of synchronous web-based collaboration with virtual and augmented reality-based visualisation and interactions. The basis is a web-based platform, which enables the remote access of dispersed actors to a working session. Tools for multi-user interaction and navigation in the same virtual environment are provided. Advanced functionality for product visualisation and customised demonstration over the web is provided through VR and AR viewers. User-friendly interfaces allow the active participation of non-experts in the design process, facilitating for example, the on-line recording of customer preferences. All product-related data may be stored in a database for managing multiple part/product versions.

6.4 Platform Design

6.4.1 Platform Architecture

The platform is designed based on an open architecture and a browser-server technology. The architecture follows the three-tier example and includes three layers namely: (a) the data layer, (b) the business layer, and (c) the presentation layer (Figure 6.4). These layers communicate through the Internet or an Intranet, depending on the type of communication.

- *Data layer* (1st tier): includes the application's database and the connections with all the other external systems, namely an external database for the recovery and storing of data. Some characteristics, such as data locking, consistency and replication, ensure the integrity of data. Oracle 9i is used for the platform's database implementation.

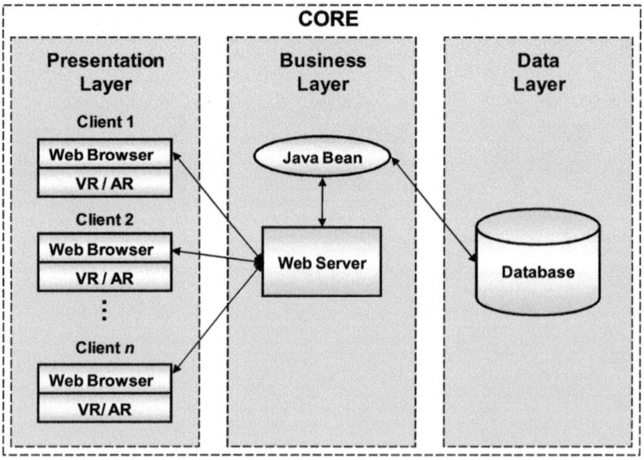

Figure 6.4. Three-tier architecture of CORE platform

- *Business layer* (2nd tier): consists of the business logic. The architecture of this level can be analysed further and divided into the connection mechanism between the mainframe PC and the application (JavaServer) as well as the Java Bean Architecture, which contains the work-division planning algorithm and the database interactions.
- *Presentation layer* (3rd tier): concerns the clients and consists of a standard web browser (*i.e.* Internet Explorer, Mozilla, *etc.*) as well as the VR and AR viewers that will be integrated into the web browser.

6.4.2 Communication

The web interface provides access to the portal. Through this portal, authorised users can upload/download the required virtual project environment and information. By the time the project-related files are uploaded, a new version of the selected project will be automatically created into the database. The communication between the front-end and the VR/AR viewers enables authorised users to open and modify a virtual project environment, realised through the XML protocol (Figure 6.5). Communication among the CORE platform and other external applications, such as databases, is also feasible.

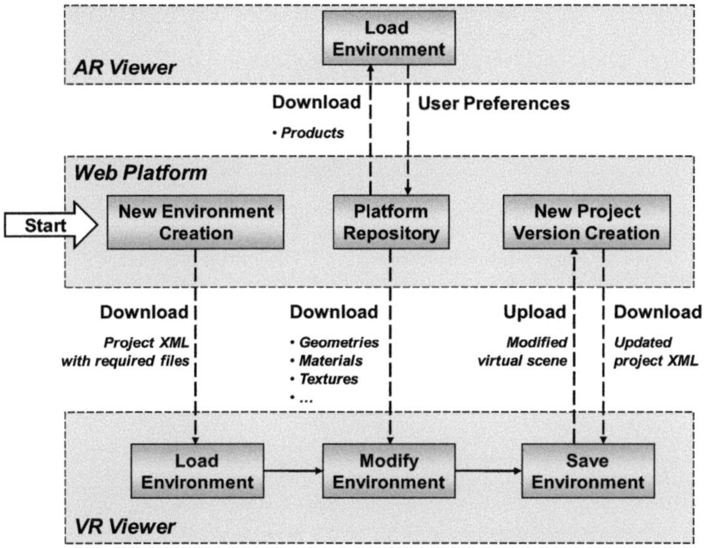

Figure 6.5. Communication interfaces between the web platform and the viewers

6.5 Platform Implementation and Functionality

The implementation of the CORE architecture is schematically shown in Figure 6.6. The Web Server allows the running of CORE through the Internet and supports the use of the underlying applications. The Database Server keeps stored all the important data in addition to the data logic code. The External Users/Clients are personal computers capable of connecting with the Internet and sending to and receiving messages from the database server.

The development of the platform has been driven by standard technologies applied to the J2EE language. Such technologies comprise the JavaServer Pages (JSP) for visualising data by creating HTML pages as well as the Servlets for data manipulation and user interaction. For the Web Server and Servlets container, Apache Jakarta Tomcat 5.5 has been used. The development has been assisted by the Eclipse, as the Integrated Development Environment (IDE) and the InterDev together with the Oracle 9i development and administration tools for the database design and creation. The development along with the installation have taken place on the Windows XP Pro operating systems but the same tools, technologies and development processes can be applied to other operating systems, such as Unix.

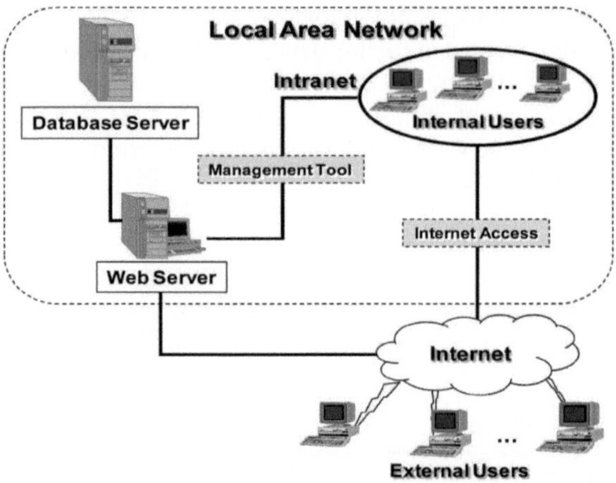

Figure 6.6. Environment of CORE infrastructure

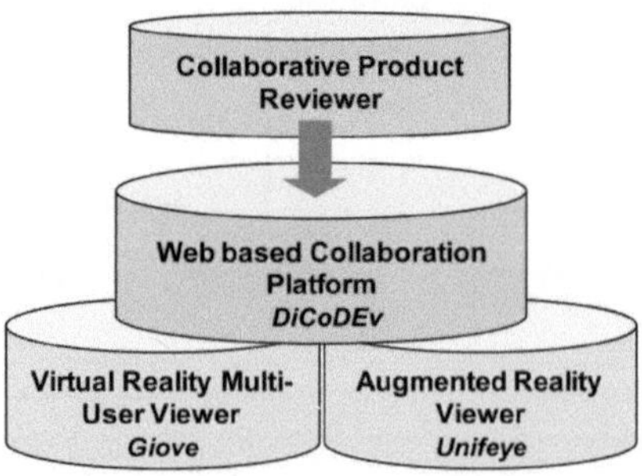

Figure 6.7. CORE functional building blocks

During a multi-user collaborative session of CORE, each participant is presented with his own copy of the graphical user interface, which provides a rendered 3D view of the virtual product, through a VR or AR-based environment. All users can interact with the virtual product at any time, with no restriction to the number of simultaneous interactions. Any changes made by a user on the virtual prototype may be seen immediately by others. Real-time chat capabilities enable the continuous communication among the online users. Remote users can join a collaborative session using TCP/IP over local or wide area networks. A 128 kbps ISDN (or DSL) line is capable of handling a simple product data load during a collaborative session. The major functional building blocks of CORE are shown in Figure 6.7.

6.5.1 Collaboration Platform

The DiCoDEv platform [6.3] served as a basis for developing the collaboration framework and functionality of CORE. A typical user workflow in the collaboration platform of CORE is shown in Figure 6.8. Major collaboration functions include:

- *User/Project Management*: Through user home page, each user can process his/her personal profile and check messages sent by the other members of the work group with reference to a specific project. The user also has the option to access an existing project or create a new one and assign the other users that will be capable of having access to its files and name their rights.

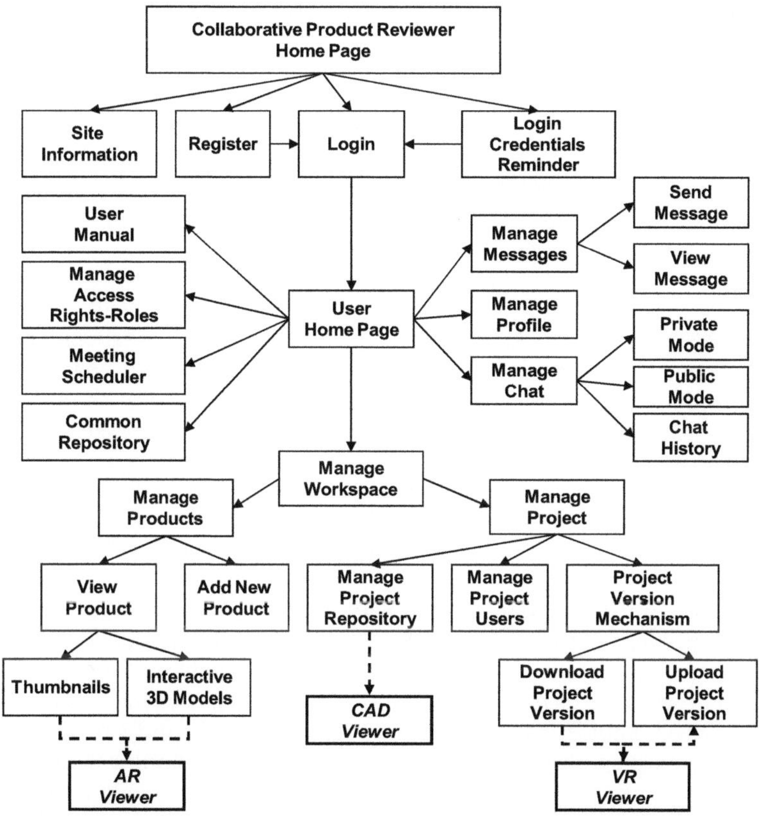

Figure 6.8. The user workflow within the collaboration platform

- *Messages/Chat*: Asynchronous communication among users is accomplished via written messages. Users can send out a message, in groups, to all the members involved directly, as well as a notification to others who are just supervising the project's progress. In order to facilitate and improve the real-time electronic collaboration among the users, the platform enables the synchronous conferencing through the exchange of instant messages among all the online users, in different chat rooms and in different modes: public/private chat. All communication may be recorded and saved so as for the user to revert to it whenever necessary after the chat session.
- *Project Versioning*: There is a mechanism for automated project file versioning. All new versions created are stored, making it easier for the users to keep track of all current modifications. This feature facilitates the review procedure, since users will have the ability to retrieve both the old and the new versions and notice the changes.
- *Calendar/Scheduler*: A calendar for meetings scheduling over the web is provided. All users have access to it, but each user has the right to see only the announcements or programmed meetings of the projects in which he/she is actively involved. There is also a mechanism to automatically send messages to the parties involved so as to remind them, in advance, of the upcoming meetings or to notify them of an announcement related to some projects.
- *File Sharing/Browser*: It is the platform's user-friendly web-based interface that allows authorised users to download and upload files. It has been implemented so as to provide rapid adoption throughout an organisation, requiring little or no training for familiarisation.

6.5.2 Virtual Reality Viewer

An OpenGL-based VR viewer, called GIOVE, has been integrated into the CORE platform to provide a shared virtual environment for collaborative product visualisation and review. The VR viewer is used for loading and displaying a product model, inside a 3D scene, and allows users to interact with objects on the scene or to change their appearance in the same shared environment. In order to integrate the viewer into the platform, a 3D scene (including the prototype and the surrounding environment) is loaded and displayed as part of the project selected. Project files include XML documents, 3D models and textures. The web platform facilitates users to select a project, download all the required files and start the viewer.

Once the virtual environment has been loaded, a context menu is available, providing functions such as session management and object editing. Both window and full screen visualisation are supported. Each user can interact with the VE while other users can notice changes in real time. The user gains exclusive control of the object selected until it is released in order to avoid conflicts among users' interactions (once an object is locked, no one can select it until it is released). The user can freely navigate so as to explore the scene or can move to a viewpoint, choosing from a set of default or user-defined viewpoints. It is possible to select objects, move and rotate them or change their appearance (*e.g.* material properties, texture, *etc.*). Stereoscopic graphics and different types of background are supported. The GIOVE viewer is built on the GIOVE library and is part of the GIOVE Toolkit [6.27]. The GIOVE library is the basic facility upon which higher level libraries and tools are built. It has a core module that provides basic common utility functionalities for the other modules: Graphics and Network. The GIOVE Toolkit is based on the GIOVE library and offers a higher level of programming interface for the tools developed on it: Viewer, Scene Editor, GUI Editor, and Application Editor.

6.5.3 Augmented Reality Viewer

An AR viewer, based on the Metaio Unifeye SDK® [6.28], has also been integrated into the CORE platform to allow for special purpose visualisation that requires the combination of virtual information with real environments. The current version of the CORE platform provides monitor-based AR visualisation.

The AR viewer offers functionality under two application modes, namely a light-weight online application and a more powerful offline application. For the online application mode, the basic AR functionalities are wrapped into a light-weight ActiveX plug-in. The 3D models are seamlessly integrated into the digital view of the real world. This is realised through an underlying marker tracking, which detects the paper marker in the image and uses this reference to place the virtual model data in the correct perspective. Next to the AR view, additional information on the product can be presented (*e.g.* size, available colours and textures, price, *etc.*). The offline application mode provides powerful interface for creating various mixed reality applications. Next to the digital images, the offline version can also use video data or real-time camera streams for visualisation and offers a large selection of configuration and measuring features. The offline version provides tools for creating and running AR-based workflows in an easy and intuitive way. For product presentation, this feature may be applied to present the composition of more complex products by visualising their step-by-step assembly.

6.6 A Textiles Industry Use Case

In order for the functionality of the CORE platform to be demonstrated, a use case from the textiles industry has been considered. This use case is related to the collaborative review, customisation and demonstration of carpets, in the context of two major activities:

- The interactive web-based product demonstration of existing carpets, by providing a virtual showroom for potential customers.
- The online collaboration during carpet design and customisation, among key actors involved in the process, such as designers, textile engineers, retailers, sales personnel and customers.

The framework of the collaborative carpet review use case is schematically shown in Figure 6.9. The information input to the design process of a carpet, is usually a market trend analysis or a specific customer demand. Within this context, a graphical environment is required that facilitates the exchange of ideas among the company's relevant departments and its customers, regarding the carpet design characteristics, such as the dimensions, materials, colours, patterns, textures, *etc*. On that basis, the collaboration platform should enable all partners to interact, assess modifications on the product characteristics, and finally make a decision.

The customised user workflow within the CORE platform for both major review activities, namely the customisation and demonstration of carpets, is shown in Figure 6.10. Figures 6.11–6.15 illustrate a number of steps of this workflow.

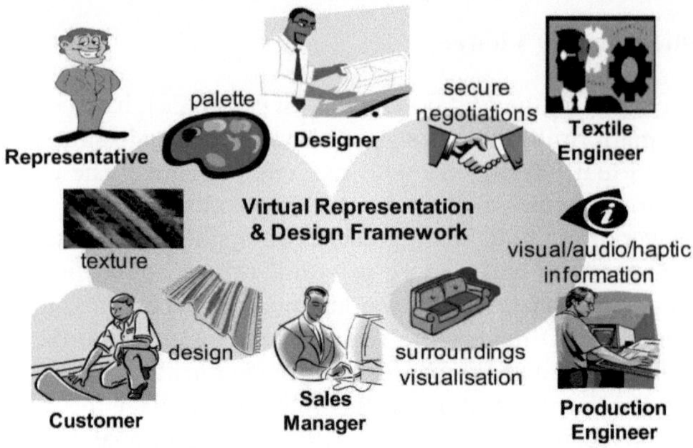

Figure 6.9. Framework of carpet review use case

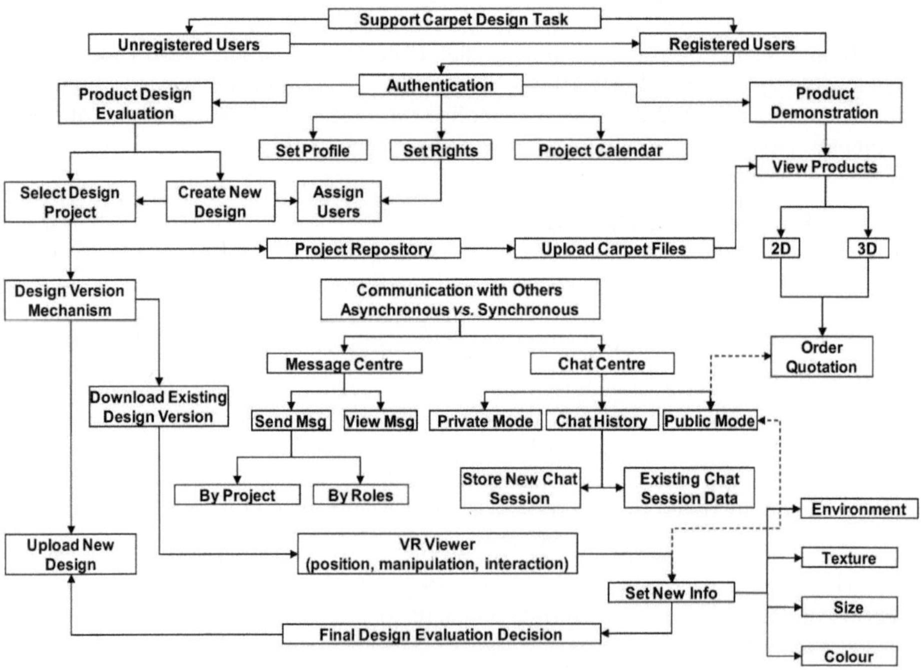

Figure 6.10. User workflow for carpet review, customisation and demonstration

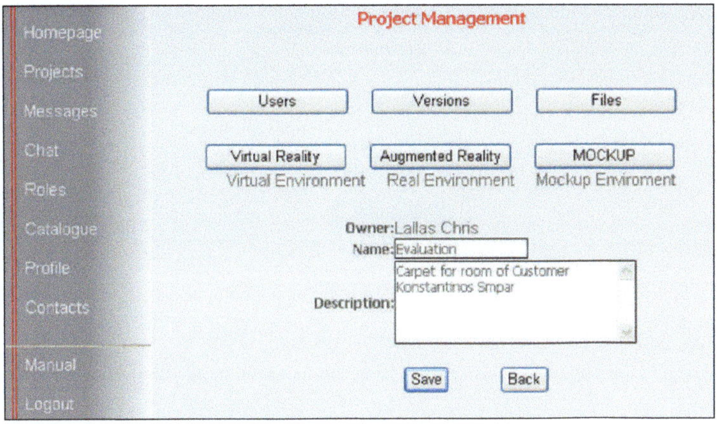

Figure 6.11. Projects management

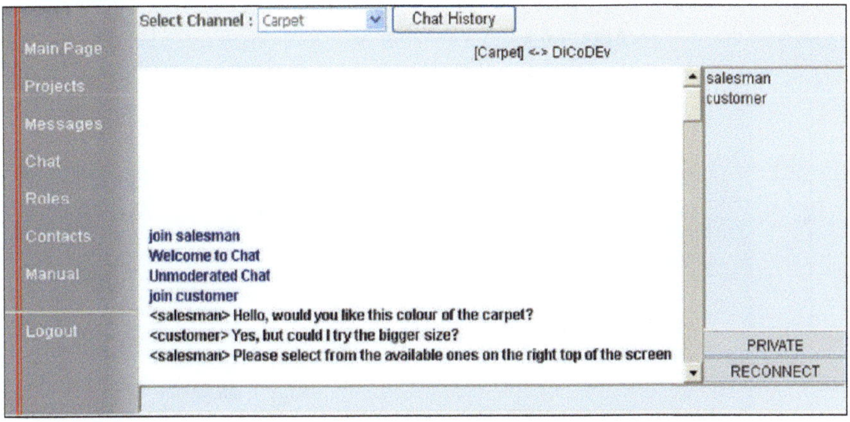

Figure 6.12. Chat between a salesman and a customer

Figure 6.13. Interacting with carpets

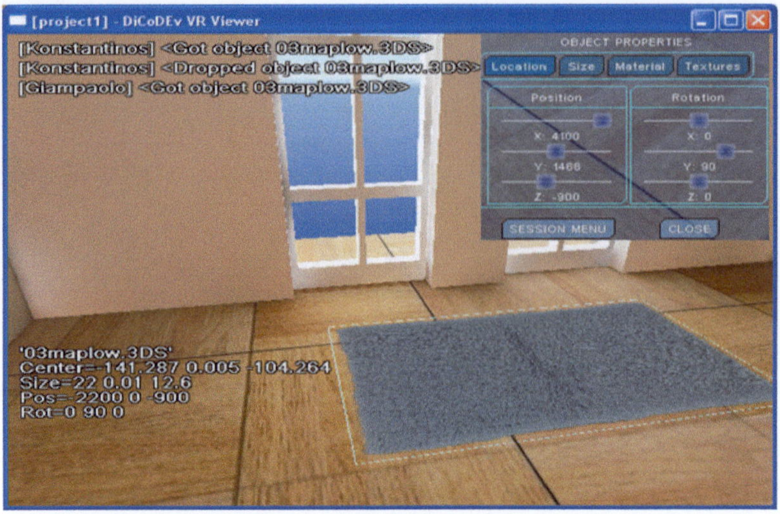

Figure 6.14. A VR-based review of carpet

Figure 6.15. An AR-based review of carpet

6.7 Conclusions

This chapter has presented a web-based collaborative platform for product review, demonstration and customisation. The CORE platform serves as a multi-user real-time collaboration tool with VR and AR integration. The potential benefits of using the proposed collaborative platform include: quick and easy exchange of design data/information among distributed users (user and data management capabilities), multi-user visualisation and interaction with the shared product prototype, real-time collaboration for online decision making on the same product design, user-friendly environment for multi-disciplinary assessment of product design outcomes, improvement on the communication among participating work groups throughout the lifecycle of the products and efficient review and evaluation of alternative virtual designs.

The CORE platform has been tested against the requirements of the use cases of non-mechanical products. Our future work includes the expansion of the platform's functionality to address use cases involving complex mechanical products.

Acknowledgement

This work is partially supported by the IST research project "DiFac – Digital Factory for Human-Oriented Production System" (FP6-2005-IST-5-035079), funded by the European Commission under IST priority 2.5.9 Collaborative Working Environments.

References

[6.1] Chryssolouris, G., 2006, *Manufacturing Systems – Theory and Practice*, 2nd Edition, Springer-Verlag, New York.

[6.2] Rezayat, M., 2000, "The enterprise-web portal for life cycle support," *Computer-Aided Design*, **32**(2), pp. 85–96.

[6.3] Pappas, M., Karabatsou, V., Mavrikios, D. and Chryssolouris, G., 2006, "Development of a web-based collaboration platform for manufacturing product and process design evaluation using virtual reality techniques," *International Journal of Computer Integrated Manufacturing*, **19**(8), pp. 805–814.

[6.4] Arangarasan, R. and Gadh, R., 2000, "Geometric modelling and collaborative design in a multi-modal multi-sensory virtual environment," In *Proceedings of the ASME 2000 Design Engineering Technical Conferences and Computers and Information in Engineering Conference*, pp. 10–13.

[6.5] Craig, D.L. and Craig, Z., 2002, "Support for collaborative design reasoning in shared virtual spaces," *Automation in Construction*, **11**(2), pp. 249–259.

[6.6] Noel, F. and Brissaud, D., 2003 "Dynamic data sharing in a collaborative design environment," *International Journal of Computer Integrated Manufacturing*, **16**(7–8), pp. 546–556.

[6.7] Xu, X.W. and Liu, T., 2003, "A web-enabled PDM system in a collaborative design environment," *Robotics and Computer-Integrated Manufacturing*, **19**(4), pp. 315–328.

[6.8] Kan, H.Y., Duffy, V.G. and Su, C.J., 2001, "An Internet virtual reality collaborative environment for effective product design," *Computers in Industry*, **45**, pp. 197–213.

[6.9] Zhan, H.F., Lee, W.B., Cheung, C.F., Kwok, S.K. and Gu, X.J., 2003 "A web-based collaborative product design platform for dispersed network manufacturing," *Journal of Materials Processing Technology*, **138**, pp. 600–604.

[6.10] Shyamsundar, N. and Gadh, R., 2002, "Collaborative virtual prototyping of product assemblies over the Internet," *Computer-Aided Design*, **34**, pp. 755–768.

[6.11] Lau, H.Y.K., Mark, K.L. and Lu, M.T.H., 2003, "A virtual design platform for interactive product design and visualisation," *Journal of Materials Processing Technology*, **139**(1–3), pp. 402–407.

[6.12] Hao, Q., Shen, W., Zhang, Z., Park, S.W. and Lee, J.K., 2006, "Agent-based collaborative product design engineering: an industrial case study," *Computers in Industry*, **57**, pp. 26–38.
[6.13] Dong, T., Tong, R., Zhang, L. and Dong, J., 2005, "A collaborative approach to assembly sequence planning," *Advanced Engineering Informatics*, **19**, pp. 155–168.
[6.14] Shen, W., 2000, "Web-based infrastructure for collaborative product design: an overview," In *Proceedings of the 6th International Conference on Computer Supported Cooperative Work in Design*, Hong Kong, pp. 239–244.
[6.15] Yang, H. and Xue, D., 2003, "Recent research on developing Web-based manufacturing systems: a review," *International Journal of Production Research*, **41**(15), pp. 3601–3629.
[6.16] CoCreate OneSpace.net, 2007, http://www.cocreate.com/
[6.17] ENOVIA SmarTeam, 2007, http://www.smarteam.com/
[6.18] ENOVIA MatrixOne, 2007, http://www.matrixone.com/
[6.19] SolidWorks eDrawings, 2007, http://www.solidworks.com/edrawings/
[6.20] TeamCenter, 2007, http://www.plm.automation.siemens.com/teamcenter/
[6.21] Chryssolouris, G., Mavrikios, D., Fragos, D., Karabatsou, V. and Pistiolis, K., 2002, "A novel virtual experimentation approach to planning and training for manufacturing processes – the virtual machine shop," *International Journal of Computer Integrated Manufacturing*, **15**(3), pp. 214–221.
[6.22] Chryssolouris, G.., Mavrikios, D., Fragos, D., Karabatsou, V. and Alexopoulos, K., 2004, "A hybrid approach to the verification and analysis of assembly and maintenance processes using virtual reality and digital mannequin technologies," In *Virtual Reality and Augmented Reality Applications in Manufacturing*, Nee, A.Y.C. and Ong, S.K. (eds), Springer-Verlag, London.
[6.23] Azuma, R., Baillot, Y., Behringer, R., Feiner, S., Julier, S. and MacIntyre, B., 2001, "Recent advances in augmented reality," *IEEE Computer Graphics and Applications*, **21**(6), pp. 34–47.
[6.24] Pentenrieder, K., Doil, F., Bade, C. and Meier, P., 2007, "Augmented reality based factory planning – an application tailored to industrial needs," In *Proceedings of the IEEE and ACM International Symposium on Mixed and Augmented Reality*, Nara, Japan.
[6.25] Dempski, K., 2000, "Context-sensitive e-Commerce," In *Proceedings of the Conference on Human Factors in Computing Systems*, Hague, The Netherlands.
[6.26] Zhang, X., Navab, N. and Liu, S.P., 2000, "E-commerce directed marketing using augmented reality," In *Proceedings of the IEEE International Conference on Multimedia and Expo (ICME 2000)*, New York, **1**, pp. 88–91.
[6.27] Viganò, G., Sacco, M., Greci, L., Mottura, S. and Travaini, E., 2007, "A virtual and augmented reality tool for supporting decisions in motorbikes design: Aprilia application case," In *The 3rd International VIDA Conference*, Poznan, Poland.
[6.28] Metaio GmbH, 2008, "Augmented reality – software, systems and consulting from Metaio," http://www.metaio.com/

7

Managing Collaborative Process Planning Activities through Extended Enterprise

H. R. Siller, C. Vila[1], A. Estruch, J. V. Abellán and F. Romero

Department of Industrial Systems Engineering and Design, Universitat Jaume I
Av. Vicent Sos Baynat s/n. 12071 Castellón, Spain
[1]Email: vila@esid.uji.es

Abstract
Nowadays, the competitive global scenario has driven companies to work within extended enterprises. In this context, the collaborative product design and development process have to take into account all the geographically dispersed manufacturing resources. In order to enable real digital manufacturing, companies are forced to share and distribute data, information and knowledge through collaborative procedures. Particularly, manufacturing process planning activities, which are the link between product development and manufacturing, become crucial to achieve global efficiency. The objective of this chapter is, for these reasons, to define a reference model for collaborative process planning, taking into account certain basic requirements in order to enable an inter-enterprise environment. For the implementation of the reference model, a workflow modelling strategy and a reference architecture are presented that could enable collaborative processes management.

7.1 Introduction

During the last few years, novel organisational structures have emerged to satisfy a global, dynamic and competitive market. Leading enterprises of different sectors innovated the way to make business with concepts like strategic joint ventures, supply chain nets, outsourcing and electronic commerce, among others.

As a result of this tendency for increasing competitiveness and transforming manufacturing business, extended enterprises have been institutionalised and consolidated. They emerged as nets of independent and geographically dispersed organisations, which work in a collaborative way to achieve common objectives, with the aid of information and communication technologies (ICT).

The concept of extended enterprise is more than just the joint of different enterprises related by a product supply chain. It is based on an organisational paradigm for satisfying not only clients' needs, but the needs of people involved in all stages of the product lifecycle, *e.g.* product design, manufacturing or recycling.

In traditional enterprises, the organisational structure is composed of isolated departments with limited functions. This leads to a rigid and sequential product lifecycle and interruptions in the flow of work and information, needed to maintain interaction and feedback across the entire enterprise. Efforts directed to overcome the resistance to simultaneous work in the new era of extended enterprises led to the emergence of collaborative engineering.

Collaborative engineering is a systematic approach that succeeds concurrent engineering and forces technical departments to consider all product lifecycle stages and to take into account all the clients' demands. This approach considers the use of ICT in the implementation of collaborative environments for the development of crucial engineering activities like design, process planning and manufacturing.

Process planning has been a relevant research topic for the past twenty years. A number of papers have been published and important advances have been achieved, especially in the development of computer-aided process planning (CAPP) systems. These systems carry out a certain level of automation in decision making and instruction sheets preparation for discrete parts manufacturing. In addition, they include reasoning mechanisms, knowledge bases and databases that help process planners to perform different procedures, from the recognition of geometric characteristics of the part to be manufactured, to the generation of numerical control programs to be executed in shop floor machines.

But in practice, the dependency on sequential and iterative work is still present in technical departments of real companies, which leads to an increase of product development cycle time and all the associated costs. Furthermore, CAPP systems have not been integrated to other enterprise functions, like conceptual design, production planning, quality control or inventory control (Figure 7.1).

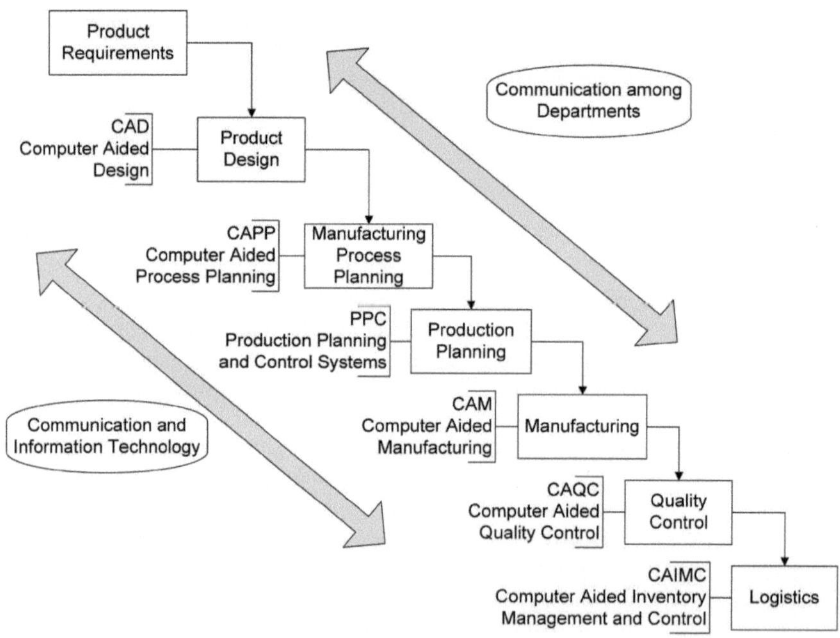

Figure 7.1. Stages of the product design, development, manufacturing and technological tools associated

In the age of the extended enterprise, the swift expansion of the Internet provides the infrastructure by which information can be made simultaneously available to all those involved in planning manufacturing processes, *i.e.* designers, planners, production managers, shop floor workers, and so forth. Yet, before this situation can be accomplished, the following problems will need to be overcome:

- Companies involved in process planning activities have different factories, information technology infrastructures and, therefore, use different data, rules and methods. Designers, even the most experienced, do not know exactly the capabilities of all manufacturing processes available in distributed plants. Furthermore, it is difficult for a process planner to guess the original design intentions of product designers.
- A part can be designed by personnel with limited skills in manufacturing engineering. Today, CAD tools allow designers to draw parts with intricate shapes, but which are sometimes difficult to manufacture. These parts must be modified later by manufacturing engineers in order to achieve manufacturability, in an iterative way, which requires resources that impact the product development cycle time.
- Decision making in all the stages of process planning is subjected to dynamic changes, due to the successive modification of production requirements and material conditions during manufacturing sequences on shop floors. In order to work in this scenario, process planning must be adaptive instead of being reactive. Furthermore, due to the diversity of processes, machines and tools for manufacturing the same part, a process planning problem can have alternative solutions. All of these factors lead to uncertainties at the time of decision making in technological departments.

These problems can be solved to some extent by distributed, adaptable, open and intelligent process planning systems within a collaborative environment [7.1]. But the implementation of these systems must be carried out by taking into account technological and economical considerations of the real industrial scenario, applying the teamwork philosophy of collaborative engineering.

In addition to these considerations, a collaborative process planning system should also help users draw up process plans at their different levels of detail. These are, according to some authors [7.2, 7.3], meta-planning, macro-planning and micro-planning.

Meta-planning is performed to determine the manufacturing process and the machines that fit the shape, size, quality and cost requirements of the parts that have been designed. In macro-planning, the equipment is selected, the minimum number of setups needed to manufacture the part is determined, and the sequence of operations is established. Micro-planning is concerned with determining the tools to be used, the tool paths to be followed during the manufacturing process (*e.g.* machining process), and the parameters associated with shop floor operations so that productivity, quality of parts and manufacturing costs can be optimised.

The state of the art presents several research works that study the problems of the integration of process planning in collaborative and distributed environments. In order to review the main contributions presented in each work and to study their technological infrastructures, a comprehensive literature review is presented in the next section.

7.2 Review of Collaborative and Distributed Process Planning

In the field of collaborative and distributed process planning, there exist research papers and prototypes diversified in terms of functionalities, communication protocols, programming languages and data structure representations. The following paragraphs include a chronological review of some of the recent works that report Web-based systems and methodologies of collaborative meta-, macro- and micro-process planning, most of them oriented to machining processes.

Van Zeir *et al.* [7.4] created a computer system for distributed process planning called Blackboard. Several expert modules of the system perform specific process planning tasks, ranging from process and machine assignment to operations sequencing and tool selection. The system generates graphs (generic Petri nets), called non-linear process plans (NLPP), which are made available to users by means of a graphic interface. The process plans that are thus generated can then be handled and modified by users, according to their own experience and skills.

Chan *et al.* [7.2] proposed a tool called COMPASS (computer oriented materials, processes and apparatus selection system) that helps designers identify potential manufacturing problems in the early stages of the product development cycle; it also helps them organise in one coherent plan all the heterogeneous technological processes involved in manufacturing. The system, which was developed in the form of a meta-planner, provides essential information about production costs, cycle times and product quality for different candidate processes, by a series of modules that receive the design information and analyse it according to the restraints of each technological process stored in databases.

Tu *et al.* [7.5] proposed a CAPP framework for developing process plans in virtual OKP (one-of-a-kind) manufacturing environment. It includes reference architecture, an IPP (incremental process planning) method, an optimal/rational cost analysis model and a database of the partner's resources. The framework was implemented for concurrently designing and manufacturing a steel frame of a rail station.

Zhao *et al.* [7.1] presented a process planning system (CoCAPP), which utilises co-operation and co-ordination mechanisms built into distributed agents with their own expert systems. Each agent has knowledge contained in databases, analytical algorithms, and conflict resolution rules for constructing feasible process plans.

Ahn *et al.* [7.3] created an Internet-based design and manufacturing tool, called CyberCut, which allows the generation, by a destructive solid geometry approach (DSG), of 3D prismatic parts from the basic machining specifications. Thus, in the design stage, the user can suggest what processes, operations, sequences and tools will be used in the actual manufacturing process.

Wang *et al.* [7.6] presented a distributed process planning (DPP) approach based on the use of function blocks, which encapsulate complete process plans for their execution in open CNC controllers. The process plans are generated through multi-agent negotiations implemented with KQML (knowledge query manipulation language).

Kulvatunyou *et al.* [7.7] described a framework for integrating collaborative process planning in which collaborative manufacturing is divided into two parts: the "design house" and the "manufacturing side". These two divisions exchange information by means of hierarchical graphs and UML (unified modelling language) models that represent the process requirements and alternatives. In this same work, a prototype application was implemented that uses the Java programming language and the data representation language XML (extensible markup language) to ensure information portability. Chung and Peng [7.8] addressed the issue of selecting machines and tools under a Web-based manufacturing environment. To ensure efficiency and functionality, they developed a selection tool using MySQL databases, Java Applets and a VRML (virtual reality modelling language) browser.

Sormaz *et al.* [7.9] created a process plan model prototype (named IMPlanner) for distributed manufacturing planning. It relies on existing CAD/CAM applications and proprietary CAPP software, and was implemented with Java and XML.

You and Lin [7.10] applied the ISO 10303 (STEP, standard for exchange of product model data) standard and Java J2EE specification, to implement a process planning platform for selecting machining operations, machines and tools, based on EXPRESS-G (object-oriented information modelling) models of the geometric characteristics of the parts to be machined.

Feng [7.11] developed a prototype of a multi-agent system that helps the user select the technological processes required to manufacture a part and the associated resources. This system is based on a platform with a knowledge base that captures design factors and classifies them in several machining features. It also integrates heterogeneous CAD, CAM and CAPP applications, databases and mathematical tools. Designers, process planners and manufacturing engineers can access it by means of web-based heterogeneous tools.

Nassehi *et al.* [7.12] examined the application of artificial intelligence techniques, as well as collaborative multi-agent systems, to design a prototype of an object-oriented process planning system, called MASCAPP (multi-agent system for computer aided process planning). This system focuses on prismatic parts and uses the STEP-NC standards (ISO 14649 and ISO 10303).

Cecil *et al.* [7.13] developed a collaborative Internet-based system to perform some of the process planning activities carried out between the partners in a virtual enterprise, based on the use of an object request broker (ORB). The distributed resources include feature identification modules, STL (stereo-lithography) files and software objects for choosing and sequencing processes, generating setups, selecting machines and tools, and also include a machining cost analysis agent.

Guerra-Zubiaga and Young [7.14] designed a manufacturing model to ensure management and storage of facility information and knowledge related to processes and resources. They developed an experimental system for the model validation using UML, Object Store databases and Visual C++ programming environment.

Mahesh *et al.* [7.15] proposed a framework for a web-based multi-agent system (Web-MAS) based on a communication over the Internet via KQML messaging. Each agent possesses unique capabilities and a knowledge base for performing different activities like manufacturing evaluation, process planning, scheduling and real-time production monitoring.

Peng *et al.* [7.16] proposed a networked virtual manufacturing system for SMEs (small and medium-sized enterprises) in which distributed users share CAD models in a virtual reality environment and contribute to the development of process plans with the aid of a system named VCAPP, implemented in Java and using VRML.

Finally, Ming *et al.* [7.17] presented a framework for collaborative process planning and manufacturing in product lifecycle management. They developed and implemented a technological infrastructure that integrates with UML information exchange, CAPP and CAM applications. They also presented an interface in which users can interact with the system developed.

Other works that complement this literature review include state of the art reviews about web-based manufacturing and collaborative systems [7.18, 7.19], in which they identify future trends of development issues like integration, security, flexibility and interoperability. The vast majority of academic works reviewed above are based on the need to integrate process planning activities across different stages of the processes of product design, development and manufacturing. Nevertheless, there are just a few studies about the integration that must exist among different organisations that collaborate in the development of process plans at their different levels of details. Table 7.1 shows the main technological characteristics (standards and information technology) of each reviewed work including different process planning levels covered, and specifying if the nature of process planning is distributed (D), collaborative (C) or agent based (A). The main findings after the literature review can be summarised as follows:

- The technological infrastructure used for distributed and collaborative process planning is predominantly based on Java, XML and VRML languages. This finding shows us the tendency to develop user-friendly interfaces that allow 3D models and document visualisation via Web applications. It is also frequent on using architectures that enable the use of distributed heterogeneous applications.
- In spite of the fast development of agent-based expert systems for process planning, the dependency on human experience is still dominant, because of the increasing demand of manufacturing flexibility that cannot be covered yet by expert systems.
- Although the above-mentioned works have established the roadmap for the next-generation commercial tools, they do not include any mechanism for the management of inter-enterprise collaborative workflow for the development of process planning activities across extended enterprises.

In order to develop a collaborative process planning model that can integrate the three hierarchical levels of process planning, and that can involve distributed participants across extended enterprises with the help of workflow coordination, we need first to recognise the ICT requirements for enabling a collaborative inter-enterprise environment.

7.3 ICT Functionalities for Collaboration

In order to explain the main functionalities that an ICT infrastructure must have for the development of process planning activities across the extended enterprise, it is necessary to describe all the requirements for the effective exchange of data, information and knowledge in an inter-enterprise environment.

Table 7.1. Literature review on collaborative and distributed process planning

Author	Prototype	Standards	ICT	Meta PP	Macro PP	Micro PP	D	C	A
Chan *et al.*	COMPASS	-	-	✓			✓	✓	
Van Zeir *et al.*	Blackboard	-	-	✓	✓		✓	✓	
Zhao *et al.*	CoCAPP	-	ABE Tool Kit, KQML, Visual C++	✓	✓		✓	✓	✓
Tu *et al.*	OKP/IPP	CSG/Brep, STEP	CORBA, Visual C++, Java	✓	✓		✓	✓	
Ahn *et al.*	CyberCut	DSG	Java	✓	✓	✓	✓		
Wang *et al.*	DPP	IEC-61499, KQML	Java	✓	✓	✓	✓	✓	✓
Kulvatunyou *et al.*	NIST IPPD/ RIOS	UML, XML	Java	✓	✓	✓	✓	✓	
Chung and Peng	WTMSS	VRML, DXF	Java, MySQL		✓	✓	✓	✓	
Sormaz *et al.*	IMPlanner	XML	Java	✓	✓	✓	✓		
You and Lin	-	STEP	J2EE	✓	✓		✓		
Feng	-	XML	ProTool Kit, CORBA, C, Java, Oracle, Matlab	✓	✓		✓	✓	✓
Nassehi *et al.*	MASCAPP	UML, OMT, STEP-NC	Java, Object Store	✓	✓		✓		
Cecil *et al.*	CHOLA	STL	CORBA, C++, Java	✓	✓		✓		
Guerra-Z. *et al.*	MKM	UML	Object Store, Visual C++	✓	✓				
Mahesh *et al.*	WebMAS	KQML	JATLite, MySQL, Java	✓	✓		✓	✓	✓
Peng *et al.*	VCAPP	VRML	Java, LDAP, MS-Access	✓	✓	✓	✓	✓	
Ming *et al.*	-	UML	Proprietary	✓	✓	✓	✓	✓	

7.3.1 Basic Requirements for Knowledge, Information and Data Management

The functionalities that an ICT infrastructure must have for the management of knowledge, information and data (KID) involve the use of highly accepted standards in order to enable communication and interoperability in heterogeneous environments. Following sections contain a review of the required functionalities emphasising the standardisation that can be applied in each one of them.

7.3.1.1 Product Data Vault and Document Storage Management

This functionality must allow a safe and controlled storage of data documents and their meta-data (attributes of those data documents), in either a proprietary or standardised formats like STEP (formally known as ISO 10303), IGES (Initial Graphic Exchange Specification) or the *de facto* standard DXF, among others in a vast catalogue of standards developed by international software consortiums, further explained in this text.

7.3.1.2 Product Data and Structure Management

According to van den Hamer and Lepoeter [7.20], the management of product data can be divided into five orthogonal dimensions: *versions*, *views*, *hierarchies*, *status* and *variants*. Each dimension plays an important role in the product structure data management, such as carrying out the iterative nature of product design, the representation of different levels of details, the division into assemblies, sub-assemblies and parts, and so forth.

With the data structure management, it is possible to create, manage and maintain bills of materials, assemblies and product configurations. The user of the technological infrastructure can navigate inside product structures, see parts and sub-assemblies, and create useful configurations, extending basic product structures to complete structures formed by options, versions and substitutes.

The system must provide a graphic visualisation of all structures and the possibility of attach documents related to the processes needed for product manufacturing. In the same way, the system administrator must be able to create and maintain authorisations for creating and modifying product structures by the participants of the collaborative environment.

Other functionality for the management of product structures is the information classification and retrieval that allow users to perform searches of product structures and attached documents. With this functionality the existing relationships between products and their associated documents can be represented in hierarchies, in a way that can be possible for the reutilisation of product structures, their classification in part families and the management of standard components.

7.3.1.3 Data Sharing and Exchange

The data sharing function must allow users to extract documents from a common repository, so that they can work with them in their private workspace. Once the tasks or modifications have been completed, the documents can be uploaded back to the shared data vault to make changes visible to other users.

This function is essential when the collaboration is carried out in heterogeneous environments, where different applications generate different file formats. Well-known standards (*e.g.* STEP, IGES, and STL) are needed for representing CAD models and product information. In addition to these consolidated standard formats, the use of XML-based formats becomes popular for the exchange of electronic data, like ebXML (modular specifications that provide support to electronic commerce) [7.21], EDIFACT (electronic data interchange for administration, commerce and transport) [7.22], STEPml (STEP markup language) [7.23] and PDML (product data markup language) [7.24].

For sharing information about manufacturing processes, there is a need of the use of languages like PSL (process specification language) and KIF (knowledge interchange format). In order to standardise the terminology that must be used in the manufacturing processes, it is needed to use proper ontology that expresses sets of terms, entities, objects and the relationships among them. Different standardised languages can be applied to achieve this purpose, such as OWL (ontology web language), RDF (resource description framework), and DAML (DARPA agent markup language) [7.25].

7.3.1.4 Visualisation and Notification Services

In order to enable communication channels needed to facilitate task assignments and interactions among distributed participants, the technological infrastructure for collaborative process planning must have electronic notification embedded services or external applications for the same purpose.

On the other hand, the functionality for data and documents visualisation must facilitate users' global access with the help of rich Internet applications (RIA) and standards for representing 3D models like VRML, U3D (universal 3D), X3D, 3DIF, DWFTM (Autodesk design web format) and JTOpen (developed by Unigraphics) among others [7.25].

Some of the requirements mentioned above can be found partially in PDM (product data management) tools, intended to store and manage engineering data and provided by 3D modelling software vendors.

7.3.2 Basic Requirements for Workflow Management

The adoption of the functionalities described previously is not a guarantee of success in the collaborative processes management. They must be complemented by extra functionalities that enable correct tasks coordination among distributed teams, in the collaborative process planning execution. Modern PLM (product lifecycle management) tools having embedded workflow management systems could support this scenario.

Workflows are useful for the coordination of interrelated activities carried out by organisation members in order to execute a business process. According to the ASQ (American Society for Quality), a business process is an organised group of related activities that work together to transform one or more kinds of inputs into outputs that are of value to the enterprise [7.26].

The term workflow is defined, according to Workflow Management Coalition (WfMC), as the automation of a business process in the course of which documents, information or tasks move from one participant to another in order to perform some actions, in accordance with a set of procedure rules [7.27]. When it is executed across an extended enterprise, it becomes a distributed workflow, in which different individuals participate in order to reach global objectives. In the work presented here, collaborative process planning can be considered as a business process that can be managed using distributed workflows.

A workflow management system (WfMS) defines, creates and manages the execution of workflows through the use of software, running on one or more workflow engines. The software components store and interpret process definitions, create and manage workflow instances as they are executed, and control their interaction with workflow participants and applications. The following sections contain a review of the required components for a WfMS to bring the basic support to coordinating collaborations among distributed participants.

7.3.2.1 Basic Workflow Components

According to WfMC, the basic items that must be represented in a workflow, to be complete and unambiguous for its execution in a WfMS, are as follows:

- *Activity*. A description of a piece of work that forms one logical step within a process. An activity may be a manual activity, which does not support computer automation, or a workflow (automated) activity.
- *Participant*. A resource which performs the work represented by a workflow activity instance. This work is normally manifested as one or more work items assigned to the workflow participant via a pending work list.
- *Role*. A mechanism that associates participants to a collection of workflow activity(s). A workflow participant assumes a role to access and process work from a workflow management system.
- *Routing*. A route defines the sequence of the steps that information must follow within a workflow. This element is fundamental for directing all the activity work to the distributed participants, in order to guarantee the success of the information flow and decisions taking.
- *Transition rule*. A transition rule is a logic expression that determines what actions need to be carried out depending on the value of logic operators. The definition of transition rules implies multiple options, variations and exceptions.
- *Event*. An occurrence of a particular condition (may be internal or external to the workflow management system) that causes the workflow management software to take one or more actions. For example, the arrival of a particular type of email message may cause the workflow system to start an instance of a specific process definition.
- *Deadline*. A time-based scheduling constraint, which requires that a certain activity work be completed by a certain time.

All of these items can be represented in a workflow model definition, that later can be interpreted and executed by a workflow engine. But first, it is needed to select a methodology to model the workflow items and their relations properly.

7.3.2.2 Workflow Modelling

There are several emerging industry standards and technologies related to workflow modelling that consider all the elements mentioned above. The Business Process Execution Language (BPEL) for web services [7.28] is emerging as a *de facto* standard for implementing business processes on top of web services technology.

Numerous WfMS support the execution of BPEL processes. However, BPEL modelling tools do not have the adequate level of abstraction required to make them usable during analysis and design phases of high complexity processes like collaborative product design, process planning and manufacturing.

On the other hand, the Business Process Modelling Notation (BPMN) [7.29] has attracted the attention of business analysts and system architects as a language for defining business process blueprints for subsequent implementation. The BPMN is a graph-oriented language in which control and action nodes can be connected almost arbitrarily. Also supported by numerous modelling tools, none of these can directly execute BPMN models, because they require the translation of BPMN to execute on a workflow enactment service with a workflow engine capable of executing BPEL directly.

7.3.2.3 Workflow Enactment Service

A workflow enactment service is a software service that may consist of one or more workflow engines for creating, managing and executing particular workflow instances. Applications may interface to this service via the workflow application programming interface (WAPI) that consists of a set of interfaces with particular components (Figure 7.2) [7.27]:

- *Interface 1*: The Process Definition Tools interface supports the exchange of process definition information. It defines a standard for interaction between a workflow process definition tool and the Workflow Enactment Service.
- *Interface 2*: This interface handles a workflow Client Application that contains various functions, for example to start and stop workflows, to write and read workflow attributes, to set up a work list for a user, among others.
- *Interface 3*: The Tool Agent application programming interface acts on behalf of a WfMS to invoke both external applications and internal customers or predefined procedures.
- *Interface 4*: This is an extensibility interface that enables communication between a WfMS and external programs. The interface enables processes and their instances to be created, managed and queried using an asynchronous request-response protocol. With this interface, it is possible to call a sub-workflow in a different system via XML and HTTP.
- *Interface 5*: This Audit Data Interface defines the format of the audit data that a conformant WfMS must generate during workflow enactment.

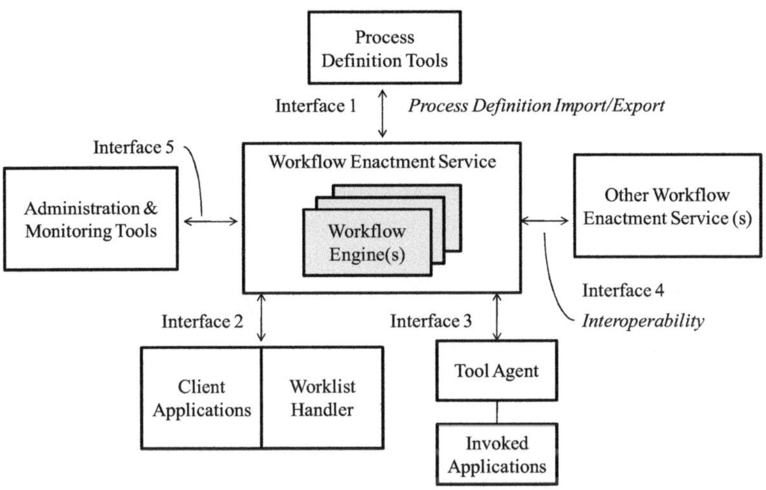

Figure 7.2. Workflow application programming interfaces (WAPI) schema

7.3.2.4 Workflow Persistence of Control Data

Workflow control data represents the dynamic state of the workflow system and its process execution instances. It may be written to persistent storage to facilitate restart and recovery of the system after a failure. The WfMC reference model identifies a number of common dynamic states that a process instance may take [7.27]:

- *Initiated*: the process instance has been created, but may not yet be running.
- *Running*: the process instance has started execution and one or more of its activities may be started.
- *Active*: one or more activities are started and activity instances exist.
- *Suspended*: the process instance is quiescent, or in other words, no further activities are started until it is resumed.
- *Complete*: the process instance has achieved its completion conditions and any post-completion system activities such as audit logging are in progress.
- *Terminated*: the execution of the process has been stopped (abnormally) due to error or user request.
- *Archived*: the process instance has been placed in an indefinite archive state (but may be retrieved for process resumption typically supported only for long lived processes).

Workflow control data may also be used to derive audit data. The audit data consists of historical records of the progress of a process instance from start to completion or termination. Such data normally incorporates information on the state transitions of the process instance.

7.3.3 Product Lifecycle Management Tools for Collaboration

Product lifecycle management (PLM) is an integrated approach that includes a series of methods, models and tools for information and process management during the different stages of a product lifecycle. PLM tools accomplish in a basic manner the aforementioned functionalities for knowledge, information and data management, and workflow management required in a collaborative ICT environment.

PLM tools are groupware tools used for storing, organising and sharing product-related data and for co-ordinating the activities of distributed teams in the progress of all product lifecycle stages like product design, manufacturing, supply, client service, recycling and other related activities [7.30]. The most immediate predecessors of PLM tools are PDM systems, which were designed to be used as databases to store engineering information such as CAD, CAE and CAM files and related documents. The basic architecture of PLM is based, according to Abramovici [7.31] and other authors, on three main functions as represented schematically in Figure 7.3:

- *KID Management*: Brings support for the identification, structuring, classification, modelling, recovery, dissemination, visualisation and storage of product and process data.
- *Workflow Management*: Brings support for modelling, structuring, planning, operating and controlling formal or semi-formal processes like engineering release processes, review processes, change processes or notification processes.

- *Application Integration*: Brings support for defining and managing interfaces between PLM and different authoring applications like CAD, CAM, CAE and integrated enterprise software such as ERP (enterprise resource planning) systems.

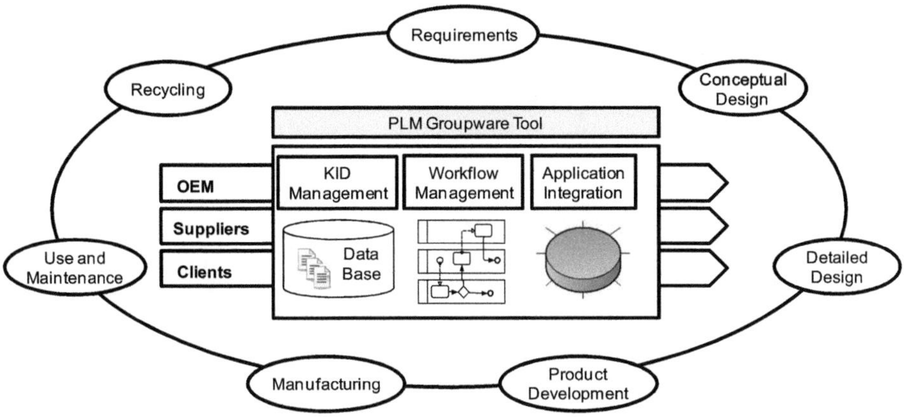

Figure 7.3. Basic components of PLM tools

Nowadays, PLM tools are consolidated in the engineering and manufacturing industry. Nevertheless, their use is concentrated in earlier stages of product lifecycle like conceptual and detailed design. But their functionalities can be exploited in other stages of product development, including process planning activities.

The current capabilities of commercial PLM tools reflect the historical evolution of their developing firms. For example, some solutions are focused on offering support to product development activities through CAD/CAM documents and engineering changes management. This is the case for PLM tools developed by Parametric Technology Corporation (PTC) [7.32], Dassault Systèmes [7.33] and Unigraphics [7.34]. Other PLM tools manage documents and data without their integration with CAD/CAM tools, but focusing on logistic processes, production control and other transactional processes. This is the case for PLM tool offered by SAP [7.35], named MySAP/PLM. Therefore, the selection of an adequate PLM tool depends on post-sale services and maintenance, integration with CAD/CAM tools already installed or other commercial considerations.

7.4 Reference Model for Collaborative Process Planning

The lack of a specific model for process planning across extended enterprises motivated us to propose a collaborative process planning reference model and an ICT architecture for its implementation, based on the required functionalities for enabling a collaborative environment, and those functionalities provided by PLM tools. The first step of the strategy needed to construct our reference model is to define a basic configuration model of an extended enterprise, useful to understand all important relationships within extended enterprises that collaborate during some

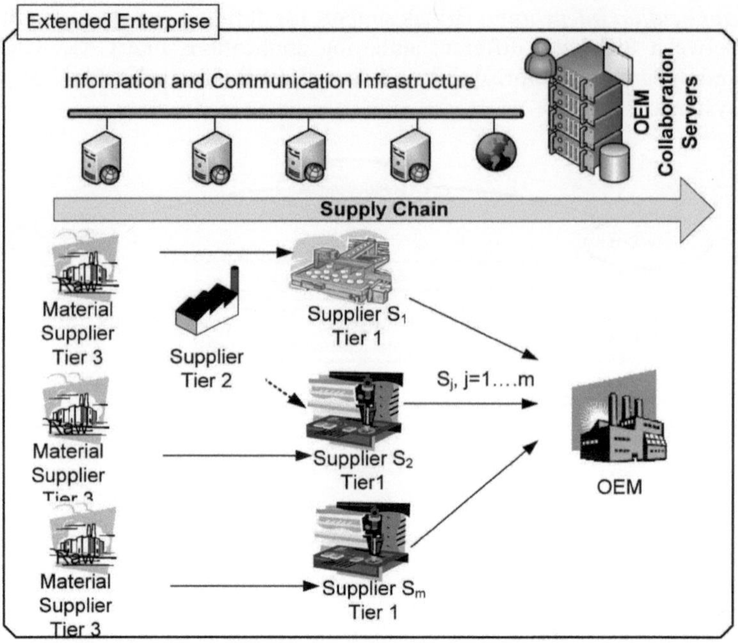

Figure 7.4. Basic configuration model of an extended enterprise

stages of the product lifecycle. The basic configuration model (Figure 7.4.) involves several distributed enterprises, partners of an extended enterprise dedicated to discrete products design and manufacturing.

The extended enterprise is constituted by an OEM (original equipment manufacturer) and by a set of enterprises; all of them are part of a supply chain. The first tier (or Tier 1) of this supply chain is composed of suppliers (S_j) of a product assembly parts. The other tiers are composed of other suppliers with less strategic value. All of these enterprises share information and data through one information and communication infrastructure provided by the OEM, specially configured to grant access to each partner in order to collaborate with them. In this particular scenario, suppliers of the first tier compete in order to obtain the manufacturing contract of a particular assembly part, solicited by the OEM.

The configuration of the extended enterprise led us to the definition of the reference model, which gives us a framework for collaborative process planning. The proposed reference model (Figure 7.5.) considers process planning as a set of activities included in the process of product design, development and manufacturing. Furthermore, an ICT reference architecture, based on functionalities that provide PLM tools (for applications integration, KID and workflow management), must be able to establish the collaborative environment.

According to the reference model, the OEM needs to pre-select manufacturing processes (meta-planning) after the stage of product design, so as to select those suppliers S_j that have the pre-selected processes in their manufacturing facilities. Selected suppliers, in order to compete for the manufacturing contract, need to

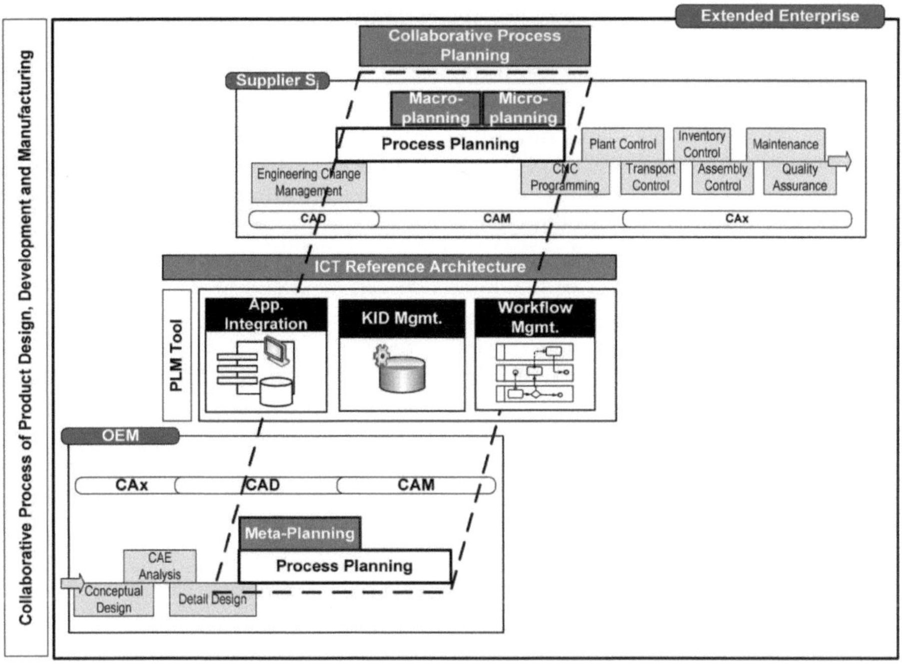

Figure 7.5. Reference model for collaborative process planning across extended enterprise

prepare the manufacturing quotes, based on the general characteristics of the assembly part to be manufactured. For this reason, they are required to elaborate a rough process plan that provides some basic information about costs and delivery times.

Once the OEM selects the best quote, the contract must be assigned to one supplier. The selected supplier personnel must elaborate a more specific and detailed process plan, performing for that purpose macro-planning and micro-planning activities aided by CAD/CAM tools.

For effective inter-enterprise sharing and exchange of data, information and knowledge and to enable the workflow management in our reference model for collaborative process planning, it is needed to identify, define and model activities, participants, roles, resources and the interactions among them. For this purpose, widely accepted techniques and methodologies are adopted for workflow modelling.

7.5 Collaborative Process Planning Activities Modelling

For the description and modelling of elements like roles, activities, deadlines, information exchange and other key factors of workflow models, the use of a single modelling methodology is not enough. According to the guidelines by OMG (Object Management Group), a combination of existing methodologies and languages that provide graphic representations understandable by users is required.

In the case of business activities modelling, the use of a business processes modelling methodology is needed, like the BPMN modelling notation. But it is first needed to identify all participants, resources and interactions. Secondly, it is also needed to revise the activities sequences performed for the execution of collaborative processes. This modelling strategy let us model adequately the workflows that reflect the collaborative process planning activities. Similar modelling strategy is also used by the OMG specification for product lifecycle management services, which defines a platform-specific model applicable to web services implementation [7.36].

UML provides two types of diagrams that can be useful to apply the proposed strategy: *use cases* and *sequence diagrams* [7.37]. The use cases diagrams describe in a graphic way the utilisation of a system by actors that can be anybody or anything. Therefore, use cases will help us describe how each participant interacts with the system (or with the environment) to achieve a specific goal in our reference model for collaborative process planning. On the other hand, the sequence diagrams allow us modelling the logic flow of interactions within the system from a chronological perspective. Figure 7.6 shows the procedure for activities modelling, using the diagrams mentioned above.

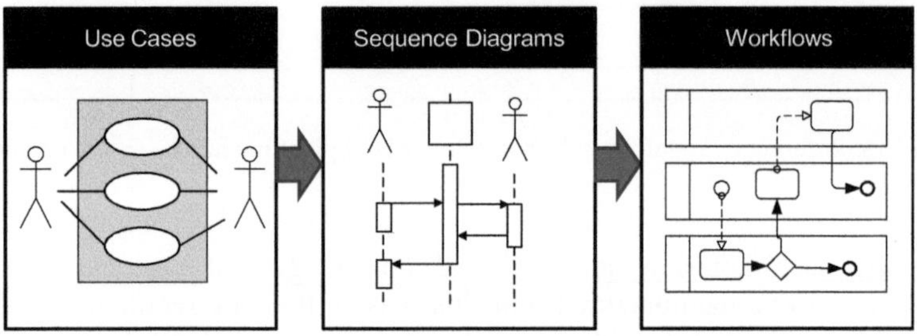

Figure 7.6. Proposed strategy for workflow modelling

7.5.1 Use Cases Modelling

Use cases diagrams in the proposed environment for process planning represent different departments of the enterprises that compose the extended enterprise. For example, Figure 7.7 represents the use case of an OEM Supply Chain Management (SCM) Department, dedicated to the manufacturing quotation management. In this diagram, a 3D model, a SCM manager and the suppliers' project managers take part in the function denominated manufacturing quotation, which includes different activities needed for its completion: execute meta-planning, send quote requests, receive quotes, assign manufacturing contract and manage engineering changes.

On the other hand, the diagram of Figure 7.8 shows the use case that represents the functions that must be performed during the manufacturing quotation in the supplier S_j Engineering Department. This diagram involves the quote request from the OEM, a database that stores former quotes and process plans, and involves a 3D model. Actors that interact in this use case diagram are the supplier project manager who performs macro-planning activities, the technical engineer who performs

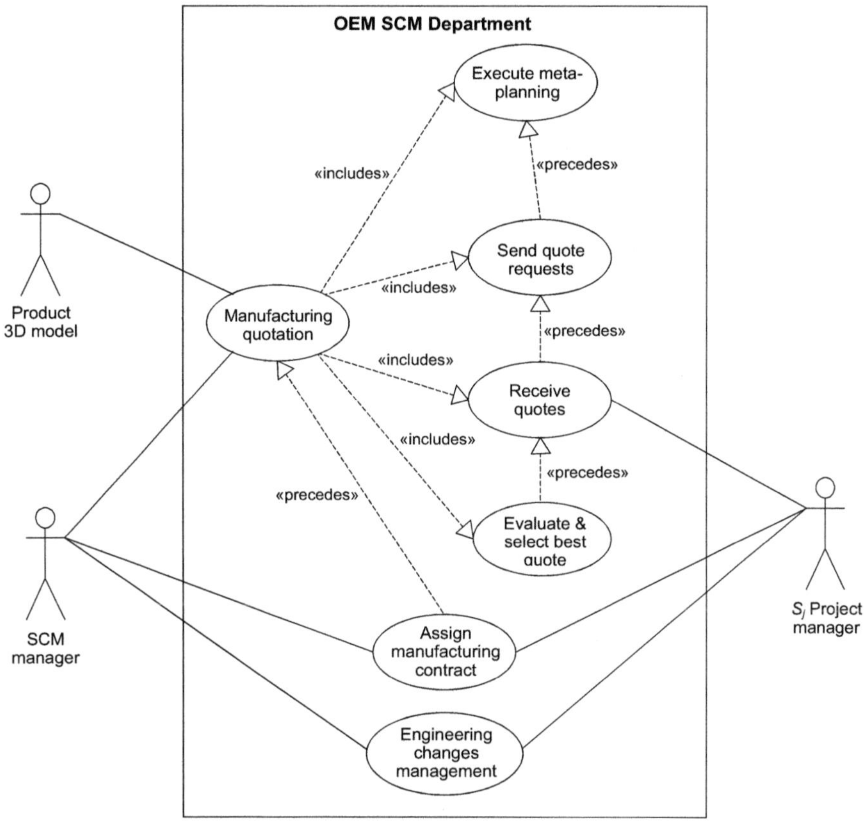

Figure 7.7. Use case of an OEM SCM department

micro-planning activities, and the OEM SCM manager who is responsible for the contract confirmation.

It can be seen in Figure 7.8 that the supplier could request engineering changes after the execution of a rough macro-planning, for enabling manufacturability. Once the contract is confirmed by the OEM SCM manager, the activities needed to execute the detailed process planning will begin.

The detailed process planning activities must be executed in a collaborative way by a technical engineer of the selected supplier S_j Engineering Department and by shop floor personnel, in order to incorporate feedback with process knowledge acquired on the shop floor, either by experimentation or by shop floor personnel experience. When the detailed process plan is finished, CNC codes and process planning documentation will be available to be used for manufacturing.

The modelling of use cases presented above has been useful for identifying and representing all the interaction among elements, participants and activities that take part in collaborative process planning across an extended enterprise. Nevertheless, it also requires the representation of this interaction in a chronological way. This is the purpose of the sequence diagrams presented in the following section.

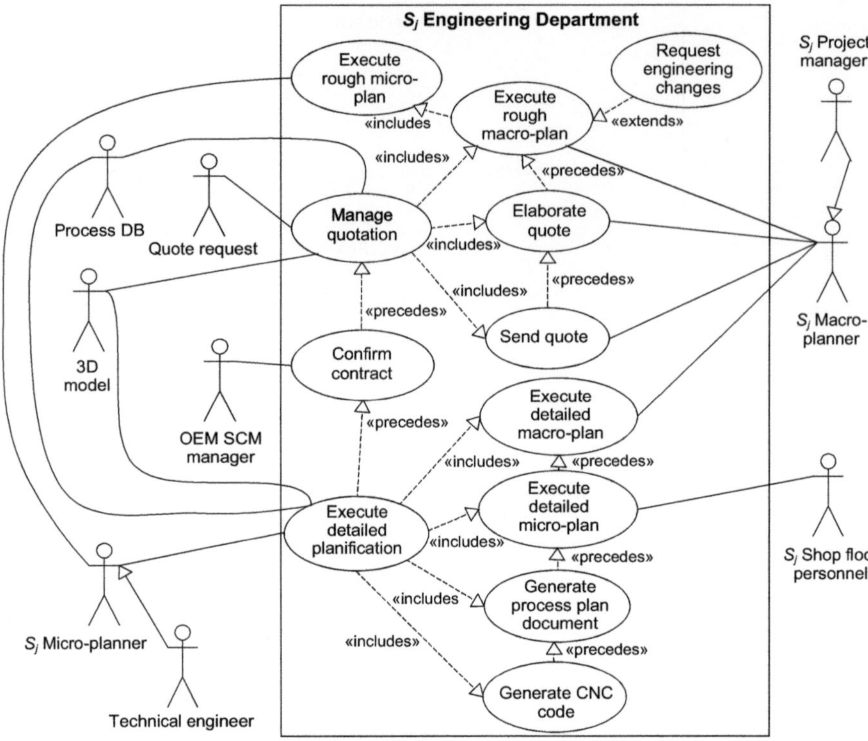

Figure 7.8. Use case of supplier S_j Engineering Department

7.5.2 Sequence Diagrams Modelling

A sequence diagram shows different processes or objects (as parallel vertical lines) that exist simultaneously and the messages exchanged between them (as horizontal arrows), in the order in which they occur. This allows the specification of simple runtime scenarios in a graphical manner. The elements of a sequence diagram are objects, lifelines (dashed lines) and message icons (horizontal arrows) among others.

The first diagram modelled is shown in Figure 7.9, in which it can be observed that the sequence of how meta-planning and the procedures of manufacturing quotation are performed by the OEM team, using product requirements and the 3D model. It also shows the request for engineering change that takes effect if the suppliers' S_j teams detect early problems for manufacturing.

The next modelled diagram (Figure 7.10) expresses the sequence of the rough process planning in the supplier S_j Engineering Department for the elaboration of the manufacturing quote. The diagram shows the chronology of rough process planning activities and management of engineering changes.

Once the sequence for rough process planning has been represented, it is necessary to represent the sequence diagram of the detailed process planning procedure, performed by the selected S_j supplier teams (Figure 7.11). As can be seen in the diagram, elaboration of the macro-plan and micro-plan depends on the process

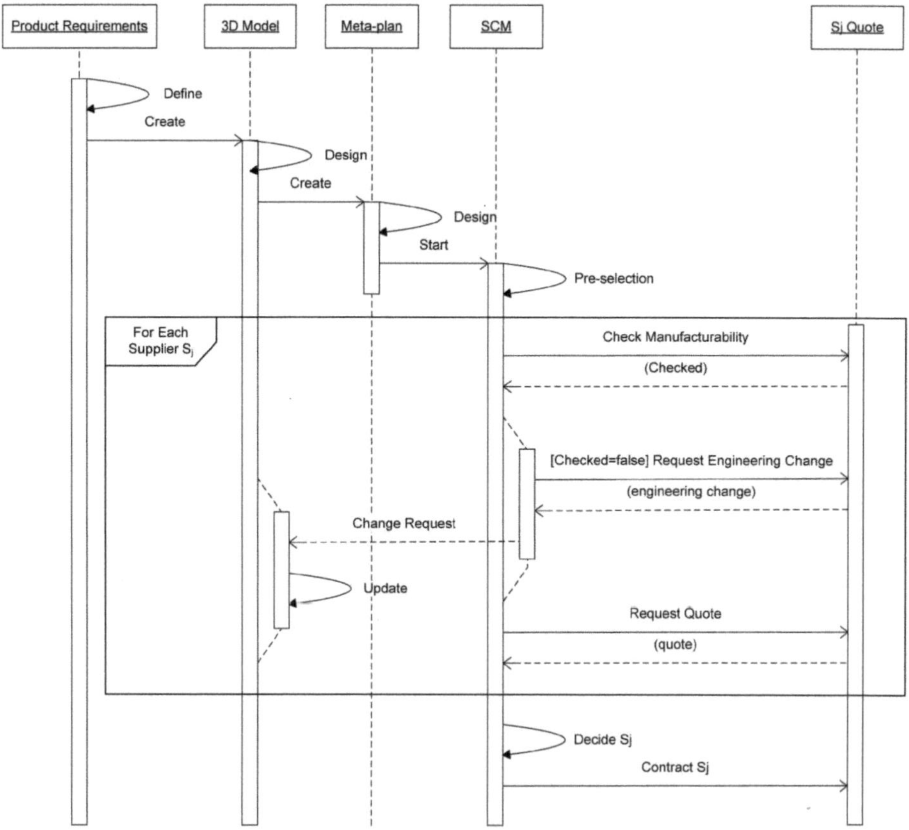

Figure 7.9. Sequence diagram of meta-planning and manufacturing quotation

information stored in databases and on feedback from the shop floor. The detailed process planning sequence is controlled by a contract execution line and ends when the CNC codes are generated for performing product manufacturing.

The set of use cases and sequence diagrams allows us to capture all relations, activities, elements, processes, sequences and products that take part in our reference model for collaborative process planning. The next step is to model the workflows to be implemented in a PLM tool with a workflow engine.

7.5.3 Workflow Modelling

A BPMN diagram is made of a set of graphical elements (some of them are shown in Figure 7.12) easy to understand by a user or a business process analyst. The particular BPMN model to represent the workflow needed for the implementation of the reference model for collaborative process planning is shown in Figure 7.13. In order to track the document history during the process planning activities and to delimit the transition between these activities, it is necessary to determine the different stages of the lifecycle of the process planning documents (coincident with

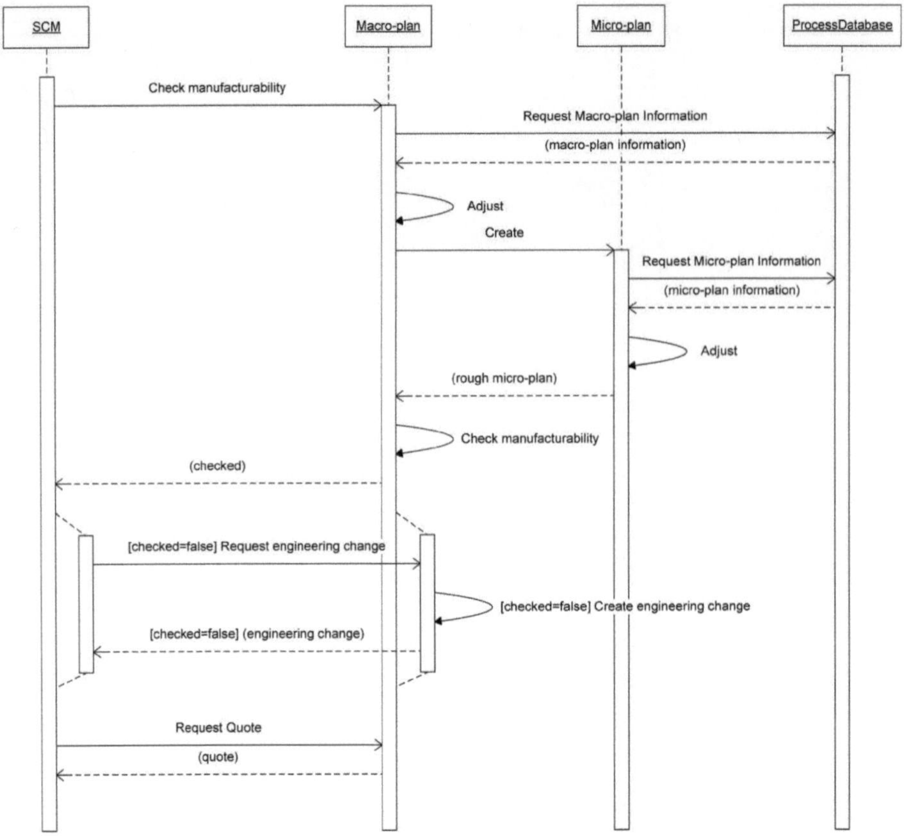

Figure 7.10. Sequence diagram of rough process planning

some product lifecycle stages), showed in the top ribbon of the BPMN model in Figure 7.13. This is also necessary for establishing access permissions to the personnel involved in each stage.

According to our BPMN model, collaborative process planning begins with the 3D modelling in the OEM Engineering Department (represented by a swim lane). This file is reviewed by the SCM team to guarantee the part manufacturability. These activities are inside the product lifecycle stage denominated *Design* in our model.

Meta-planning activities are performed in the SCM Department for pre-selecting the manufacturing processes that must be performed and the suppliers that have the pre-selected processes in its facilities. A quote request is sent to the pre-selected suppliers and with this activity the product lifecycle stage denominated Meta-Planning is finished.

The pre-selected suppliers (S_1 and S_2) receive the quote request together with a process planning file, which contain the 3D model and other product data. With this file, each project manager and technical engineer performs a rough macro-planning and a rough micro-planning, respectively, in order to elaborate an adequate quote or

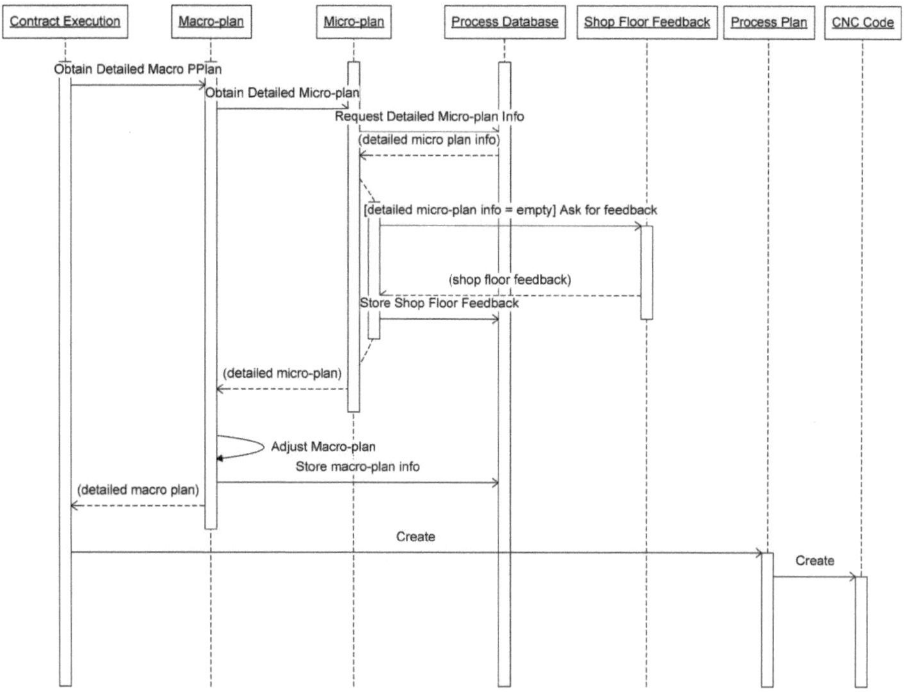

Figure 7.11. Sequence diagram of detailed process planning

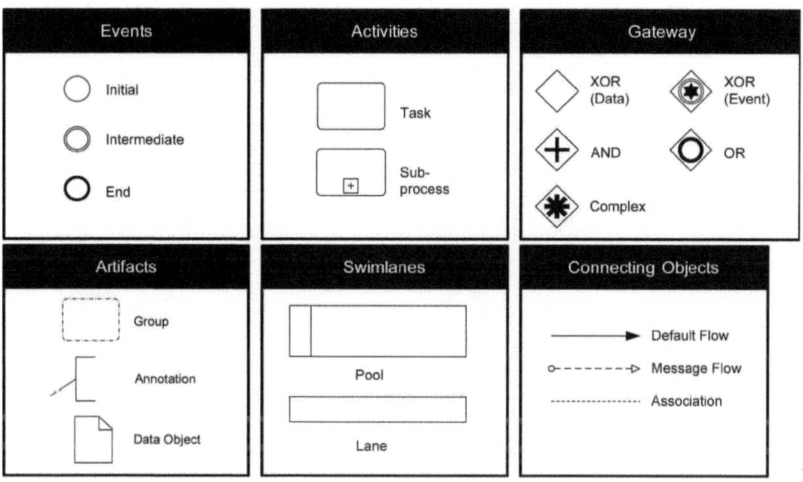

Figure 7.12. Most common modelling objects in BPMN

to request an engineering change to the OEM designer. This redesigns the 3D model and with this activity the product lifecycle stage denominated Rough Process Planning ends.

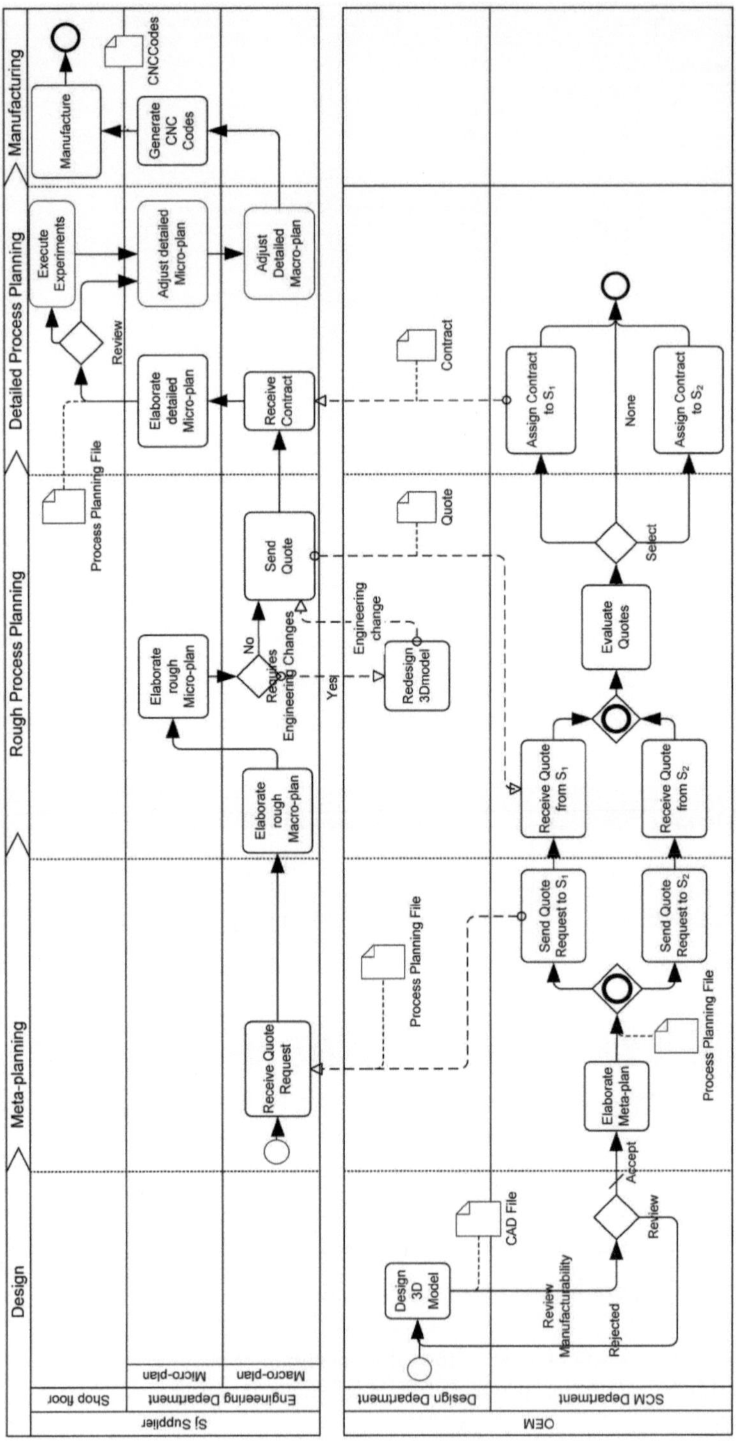

Figure 7.13. BPMN workflow model for collaborative process planning across an extended enterprise

The next stage of the product lifecycle is denominated Detailed Process Planning and consists of a set of activities for the execution of detailed macro-planning and micro-planning, once the contract has been assigned to one supplier (S_1 or S_2). These activities are performed in a collaborative way, as it is modelled in the use cases and in the sequence diagrams, between the technical department and shop floor personnel, in order to enrich the process plan with shop floor knowledge that impacts quality, productivity and costs (*e.g.* cutting parameters and cutting tool paths). This knowledge must be incorporated in the definitive process plan, by adjusting the detailed macro- and micro-plans.

The last stage of the product lifecycle represented in our workflow model is denominated Manufacturing. In this stage, the CNC codes are generated and are sent to shop floor machines. Thus, product lifecycle stages related to process planning activities ends.

The modelling of process planning activities with the strategy presented above can be useful for its implementation in a PLM tool proprietary workflow engine. In order to implement this collaborative environment in a real industrial scenario, we present the implementation of ICT reference architecture in the next section.

7.6 Implementation of ICT Reference Architecture

As mentioned previously, an ICT reference architecture is required in order to implement a platform that enables the collaborative process planning across an extended enterprise. This architecture must present the required ICT infrastructure to implement each of the previously identified components and information exchange and communications needs. The implementation of this ICT reference architecture (Figure 7.14) is intended to be the backbone of the collaborative processes management.

From a functional perspective, the architecture that gives support to the collaborative process planning model must be of client-server architecture. At the server side, a shared web application server should contain at least software components for managing product data, product lifecycle, related documents, and support for workflows and visualisation services. To support collaborative data storage and retrieval, it must have a shared file vault repository and connectivity with other enterprise systems and databases implicated. On the other hand, at client side, the collaborative platform clients must have connectivity with the application server through web browsers and CAD/CAM tools with integrated plug-ins to access and modify shared data.

The required communication between clients and server through the Internet must deal with the use of common communication standards like HTTP/S, SOAP (simple object access protocol), XML, WSDL (web service description language) and RMI (remote method invocation) among others. Also, the platform must be enabled for the exchange of pre-arranged standardised formats like STEP, IGES, DXF, STL, VRML, and other design, manufacturing and office document formats. By the use of standardised communication protocols and interchange formats, this platform is ready to offer interoperability across the extended enterprise with each organisation system like CAPP and other CAx, ERP, SCM and MES (manufacturing execution systems).

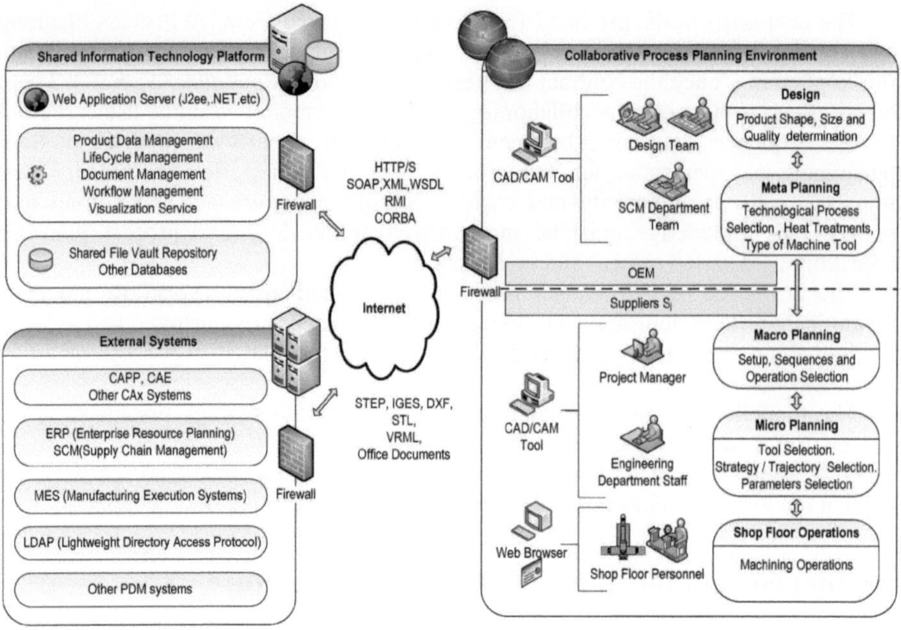

Figure 7.14. Implementation ICT architecture for collaborative process planning across an extended enterprise

Another important issue to be accomplished by the platform is the security management in the communication level ensuring the user authentication and the secure information transport. Firewalls and encrypting protocols like TLS (Transport Layer Security) or its predecessor SSL (Secure Sockets Layer) must be used to avoid security attacks to the platform or undesired access to data like, for example, packet sniffing. Directory services like LDAP (Light-weight Directory Access Protocol) are needed for storing user information of participants and for providing authentication methods.

Almost all of the above technical requirements can be found in modern PLM tools available in the market today. However, commercial PLM systems are still limited for working in heterogeneous environments, due to their lack of interoperability with other enterprise systems. An appropriate study of the tools and systems used across the extended enterprise is necessary to identify the interoperability issues and to select a PLM tool that best fit the environment.

From an operational point of view, in the collaborative process planning framework, participants from different teams must interact with the application server using client tools like CAD/CAM tools and web browsers.

These interactions must be managed with the help of the workflow management features presented in this architecture. It must execute instances of the workflow model presented previously, to ensure the correct coordination of the OEM and suppliers participants for enabling or disabling dynamic access to the documents or product data information. Thus, the process planning activities can be performed in a collaborative way between the OEM teams and each supplier team in order to carry out business processes associated with process planning tasks, like budgeting and engineering changes approvals.

7.7 Case Study

In the same way of product design, manufacturing process plans can also be created, modified and reviewed, collaboratively, using PLM tools. In order to validate the previously proposed methodology, a case study was carried out to model the activities involved in the collaborative process planning.

This work was divided in three stages: (1) setup of a collaborative environment, (2) creation of the lifecycle phases in a manufacturing process plan, and (3) implementation of the required workflow.

7.7.1 Setup of a Collaborative Environment

The environment chosen for the case study is a scenario of an extended enterprise dedicated to the manufacture of moulds for ceramic tiles. Within this context, three geographically distributed companies interact. One of them (A) is the owner of the mould products and the other two (B and C) are members of the supply chain providing parts for the moulds (Figure 7.15), more specifically, parts that require certain machining operations. The described scenario fits well into the reference model for collaborative process planning.

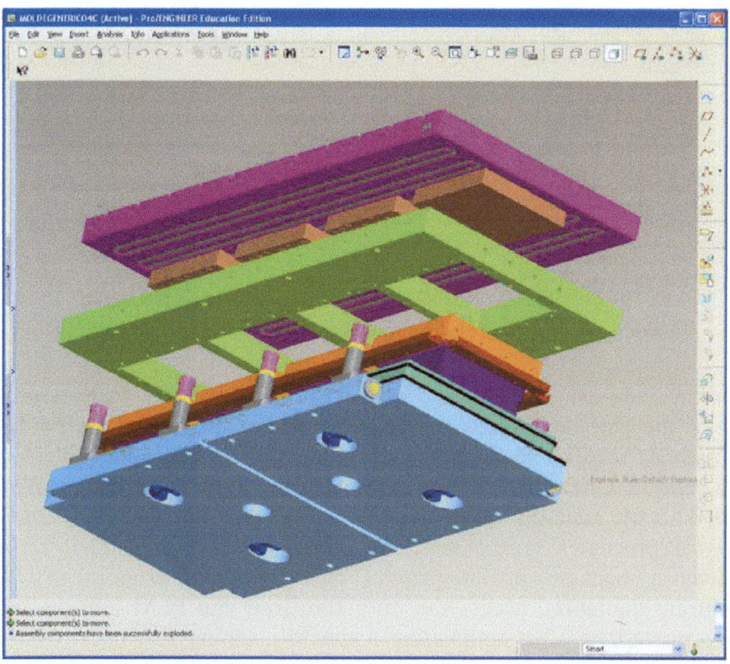

Figure 7.15. Exploded view of parts of a mould for forming ceramic tiles

The collaborative environment was set up using a PLM tool (Windchill by PTC), a CAD/CAM application (ProEngineer Wildfire 3.0) and Internet tools. The commercial software and matching hardware facilitated the implementation of the proposed ICT reference architecture.

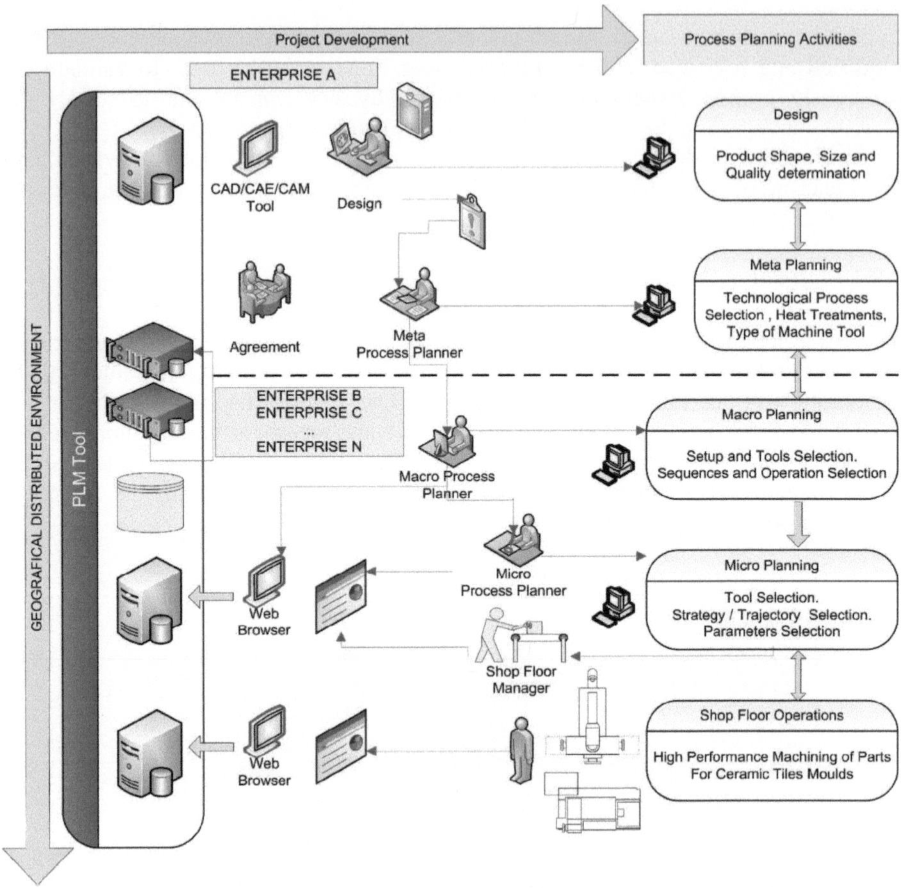

Figure 7.16. Proposed environment for collaborative process planning

The case study can be described as follows (Figure 7.16). In Enterprise A, the designer of the part determines the shape, size and quality of the product and a process planner reviews the manufacturability of the design and then uses a 3D model to create a CAD/CAM file in manufacturing process format (*.mfg) for Pro-Engineer. This file contains the process plan at the meta-planning level (selection of technological processes, type of machines, thermal treatments, and so forth). Later, the engineering managers at Enterprises B and C receive a proposal to draw up the manufacturing quotation, access the .mfg file using the PLM tool and send their respective quotations after completing a macro-planning (selection of setups and equipment, and selection and sequencing of operations). Once a quotation has been approved or rejected by Enterprise A, after a due date, a member of the team in the enterprise that wins the contract (B or C) undertakes the micro-planning (selection of machining parameters, tools and strategies).

The plant manager at that enterprise receives the preliminary process plan, through the PLM tool (by means of a viewer in their web browser), and determines the real operating conditions in accordance with the performance and the capacity of the manufacturing process. The process micro-planner integrates them into the final process plan and sends them to the shop floor in the form of CNC files and instruction sheets. The files are saved in the PLM tool database so that they can be retrieved for planning in the future.

7.7.2 Creation of Lifecycle Phases in a Manufacturing Process Plan

This case study involved the following process stages: initial design of the process plan and the manufacturing proposal (PROPOSAL), quotation activities for macro-planning (QUOTATION), micro-planning (PLANNING), and final development and manufacturing (MANUFACTURING) (Figure 7.17). Each of the stages is delimited by approval "gates" that determine the transition from one stage to another, as the tasks set out in the workflow are completed. In Figure 7.17, a simplified view of the lifecycle with the four activities and the gates is shown; a trigger workflow has been included for a better understanding of the collaborative activities management with the workflow module of the PLM.

7.7.3 Implementation of Required Workflow

The methodology for modelling the collaborative process planning activities for the reference model presented previously was applied in this case study. As a result, the BPMN reference model was translated to the proprietary PTC workflow notation. The final workflow is shown in Figure 7.17 and it is used to automate the running of tasks that allow the process planning to move on from one stage to another during its lifecycle. It includes particular workflows (Figure 7.18) for the co-ordination of specific tasks to be executed accordingly to the collaborative interactions identified previously.

All the processes start with the assignment of a lifecycle and a workflow to the manufacturing file (*.mfg) that stores all the required manufacturing information. It is then made available (according to a prior assignation of roles and authorisations) to all those involved in the collaborative process planning so that it can be modified or viewed as needed.

7.7.4 Results and Discussions

The case study has achieved the aim of validating some aspects of the general proposal for managing collaborative process planning activities. In this case, the development of a prototype project of a ceramic tile mould, with the proposed methodology and a PLM system, has helped to match the theoretical proposal to a real case. Figure 7.19 shows the process plan for machining a mould part, and the simulation of its execution in Vericut software within ProEngineer.

In order to achieve this, a modification suggested by the plant manager had to be incorporated because the machining parameters calculated by the micro-planner

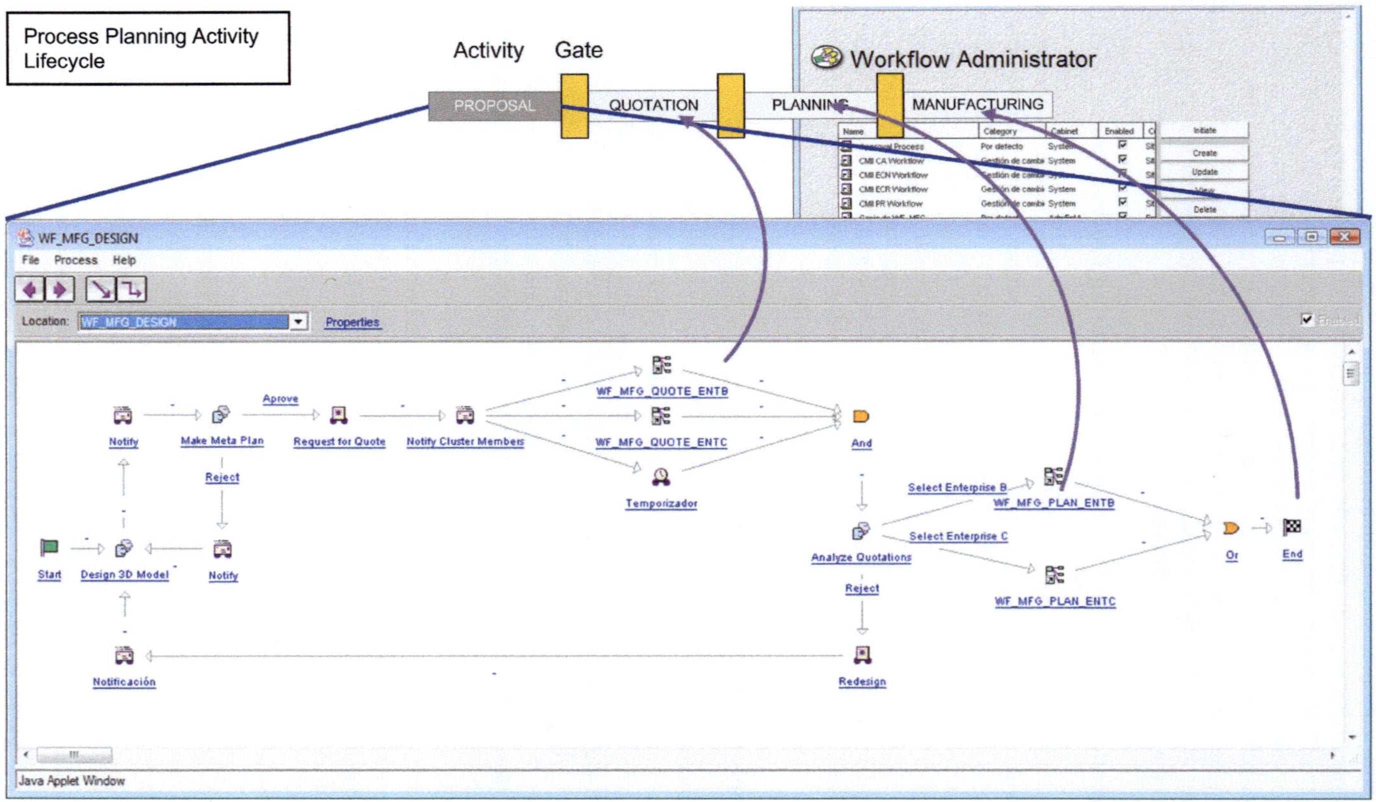

Figure 7.17. Collaborative process planning lifecycle and workflow

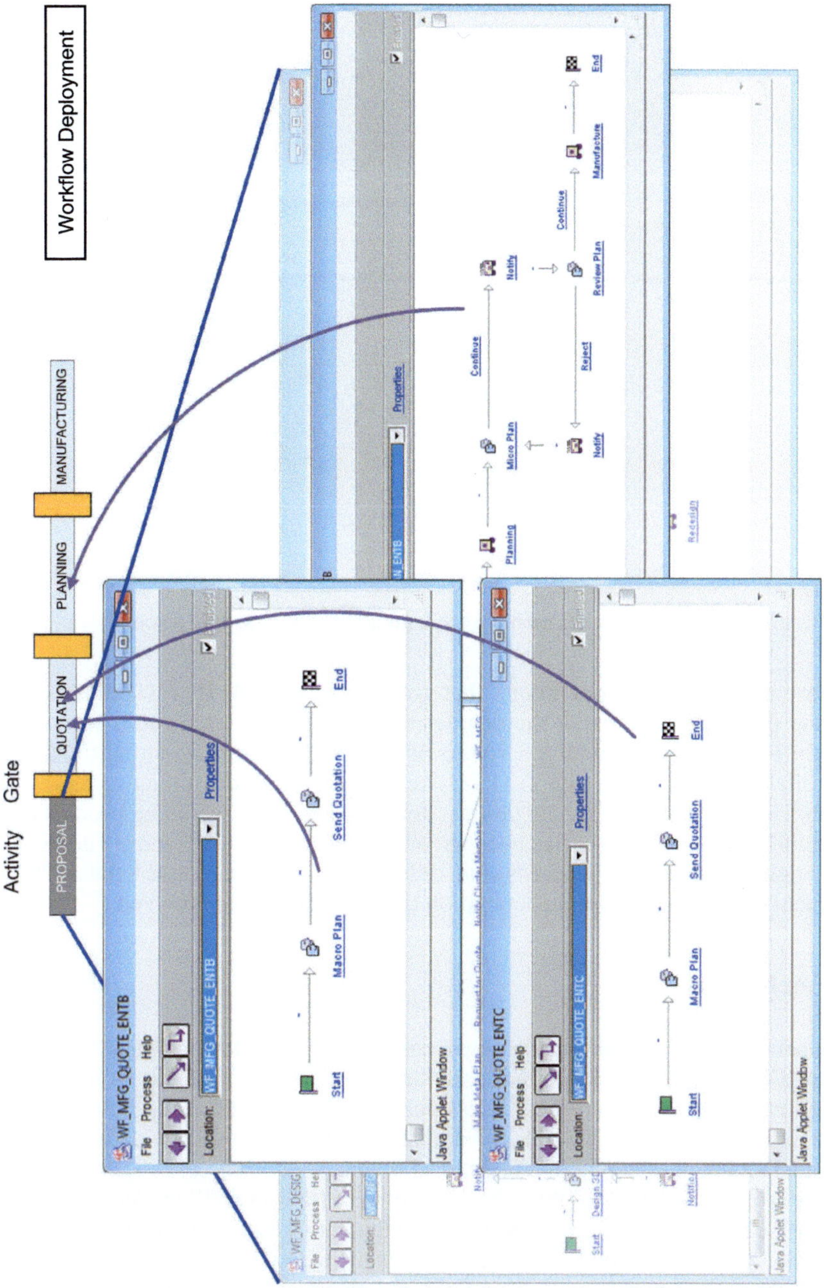

Figure 7.18. Workflow deployment across collaborative process planning

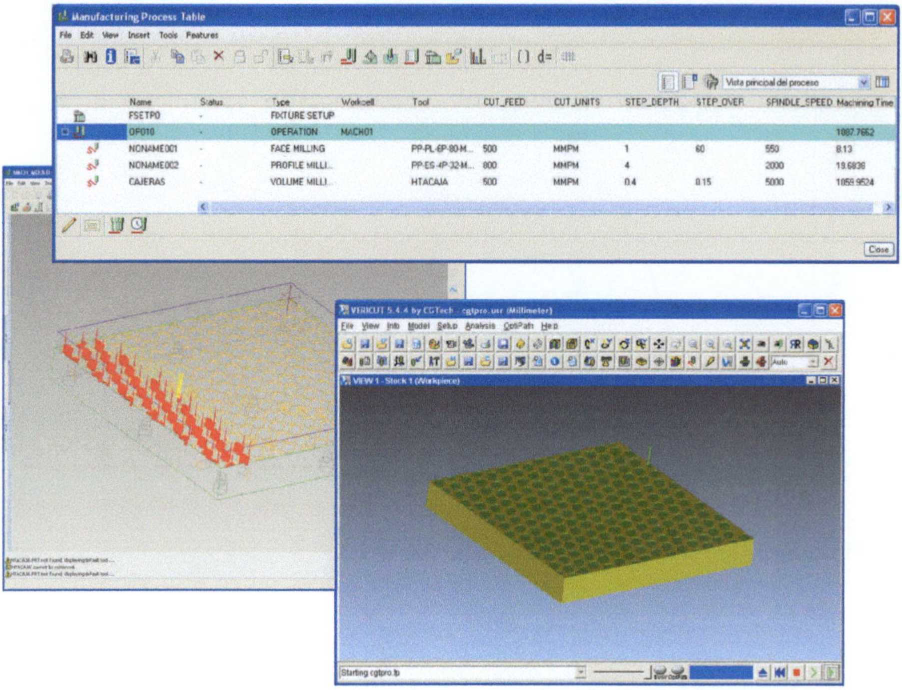

Figure 7.19. Simulation of machining of a mould part and its final process plan

were not suitable for the operations that were programmed. It was necessary to adjust the parameters to match the characteristics of the material (cold work tool steel DIN 1.2080/AISI D3 with a hardness of 60 HRC) and of the cutting tools that were available (tool holders with TiAlN-coated tungsten carbide milling cutters).

With the experience, the whole product development time, including product and manufacturing process design, has been slightly reduced but what is more interesting is that engineering change requests and orders have been minimised thanks to a more fluent communication and cooperation between the people involved.

7.8 Conclusions

The consolidation of virtual organisations, particularly extended enterprises focused on product design, development and manufacturing, has motivated us to propose a reference model for collaborative process planning. This model could be useful for identifying the collaborative process planning activities that must be carried out in this context and their mapping in a real-world scenario.

The strategy presented in this chapter for workflow modelling, including use cases and sequencing diagrams, should be taken into account as a guide to help achieve the co-ordination of activities and exchange of information during the collaborative manufacturing activities of process planning that could involve several companies of an extended enterprise.

The implementation of this reference model can be technically achieved, thanks to the use of ICT available today, which has been presented in the ICT reference architecture. Companies belonging to a supply chain should exploit this kind of tools in order to co-operate in the product development process, shortening lifecycles of new products introduction to market.

Nevertheless, they must consider manufacturing planning activities as one of the key product development stages, taking especial attention when it must be done abroad or it should be outsourced. In this case, collaborative engineering ought to be transferred to effective process planning activities beyond the frontiers of the company and complementing the concurrent design.

The literature review has been valuable to detect future technology trends for enabling distributed process planning. It is important to point out that PLM tools need to interoperate with CAPP systems to bridge the gap between concurrent product design and collaborative manufacturing within the product lifecycle.

It is expected to work in the direction of integrating not only CAD/CAM tools in PLM systems but also CAPP tools, because PLM must act as a backbone to provide all the required product information to other applications. Therefore, future research should integrate next generation CAPP systems with PLM tools through automated workflows that will enable the embedded execution of CAPP applications for an entirely automated collaborative process planning.

Acknowledgements

This research was funded by the Fundación Caja Castelló-Bancaixa and the Universitat Jaume I through the project entitled "Integration of Process Planning, Execution and Control of High Speed Machining in Collaborative Engineering Environment – Application to Ceramic Tiles Moulds Manufacturing". We are also grateful for support from the European Union's Alβan Programme of Scholarships for Latin America, grant number E04D030982MX.

References

[7.1] Zhao, F.L., Tso, S.K. and Wu, P.S.Y., 2000, "A cooperative agent modelling approach for process planning," *Computers in Industry*, **41**, pp. 83–97.

[7.2] Chan, K., King, C. and Wright, P., 1998, "COMPASS: Computer oriented materials, processes, and apparatus selection system," *Journal of Manufacturing Systems,* **17**, pp. 275–286.

[7.3] Ahn, S.H., Sundararajan, V., Smith, C., Kannan, B., D'Souza, R., Sun, G., Mohole, A., Wright, P.K., Kim, J., McMains, S., Smith, J. and Sequin, C.H., 2001 "CyberCut: an Internet-based CAD/CAM system," *Transactions of the ASME, Journal of Computing and Information Science in Engineering*, **1**, pp. 52–59.

[7.4] Van Zeir, G., Kruth, J.P. and Detand, J., 1998, "A conceptual framework for interactive and blackboard based CAPP," *International Journal of Production Research,* **36**, pp. 1453–1473.

[7.5] Tu, Y.L., Chu, X.L. and Yang, W.Y., 2000, "Computer-aided process planning in virtual one-of-a-kind production," *Computers in Industry,* **41**, pp. 99–110.

[7.6] Wang, L., Feng, H.-Y. and Cai, N., 2003, "Architecture design for distributed process planning," *Journal of Manufacturing Systems,* **22**, pp. 99–115.

[7.7] Kulvatunyou, B., Wysk, R.A., Cho, H.B. and Jones, A., 2004, "Integration framework of process planning based on resource independent operation summary to support collaborative manufacturing," *International Journal of Computer Integrated Manufacturing,* **17**, pp. 377–393.

[7.8] Chung, C.H. and Peng, Q.J., 2004, "The selection of tools and machines on web-based manufacturing environments," *International Journal of Machine Tools & Manufacture,* **44**, pp. 317–326.

[7.9] Sormaz, D.N., Arumugam, J. and Rajaraman, S., 2004, "Integrative process plan model and representation for intelligent distributed manufacturing planning," *International Journal of Production Research,* **42**, pp. 3397–3417.

[7.10] You, C.F. and Lin, C.H., 2005, "Java-based computer-aided process planning," *International Journal of Advanced Manufacturing Technology,* **26**, pp. 1063–1070.

[7.11] Feng, S.C., 2005, "Preliminary design and manufacturing planning integration using web-based intelligent agents," *Journal of Intelligent Manufacturing,* **16**, pp. 423–437.

[7.12] Nassehi, A., Newman, S.T. and Allen, R.D., 2006, "The application of multi-agent systems for STEP-NC computer aided process planning of prismatic components," *International Journal of Machine Tools & Manufacture,* **46**, pp. 559–574.

[7.13] Cecil, J., Davidson, S. and Muthaiyan, A., 2006, "A distributed internet-based framework for manufacturing planning," *International Journal of Advanced Manufacturing Technology,* **27**, pp. 619–624.

[7.14] Guerra-Zubiaga, D. and Young, R.I.M., 2007, "Design of a manufacturing knowledge model," *International Journal of Computer Integrated Manufacturing*, Published Online: DOI: 10.1080/09511920701258040.

[7.15] Mahesh, M., Ong, S.K. and Nee, A.Y.C., 2007, "A web-based multi-agent system for distributed digital manufacturing," *International Journal of Computer Integrated Manufacturing,* **20**, pp. 11–27.

[7.16] Peng, Q., Chung, C. and Luan, T., 2007, "A networked virtual manufacturing system for SMEs," *International Journal of Computer Integrated Manufacturing,* **20**, pp. 71–79.

[7.17] Ming, X.G., Yan, J.Q., Wang, X.H., Li, S.N., Lu, W.F., Peng, Q.J. and Ma, Y.S., 2008, "Collaborative process planning and manufacturing in product lifecycle management," *Computers in Industry*, **59**, pp. 154–166.

[7.18] Yang, H. and Xue, D., 2003, "Recent research on developing Web-based manufacturing systems: a review," *International Journal of Production Research,* **41**, pp. 3601–3629.

[7.19] Li, W.D. and Qiu, Z.M., 2006, "State-of-the-art technologies and methodologies for collaborative product development systems," *International Journal of Production Research,* **44**, pp. 2525–2559.

[7.20] Van den Hamer, P. and Lepoeter, K., 1996, "Managing design data: the five dimensions of CAD frameworks, configuration management, and product data management," In *Proceedings of the IEEE*, **84**, pp. 42–56.

[7.21] United Nations Centre for Trade Facilitation and Electronic Business, http://www.unece.org/cefact/tc154_h.htm.

[7.22] United Nations Directories for Electronic Data Interchange for Administration, Commerce and Transport, http://www.unece.org/trade/untdid/welcome.htm.

[7.23] PDES, http://pdesinc.aticorp.org/

[7.24] Organisation for the Advancement of Structured Information Standards, http://xml.coverpages.org/pdml.html.

[7.25] Subrahmanian, E., Rachuri, S., Fenves, S.J., Foufou, S. and Sriram, R.D., 2005, "Product lifecycle management support: a challenge in supporting product design and

manufacturing in a networked economy," *International Journal of Product Lifecycle Management,* **1**, pp. 4–25.
[7.26] American Society for Quality, http://www.asq.org/
[7.27] Workflow Management Coalition, http://www.wfmc.org/
[7.28] WS-BPEL Committee, *Web Services Business Process Execution Language*, http://www.oasis-open.org/committees/tc_home.php?wg_abbrev=wsbpel.
[7.29] Object Management Group, *Business Process Management Initiative*, available at http://www.bpmn.org/
[7.30] Contero, M. and Vila, C., 2004, "Collaborative engineering," In *Advances in Electronic Business*, **I**, Li, E. and Du, T.C. (eds), Idea Group Publishing, Hershey Pennsylvania, pp. 87–122.
[7.31] Abramovici, M.A., 2007, "Future trends in product lifecycle management," In *The Future of Product Development*, Krause, K-L. (ed.), Springer, Berlin, pp. 675–674.
[7.32] Parametric Technology Corporation, http://www.ptc.com/
[7.33] Dassault Systèmes, http://www.dsdsf.com/
[7.34] Unigraphics, http://www.ugs.com/
[7.35] SAP Product Lifecycle Management, http://www.sap.com/solutions/business-suite/plm/index.epx.
[7.36] Product Lifecycle Management Specification 2.0 – OMG Adopted Specification dtc/07-05-01, http://www.omg.org/
[7.37] Object Management Group – Unified Modelling Language, http://www.uml.org/

8

Adaptive Setup Planning for Job Shop Operations under Uncertainty

Lihui Wang[1], Hsi-Yung Feng[2], Ningxu Cai and Ji Ma[2]

[1] Centre for Intelligent Automation
University of Skövde, PO Box 408, 541 28 Skövde, Sweden
Email: lihui.wang@his.se

[2] Department of Mechanical Engineering
The University of British Columbia, Vancouver, BC V6T 1Z4, Canada
Email: feng@mech.ubc.ca

Abstract
This chapter presents a novel decision-making approach toward adaptive setup planning that considers both the availability and capability of machines on a shop floor. It loosely integrates scheduling functions at the setup planning stage, and utilises a two-step decision-making strategy for generating machine-neutral and machine-specific setup plans at each stage. The objective of this research is to enable adaptive setup planning for dynamic job shop machining operations through collaborations among multiple system modules residing in different resources and interactions with human operators. Particularly, this chapter covers basic concepts and algorithms for one-time generic setup planning, and run-time final setup merging for specific machines. The decision-making process and algorithms validation are further demonstrated through a case study. It is expected that the proposed approach can largely enhance the dynamism of fluctuating job shop operations.

8.1 Introduction

With increased product diversification today, companies must be able to profitably produce in small quantities and make frequent product changeovers. This leads to dynamic job shop operations that require a growing number of setups in a machine shop. How to come up with effective and efficient setup plans in such a changing environment is highly in demand.

Setup planning is the critical link between general process planning and detailed operation planning in a machine shop; it is also the intimate upstream of fixture planning [8.1]. The purpose of setting up a part is to ensure its stability during machining, and more importantly, to guarantee the precision of the machining process. The task of setup planning is to determine the number and sequence of setups, the machining features in each setup, and the locating orientation of each setup. It is also closely relevant to process planning and scheduling.

For many years, process planning and scheduling have been treated as separate systems, each of which has its own targets and optimisation rules. On one hand, process planning systems usually focus on analysing design requirements and often overlook the potential of integration with scheduling functions. Limited attention has been paid to the effect that changing shop floor conditions may have on the desirability of process plans. On the other hand, research on scheduling has been primarily focused on the construction of efficient algorithms to solve different types of scheduling problems [8.2]. The result of the separation is a gap between process planning and scheduling, with little flexibility of a process plan at scheduling stage. Up to 30% of process plans have to be modified due to the changing shop floor conditions [8.3]. Although various approaches and algorithms have been reported in the literature over the last decade attempting to integrate process planning and scheduling functions, the integration is limited in functionality or compromised in computational efficiency due to the complexity of this NP-hard problem.

Usually, setup planning is done after process planning but before scheduling. Optimisation of a process plan after setup planning is limited to a specific machine determined by the scheduling system. Moreover, in a machine shop, setup is the most commonly used dispatching unit of machining jobs to the assigned machines. The changing shop floor conditions including machine availability and capability should be considered at the setup planning stage. Therefore, it is our view that setup planning can play an important role in adaptive decision making toward process planning and scheduling integration.

The objective of this research is to develop adaptive setup planning algorithms according to the changing scheduling requirements (machine availability, machining cost, make span, and machine utilisation), so as to bridge the gap between process planning and scheduling. The rest of the chapter is organised as follows. Following a literature review in Section 8.2, the principles of the adaptive setup planning, including generic setup planning and adaptive setup merging, are introduced in Section 8.3. A prototype system implemented in Java and supported by MATLAB® is presented in Section 8.4, together with a case study for system validation. Finally, research findings and contributions are summarised in Section 8.5.

8.2 Literature Review

Setup planning encounters two major constraints coming from design specifications and manufacturing resources. Most existing setup planning methods attempted to satisfy the first constraint that involves one or more of the following considerations: tolerance, precedence, machining knowledge, and fixturing.

Tolerance analysis in setup planning aims at minimising locating error and calculating machining error stack-up, which will influence the locating datum, direction and machining sequence of a workpiece. Zhang and Lin [8.1] introduced the concept of a hybrid graph and used tolerance as the critical constraint for setup planning. A tolerance analysis chart was employed for tolerance control in setup planning by Zhang *et al.* [8.4], and a graphical approach was proposed to generate optimal setup plans based on design tolerance specification. Wu and Chang [8.5] reported an approach that analyses the tolerance specification of a workpiece to generate feasible setup plans with explicit datum element. The optimal setup plan that has a minimum number of setups and the most accurate position relationship between features is ranked together with other feasible plans by those two criteria.

Precedence constraint analysis is to find the optimal setup plan by looking at the problem caused by the conflict between the precedence relationships of design specification and those of machining process. These precedence relationships were classified by Zhang *et al.* [8.6] as geometric constraints, datum and/or reference requirement, feature interaction, good manufacturing practice and indirect links.

Machining knowledge analysis considers operation type, tool type, and best practice knowledge that are associated with certain machining features and parts. They are important factors for feature grouping and process sequencing. In the system proposed by Ferreira and Liu [8.7], descriptive features of machines, tools and operations are represented together with workpiece description, then rule and strategies are applied for reasoning. Contini and Tolio [8.8] investigated the manufacturing requirements of a workpiece before generating setup plans for different types of machining centre configurations, which includes tool machining direction and precedence constraints. Feature technology and associated machining information were represented as a precedence matrix for setup planning and fixture design by Öztürk [8.9].

Fixturing analysis during setup planning addresses on locating and clamping region selection, interference checking, stability and load distribution study. Sakurai [8.10] analysed the requirements for setup planning and fixture configuration that include accurate locating, total restraint during machining, and limited deformation. Detailed algorithm and considerations about setup planning and fixturing were also given by Sundararajan and Wright [8.11]. Joneja and Chang [8.12] considered more fixture planning during setup planning. A generalised representation scheme for a variety of fixture elements using geometric and functional properties was developed. A methodology was described to build up assemblies of fixture elements completing with the workpiece. Lin *et al.* [8.13] treated setup planning and fixture design as two stages of conceptual fixture design: pre-setup and post-setup conceptual design. Post-setup conceptual design takes place when the setup plan is available. After the best ranked surface for each fixturing datum is selected, the system then performs layout design by generating all the supporting, locating and clamping points for each setup, and various constraints are also considered.

Some other proposed setup planning approaches targeted the integration of more than one consideration or optimisation of feasible setup plans. Zhang *et al.* [8.14] considered tolerance decomposition, fixture design and manufacturing resource capability complying with setup planning. A graph-based method was employed for expressing tolerance and datum relationships. Ong *et al.* [8.15] proposed a hybrid approach to setup planning optimisation using genetic algorithms, simulated annealing, and a precedence relationship matrix using six cost indices. An integrated approach to automatic setup planning was also presented by Huang and Xu [8.16], which tried to systematically consider various components: geometry, precedence constraint, kinematics, force and tolerance. More recently, Gologlu [8.17] applied component geometry, dimensions and tolerances to extract constraint-imposed precedence relations between features, and took fixturing strategies into account.

The second constraint from manufacturing resources was normally considered at optimisation stage, in terms of cost, quality, lead time, and agility, but under an assumption that machines are given. In other words, the two constraints of setup planning are treated separately; thus, the search space is narrowed down before the search begins [8.15]. Recent efforts include setup grouping strategies for make span minimisation [8.18] and automated setup planning at both single part level and machine station level [8.19].

From the literature, it is evident that the flexibility of setup planning has not been fully addressed, especially from the process planning and scheduling integration point of view. Targeting this problem, machine availability and capability as well as other scheduling requirements on make span, machining cost and machine utilisation have to be considered during setup formation, sequencing, and optimisation. As the first step, a tool accessibility examination approach was developed by the authors for adaptive setup planning (ASP) [8.20], which focused on kinematic analysis of different machine tools, as well as setup locating and grouping algorithms for machines with varying configurations. The work reported in this chapter extends the ASP algorithms to further cover multi-machine setup planning or cross-machine ASP problems (*i.e.* a part is to be fabricated on more than one machine), by applying appropriate AI heuristics.

In the literature, various AI techniques have been popularly applied for decision making in setup planning and sequencing, *etc*. For example, iterative optimisation of machining sequence was done using a GA (genetic algorithm) in the work of Singh and Jebaraj [8.21], whereas simulated annealing, Hopfield net, neural networks, fuzzy logic and rule-based techniques were attempted by other researchers [8.22, 8.23]. Among these approaches, GA has been found to be effective in solving large-size optimisation problems. Although optimal solutions cannot be guaranteed by GA, computational complexity is reduced. Since a so-generated near-optimal result can still satisfy in real shop floor conditions, GA is adopted for adaptive setup planning in our research. We propose to use ASP as the main thread, where setup plans are generated adaptively according to machines' availability and capacity. More importantly, its optimisation considers cost, quality, make span and machine utilisation, individually or in combination.

8.3 Adaptive Setup Planning

8.3.1 Research Background

Since 2000, the authors have been working on a distributed process planning (DPP) [8.24] project targeting uncertainty issues in job shop machining operations. As shown in Figure 8.1, the DPP approach is realised by a two-layer structure of shop-level Supervisory Planning and machine-level Operation Planning. Its ultimate goal is to enable adaptive decision making by separating generic decisions from those machine-specific ones. Thus, only the generic decisions are to be made in advance, leaving the machine-specific ones to be decided at the last minute to accommodate any possible changes, adaptively.

Among the decision-making modules in Figure 8.1, adaptive setup planning is embedded in DPP in two steps during Process Sequencing and Execution Control. A generic setup plan can be generated at the stage of feature grouping, and final setups are determined through setup merging once available machines are given.

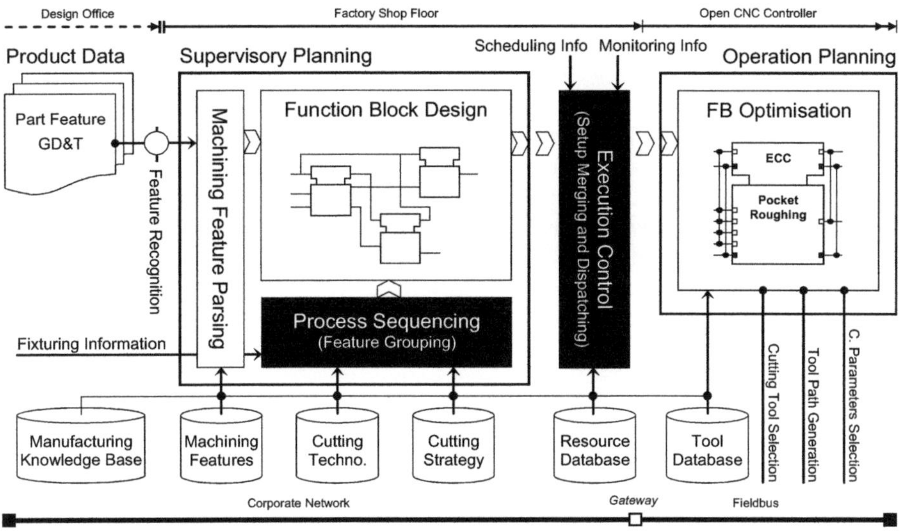

Figure 8.1. DPP architecture for adaptive decision making

Since process planning is beyond the scope of this chapter, interested readers are referred to [8.24] for more details on DPP.

8.3.2 Generic Setup Planning

In DPP, a part design is represented by machining features either through feature-based design or via a third-party feature recognition solution. Its machining process is equivalent to the machining features fabrication in proper setups and sequence. As shown in Figure 8.2, machining features are those shapes such as step, slot, pocket, and hole that can be easily achieved by the defined machining technologies.

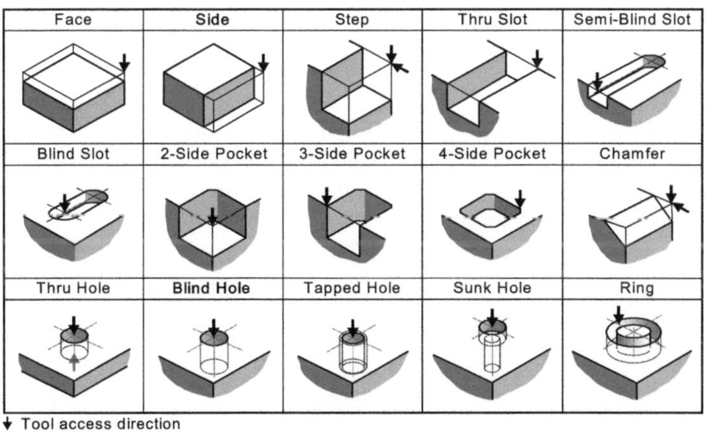

Figure 8.2. Typical machining features for prismatic part design and machining

Before a machining feature can be machined, it must be grouped into a setup for the ease of fixturing. The basic idea of generic setup planning (or feature grouping) is to determine a primary locating direction of a setup, and group the appropriate features into the setup according to their tool access directions (TADs, shown as unit vectors in Figure 8.2). This process is repeated for a secondary locating direction and so on until all the features are properly grouped. This TAD-based setup planning implicitly considers 3-axis machines only, each having a fixed tool orientation. As a 3-axis machine possesses the basic configuration of other types of machine tools (3-axis machine with an indexing table, 4-axis machine, and 5-axis machine, *etc.*), a so-generated setup plan becomes generic and applicable to all machines.

Within the context, a primary locating direction is the surface normal $\vec{V}$ of the primary locating surface (*LS*). It can be determined by the following equations:

$$LS = \left\{ f\left(A^*, T^*\right) \middle| W_A \times \frac{A^*}{A_{\max}} + W_T \times \frac{T^*}{T_{\max}} = \max\left(W_A \times \frac{A}{A_{\max}} + W_T \times \frac{T}{T_{\max}} \right) \right\} \tag{8.1}$$

$$\vec{V} = \left[\frac{\partial f}{\partial x}, \frac{\partial f}{\partial y}, \frac{\partial f}{\partial z} \right] \tag{8.2}$$

where, A^* and T^* are the surface area and the generalised accuracy grade of an *LS*; W_A is the weight factor of A^*; W_T is the weight factor of T^*; $A_{\max}$ and $T_{\max}$ are the maximum values of A^* and T^* of all candidate locating surfaces. A generalised accuracy grade *T* can be obtained by applying the algorithms described in [8.25–8.27]. Based on the primary locating direction $\vec{V}$, those machining features (*MF*) whose tool access directions $\vec{T}_{MF}$ are opposite to $\vec{V}$ are grouped into setup $ST_{\vec{V}}$, as denoted below:

$$ST_{\vec{V}} = \left\{ MF \mid \vec{T}_{MF} = -\vec{V} \right\} \tag{8.3}$$

It is worth of mentioning that the remaining features are grouped in the same way but based on the secondary locating direction and so on. The setups at this stage are planned for 3-axis machines. An adaptive setup merging is required for 4- or 5-axis machines, should they be selected for the part machining.

8.3.3 Setup Merging on a Single Machine

8.3.3.1 Tool Orientation Space of Machine Tools

Setup merging is based on the cutting tool accessibility analysis of machine tools. Although both part geometry and cutter geometry (including length and diameter) may play important roles in a complete tool accessibility examination, only feasible tool orientations are considered in this study. Compared with the workspace of a machine, most workpieces are quite small in size. Therefore, the tool orientation space (TOS) becomes the major constraint affecting the capacity of a machine. In other words, the setup merging considered here is under an assumption that there are no constraints of translational limits for small workpieces. The TOS of a machine is determined by its configuration and the motion ranges of its rotational axes. A unit spherical surface representation of TOS is adopted. Figure 8.3 shows a generic kinematic model of the rotational axes *A*, *B*, and *C* for TOS calculation.

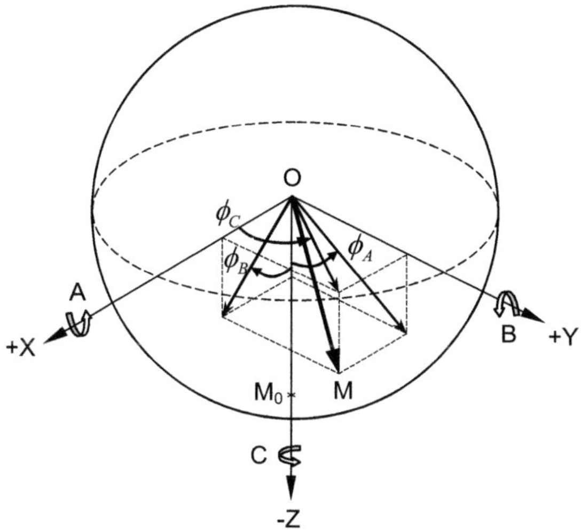

Figure 8.3. Kinematic model of rotational axes

As shown in Figure 8.3, OM_0 is the original (or home) orientation of the cutter. When the cutter is at position *OM*, its component motion angles of *A*, *B*, and *C* are ϕ_A, ϕ_B, and ϕ_C, respectively, where,

$$\begin{aligned} \Phi_A^- \le \phi_A \le \Phi_A^+ \\ \Phi_B^- \le \phi_B \le \Phi_B^+ \\ \Phi_C^- \le \phi_C \le \Phi_C^+ \end{aligned} \tag{8.4}$$

$[\Phi_A^-, \Phi_A^+]$, $[\Phi_B^-, \Phi_B^+]$, and $[\Phi_C^-, \Phi_C^+]$ denote the motion ranges of the three rotational axes. Together with the three translational motions along *X*, *Y*, and *Z* axes, if any two of the three rotational motion ranges are non-zero, the machine is of 5-axis. Similarly, any one non-zero rotational motion may result in a 4-axis machine. In a special case, if $|\Phi_A^+ - \Phi_A^-|$, $|\Phi_B^+ - \Phi_B^-|$ and $|\Phi_C^+ - \Phi_C^-|$ are all zero, the machine only has three translational axes. For prismatic part machining, since simultaneous axis movement is not mandatory, a 3-axis machine with an indexing table is also considered in our approach for setup merging. Because the original position of the cutter is along the Z-axis in the kinematic model, 5-axis machine tools are classified into two types by looking at whether a C-motion is involved: AC/BC type with C-motions and AB type without a C-motion. Since AC and BC types are similar, only AC and AB types are analysed in the following discussions. The TOS of these two types of machines are derived as follows:

$$\begin{cases} x^2 = \dfrac{(\cos\phi_C)^2}{1+(\sin\phi_C/\tan\phi_A)^2} \\ y^2 = \dfrac{(\sin\phi_C)^2}{1+(\sin\phi_C/\tan\phi_A)^2} \\ z^2 = \dfrac{(\sin\phi_C/\tan\phi_A)^2}{1+(\sin\phi_C/\tan\phi_A)^2} \end{cases} \quad \text{(AC type)} \qquad (8.5)$$

$$\begin{cases} x^2 = \dfrac{(\sin\phi_B)^2}{1+(\cos\phi_B * \tan\phi_A)^2} \\ y^2 = \dfrac{(\cos\phi_B * \tan\phi_A)^2}{1+(\cos\phi_B * \tan\phi_A)^2} \\ z^2 = \dfrac{(\cos\phi_B)^2}{1+(\cos\phi_B * \tan\phi_A)^2} \end{cases} \quad \text{(AB type)} \qquad (8.6)$$

According to Equations (8.4–8.6), the TOS of a 3-axis machine, a 3-axis machine with an indexing table (or a 4-axis machine), and a 5-axis machine can be represented on a unit spherical surface as a point, a curve, and a spherical surface patch, respectively. Consequently, setup merging is performed against the TOS.

8.3.3.2 Setup Merging on a 5-axis Machine

As mentioned earlier, a generic setup plan is created for 3-axis machines. In the case that a 4-/5-axis machine is chosen, proper setup merging is required for the best utilisation of the machine's capacity. This is explained through an example.

By applying Equations (8.1–8.3) for feature grouping and other five DPP reasoning rules [8.28] for feature sequencing, a generic setup plan of a test part with 26 machining features (Figure 8.4(a)) can be generated. It consists of five generic 3-axis-based setups, each of which contains a set of partially-sequenced machining features, as shown in Figure 8.4(b). The light gray areas indicate setups and the dark gray areas are the groups that share the same tool types. Each 3-axis-based setup can be represented by a unique unit vector $\boldsymbol{u}$ indicating its tool access direction (identical to the TAD of machining features in the setup). When a 5-axis machine tool {X, Y, Z, A (around X), B (around Y)} is selected, more than one of the 3-axis-based setups of the test part may have a chance to be machined in one *final* setup through setup merging.

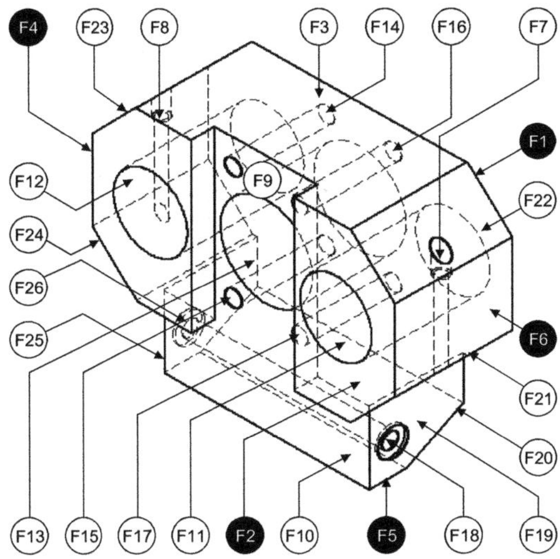

(a) A test part with 26 machining features

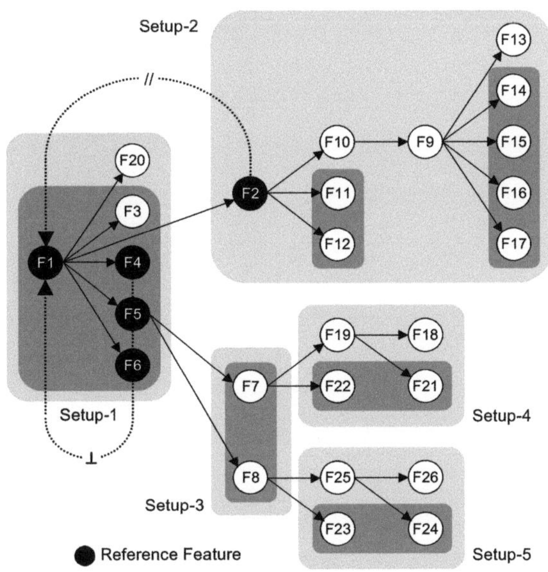

(b) A generic setup plan with five 3-axis-based setups

Figure 8.4. Results of generic setup planning of a test part

The setup merging on a single machine (a 5-axis machine in this case) examines whether other 3-axis-based setups can be included in a final setup by checking the unit vector $\boldsymbol{u}$ of each of the 3-axis-based setups against the TOS of the machine. The procedure is straightforward by following two steps and their iterations, *i.e.* (1) aligning the locating direction of a final setup to the spindle axis Z, and (2) searching for an orientation that includes a maximum number of 3-axis-based setups by rotating the part around the spindle axis *Z*. This merging process is repeated for all setups until a minimum number of 5-axis-based final setups can be reached. Since the first step can be done easily by using matrix transformation, we only provide details on the second step due to page limitation.

Figure 8.5(a) shows a typical scenario, where a setup has been aligned with -*Z* axis and another 3-axis-based setup with a TAD $\boldsymbol{u_i}$ (x_i, y_i, z_i) is under consideration. The goal is to rotate the vector $\boldsymbol{u_i}$ (or the test part) around *Z* and at the same time determine a mergable range (or ranges) within 2π, where $\boldsymbol{u_i}$ can fit in the TOS of the machine. The TOS of a 5-axis machine is represented as a spherical surface patch denoted by EFGH in Figure 8.5(a).

As shown in Figure 8.5(a), the spherical coordinates of $\boldsymbol{u_i}$ are (1, γ_i, θ_i). By rotating $\boldsymbol{u_i}$ around *Z*, a circle $\boldsymbol{C_i}$ on the unit sphere is obtained.

$$\begin{cases} x_i = \sin\theta_i \cos\gamma_i \\ y_i = \sin\theta_i \sin\gamma_i \\ z_i = -\cos\theta_i \end{cases} \tag{8.7}$$

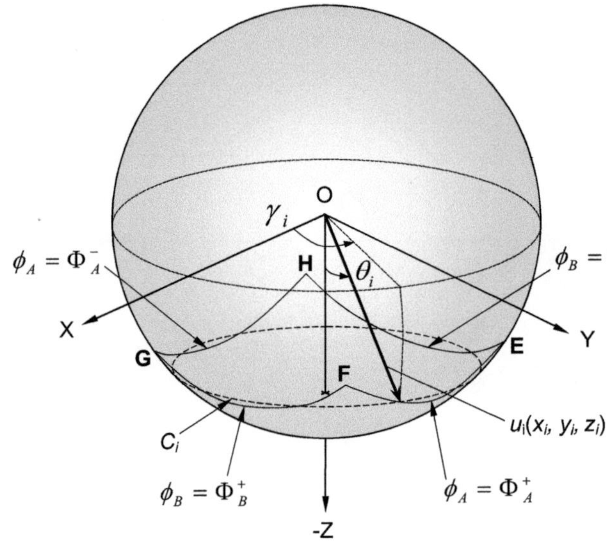

(a) Searching for setup mergability in TOS

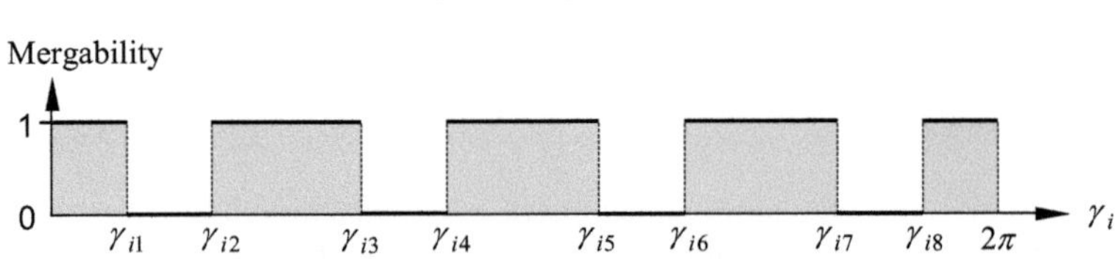

(b) Mergable range of a 3-axis-based setup with TAD $\boldsymbol{u_i}$

Figure 8.5. Setup merging on a 5-axis machine

where, θ_i is a constant and $\gamma_i \in [0, 2\pi]$. The $\boldsymbol{C_i}$ may intersect with the spherical surface patch EFGH defined by

$$\text{EF:} \quad \phi_A = \Phi_A^+, \quad \phi_B \in [\Phi_B^-, \Phi_B^+] \tag{8.8}$$

$$\text{FG:} \quad \phi_B = \Phi_B^+, \quad \phi_A \in [\Phi_A^-, \Phi_A^+] \tag{8.9}$$

$$\text{GH:} \quad \phi_A = \Phi_A^-, \quad \phi_B \in [\Phi_B^-, \Phi_B^+] \tag{8.10}$$

$$\text{HE:} \quad \phi_B = \Phi_B^-, \quad \phi_A \in [\Phi_A^-, \Phi_A^+] \tag{8.11}$$

For the segment EF expressed in Equation (8.8),

$$|z| = \sqrt{\frac{(\cos(\Phi_A^+))^2}{1+(\cos(\Phi_A^+)\tan(\phi_B))^2}}, \quad \phi_B \in [\Phi_B^-, \Phi_B^+] \tag{8.12}$$

If $|z_i| < |z|_{\min}$, the segment EF:$\{\phi_A = \Phi_A^+, \ \phi_B \in [\Phi_B^-, \Phi_B^+]\}$ and the circle $\boldsymbol{C_i}$ has no intersection. If $z_i < 0$ and $|z_i| > |z|_{\max}$, the segment EF and circle $\boldsymbol{C_i}$ intersect over the entire range of [0, 2π]. Otherwise, if $z_i < 0$ and $|z|_{\min} < |z_i| < |z|_{\max}$, EF and $\boldsymbol{C_i}$ intersect with each other along the edge of the TOS. Figure 8.5(b) gives the mergable range of the case shown in Figure 8.5(a), which can be calculated for every 3-axis-based setup. As shown in Figure 8.6, a pose (position and orientation) of the test part that provides the most overlapping mergable range determines the 5-axis-based setup.

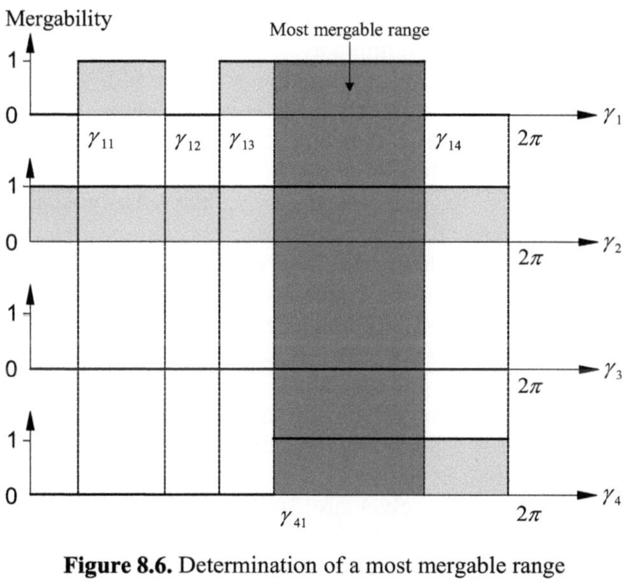

Figure 8.6. Determination of a most mergable range

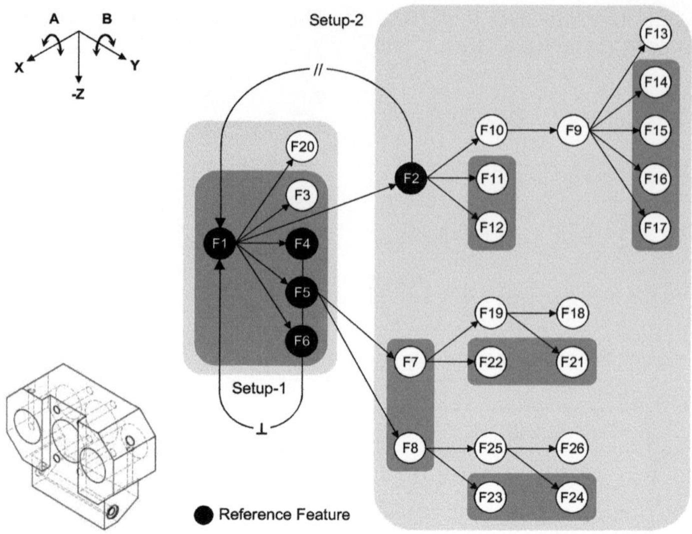

Figure 8.7. Results of setup merging on a 5-axis machine

Figure 8.7 depicts the result of setup merging of the test part after the five 3-axis-based generic setups in Figure 8.4(b) have been merged to two final setups (the light grey areas) for the 5-axis machine.

8.3.4 Adaptive Setup Merging across Machines

In practice, a part to be machined may need to be decomposed into several setups and finished on more than one machine. Owing to the combinations of machining features, setups and different types of machines, setup merging becomes complex. Adaptive setup merging across multiple available machines (or cross-machine ASP) is in fact a typical combinatorial optimisation problem and NP-hard. For a cross-machine ASP problem with a total of I 3-axis-based setups, a total of K primary locating surfaces and a total of L machine tools available on a machine shop, the maximum solution space would be ($I!\times K!\times L!$). For example, for the case study given in Section 8.4 (I = 10, K = 6, and L = 3), the solution space is as huge as (10!×6!×3!) or 15,676,416,000.

The goal of the cross-machine ASP is to generate an optimal (or near optimal) setup plan for a part based on the capability and configuration of each available machine, machine combinations, and other requirements from a scheduling system (*e.g.* machine utilisation and make span). It is evident from the literature that GA is capable of solving large-size optimisation problems like this. In order to find an optimal or near-optimal solution effectively in the huge solution space, an extended GA approach (denoted GA^+) is proposed to handle setup-specific issues.

The basic idea of GA^+ in the cross-machine ASP is illustrated in Figure 8.8, where the inputs to the GA^+ solver are 3-axis-based setups and the TOS of all available machines. The output of the GA^+ is the optimal or near-optimal setup plan corresponding to a chosen objective through optimisation.

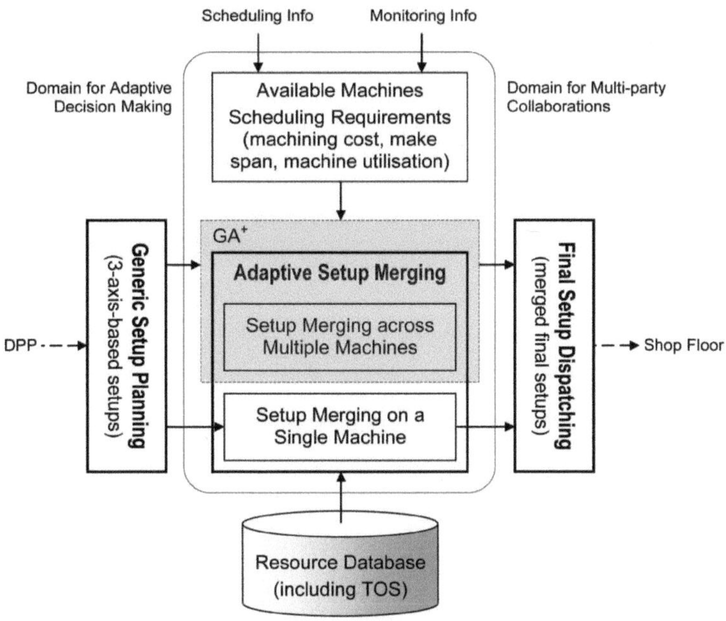

Figure 8.8. An extended GA in adaptive setup planning

8.3.4.1 Objective Function

The GA$^+$ based optimisation of the cross-machine ASP considers six objectives that can be summarised as: (1) to locate a part as stably and accurately as possible, (2) to group as many 3-axis-based setups as possible into a merged final setup, (3) to minimise the total number of final setups, (4) to minimise the machining cost of the part, (5) to minimise the make span of the part, and (6) to maximise the machine utilisation. Since Objective (2) is guaranteed by the search algorithm presented in Section 8.3.3.2 for tool accessibility analysis in single-machine setup merging, the GA$^+$ based optimisation only considers the remaining five objectives.

For Objective (1), the locating factor LF_p of a setup plan p can be calculated based on the locating factor lf_p of individual setups in the setup plan p and the number of machining features Q_p in each individual setup:

$$LF_p = \frac{Q_p^1 \times lf_p^1 + Q_p^2 \times lf_p^2 + ... Q_p^{S_p} \times lf_p^{S_p}}{Q_p^1 + Q_p^2 + ... + Q_p^{S_p}} = \frac{\sum_{s=1}^{S_p} \left(Q_p^s \times lf_p^s \right)}{\sum_{s=1}^{S_p} Q_p^s} \tag{8.13}$$

$$lf_p^s = \frac{\left(W_A \times \frac{A_s}{A_{\max}} + W_T \times \frac{T_s}{T_{\max}} \right)}{\max\limits_{s \in [1, S_p]} \left(W_A \times \frac{A_s}{A_{\max}} + W_T \times \frac{T_s}{T_{\max}} \right)}$$

where S_p is the total number of setups included in the setup plan p; W_A and W_T are the weight factors of the surface area A_s and generalised surface accuracy grade T_s of a candidate locating surface, respectively; $A_{\max}$ and $T_{\max}$ are the maximum values of all A_s and T_s. The concept of a generalised accuracy grade T_s is adopted to evaluate locating quality of a surface feature by integrating different types of tolerances (dimensional tolerance and geometrical tolerance) into a comparable format. It can be obtained by applying the algorithms described by Boerma and Kals [8.25] and Ma *et al.* [8.27].

To represent Objective (3), a grouping factor GF_p is defined as

$$GF_p = \frac{S_{\min}}{S_p},\ S_{\min} \xrightarrow{theorically} 1 \tag{8.14}$$

where $S_{\min}$ is the minimum value of S_p among all alternative setup plans; $S_{\min} \to 1$, if all machining features can be grouped into one setup on a *perfect* machine. Thus, the grouping factor GF_p can provide a relative rate for evaluating setup plan p in terms of setup number minimisation.

For Objectives (4), (5) and (6), the machining cost MC_p of setup plan p can be accumulated based on the unit cost UC_p of a machine and the estimated machining time MAT_f of each machining feature f to be machined; the make span MS_p is the time difference of the starting time and the finishing time in order to produce a part (if a multi-machine setup plan is given, it is equivalent to the longest machining time of each machine MT_l ($l \in [1, L]$) plus non-machining time T^0, including setup time, tool change time, machine idle time, *etc.*); whereas the machine utilisation MU_l can simply be represented by the total machining time of this machine, as denoted below

$$MC_p = \sum_{s=1}^{S_p}\left(\sum_{f=1}^{Q_p^s} MAT_f \times UC_p^s\right) \tag{8.15}$$

$$MS_p = \max_{MT_l | l\in[1,L]}\left(\sum_{s=1}^{S_l}\sum_{f=1}^{Q_p^s} MAT_f\right) + T^0 \tag{8.16}$$

$$MU_l = \sum_{s=1}^{S_l}\sum_{f=1}^{Q_p^s} MAT_f \tag{8.17}$$

where Q_p^s is the number of machining features in setup s, S_l the number of setups on machine l, and L the number of all available machines. As T^0 is assumed to be a constant for all machines, it can be scaled down to zero without affecting the make span minimisation. It is therefore neglected in the following make span calculations. The overall objective function for achieving an optimal setup plan OSP is

$$OSP = \max_{p\in[1\ldots P]} (W_L \times LF_p + W_G \times GF_p + W_C \times CF_p + W_M \times MF_p + W_U \times UF_p) \tag{8.18}$$

where CF_p and MF_p are the cost and make span factors, respectively, defined as relative ratios for evaluating setup plan p:

$$CF_p = \frac{MC_{\min}}{MC_p} \tag{8.19}$$

$$MF_p = \frac{MS_{\min}}{MS_p} \tag{8.20}$$

where $MC_{\min}$ is the minimum machining cost on a low-end machine with a cheap unit cost, and $MS_{\min}$ is the shortest machining time if the job can be distributed evenly among all available machines. Machine utilisation factor UF_l for machine MT_l is obtained by comparison of the total machining time and the available time AT_l of this machine. Machine utilisation factor UF_p of the setup plan p is the average of all UF_l, $l \in [1, L]$. However, if any UF_l is smaller than 0, then $UF_p = 0$, meaning that the setup plan p under evaluation is not acceptable.

$$UF_l = \frac{MU_l}{AT_l} \tag{8.21}$$

$$UF_p = \frac{1}{L} \times \left(\sum_{l=1}^{L} UF_l \right) \tag{8.22}$$

Substituting Equations (8.13–8.17) and (8.19–8.22) into Equation (8.18), yields

$$\begin{aligned} OSP = \max_{p \in [1 \ldots P]} & \left(W_L \times \frac{\sum_{s=1}^{S_p} \left(Q_p^s \times lf_p^s \right)}{\sum_{s=1}^{S_p} Q_p^s} + W_G \times \frac{S_{\min}}{S_p} \right. \\ & + W_C \times \frac{MC_{\min}}{\sum_{s=1}^{S_p} \left(\sum_{f=1}^{Q_p^s} MAT_f \times UC_p^s \right)} + W_M \times \frac{MS_{\min}}{\max\limits_{MT_l | l \in [1,L]} \left(\sum_{s=1}^{S_l} \sum_{f=1}^{Q_p^s} MAT_f \right)} \\ & \left. + W_U \times \left(\frac{1}{L} \times \left(\sum_{l=1}^{L} \frac{\sum_{s=1}^{S_l} \sum_{f=1}^{Q_p^s} MAT_f}{AT_l} \right) \right) \right) \end{aligned} \tag{8.23}$$

where, W_L, W_G, W_C, W_M and W_U are the weight factors of LF_p, GF_p, CF_p, MF_p, and UF_p, respectively. The weighted-sum multi-objective function in nature is used to obtain a compromised solution. For the setup plan p under evaluation, the fitness of

Equation (8.23) can be tuned accordingly by a set of appropriate weights. The assigned weights give generic algorithms a tendency to sample the area towards a fixed point in the criteria space, and further to meet the need of a specific application so as to offer flexibility for the cross-machine ASP.

8.3.4.2 Constraints and Assumptions

The objective function described in Equation (8.23) bears the following constraints and assumptions:

(1) Setups in a final setup plan can be formed for one machine or multiple machines.
(2) All machining features are grouped into 3-axis-based setups first; and each machining feature can only be grouped into one 3-axis-based setup.
(3) All 3-axis-based setups can be merged into fewer setups; and each 3-axis-based setup can only be merged into one final setup.
(4) A qualified surface can be used as a primary locating surface only once on each machine.

In order to solve cross-machine ASP problems, feasible setup plans generated by considering tool accessibility in terms of TOS of the available machines and the combinatorial optimisation of varying scheduling requirements on cast, make span and machine utilisation are built into an extended GA approach.

8.3.4.3 An Extended GA Approach

Figure 8.9 illustrates the decision-making process of our extended GA approach (GA^+) based on the available machines in a machine shop. This approach includes the following steps:

(1) Define constraints and optimisation criteria
(2) Encode the problem in a chromosome
(3) Choose a fitness function to evaluate the performance of each chromosome
(4) Construct the genetic operators
(5) Run the GA^+ algorithm and tune its parameters

The problem definition of the cross-machine ASP and its fitness evaluation using a weighted-sum multi-objective optimisation function are explained in Sections 8.3.4.1 and 8.3.4.2. The major challenges here in GA^+ are how to embed the tool accessibility examination algorithms into a chromosome and how to form feasible setup plans. This is also referred to as problem encoding. As shown in Figure 8.9, a potential setup plan is digitised as index numbers in a chromosome, representing 3-axis-based setups on all available machines. At the same time, a pre-processing is preformed based on the tool accessibility analysis for gene pool generation. After each GA operation, a *PLS* (primary locating surface) and $ST_{3\text{-}axis}$ (3-axis-based setup) based post-processing is carried out for fitness evaluation. Iteratively, an optimal or near-optimal setup plan can be obtained.

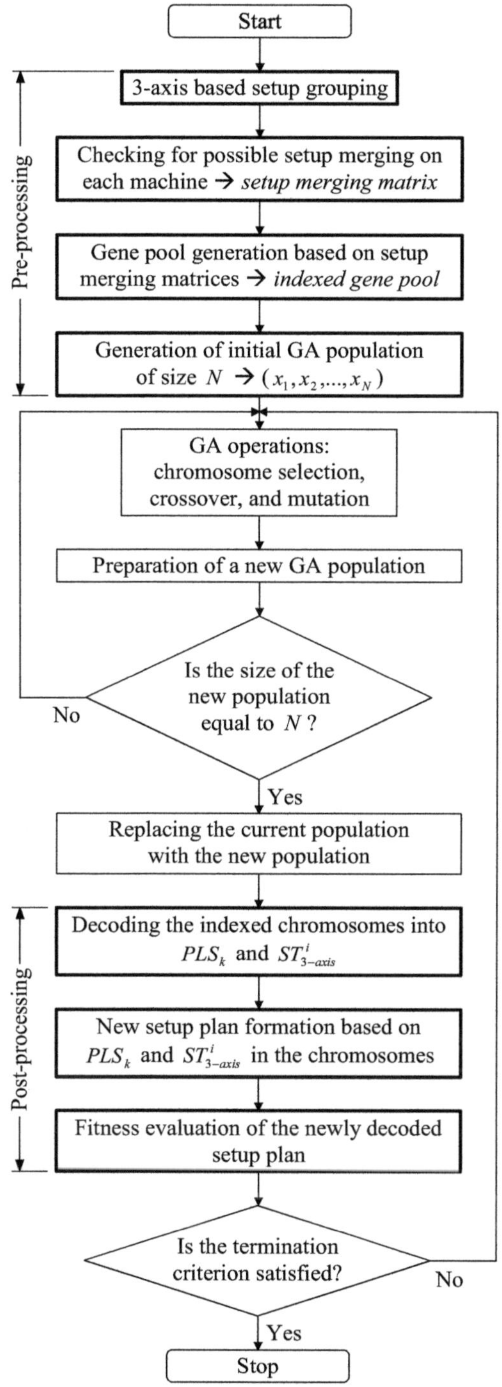

Figure 8.9. Decision-making process of an extended GA approach

Problem Encoding

A simplified example of a part with four 3-axis-based setups to be machined on three machines is used for chromosome encoding as shown in Figure 8.10. It also represents a typical setup plan. The order of genes on each machine indicates the order of the 3-axis-based setups. The value coded in each gene is the index number of a primary locating surface. Because each primary locating surface is unique in its orientation, it is used in GA^+ to represent one setup by its normal vector. The 3-axis-based setups relying on the same primary locating surface will be merged into one setup. This is also crucial for a setup plan formation in the post-processing.

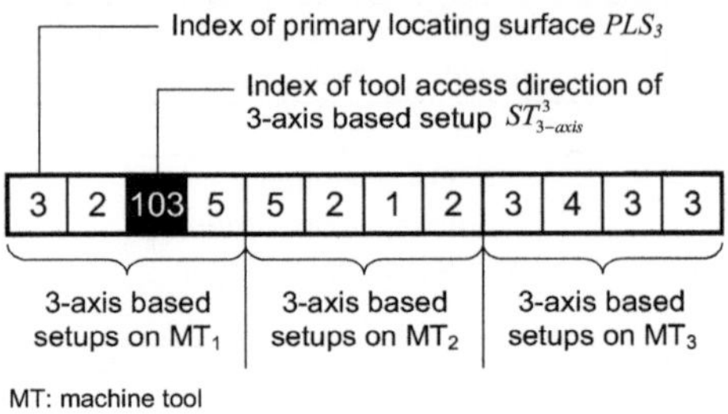

Figure 8.10. An encoded chromosome

The total number of genes in a chromosome equals the total number I of 3-axis-based setups multiplied by the total number L of available machines, *i.e.* $I \times L$. In the case shown in Figure 8.10, the chromosome is 4×3 in length. In other words, each chromosome holds 12 independent genes. In order to encode a setup plan into a fixed-length chromosome, an index-based gene pool must be generated first before chromosome constructions. This is a crucial task in the cross-machine ASP and is called pre-processing in GA^+ (Figure 8.9). Another important task is the post-processing after each generation for chromosomes decoding and fitness evaluation against the defined objective function expressed in Equation (8.23).

Pre-processing

Gene pool generation is partially based on the tool accessibility analysis in Section 8.3.3, which results in a set of 3-axis-based setups for a given part. As shown in Figure 8.9, the next step in pre-processing is to prepare a *setup merging matrix* SM_l for each available machine MT_l, $l \in [1, L]$ by checking the possibility of merging those 3-axis-based setups on each primary locating surface PLS_k, $k \in [1, K]$. This searching process is repeated for all available machines so as to generate a total of L setup merging matrices.

$$\underset{l\in[1,L]}{SM_l} = \begin{bmatrix} a_{11} & a_{12} & .. & a_{1I} \\ a_{21} & a_{22} & .. & a_{2I} \\ : & : & a_{ki} & : \\ a_{K1} & a_{K2} & .. & a_{KI} \end{bmatrix} \tag{8.24}$$

where $a_{ki} = 1$ when the ith 3-axis-based setup can be merged to the kth primary locating surface; otherwise, $a_{ki} = 0$. From the matrix (8.24), a gene pool indicating all applicable *PLS* to each $ST_{3\text{-}axis}$ can be generated for each machine by replacing the non-zero a_{ki} with the index k of the corresponding PLS_k. Taking the third column (or the third $ST_{3\text{-}axis}$) of matrix (8.24) as an example, if

$$\begin{bmatrix} a_{13} \\ a_{23} \\ a_{33} \\ a_{43} \end{bmatrix} = \begin{bmatrix} 1 \\ 0 \\ 1 \\ 1 \end{bmatrix}, \text{ then } gp_{l3} = [1, 2, 3] \text{ on } MT_l$$

where gp_{l3} is a part of the gene pool for ST^3_{3-axis} on MT_l. In a special case if $gp_{li} = \varnothing$, there is no *PLS* available for merging the ST^i_{3-axis} on MT_l, meaning that a special fixture is needed to hold the setup in position and orientation. In this case, the TAD of the ith 3-axis-based setup is used as the locating direction. "100" is purposely added to the index of ST^i_{3-axis}, denoted as $gp_{li} = 100+i$ as shown in Figure 8.10, so that the GA$^+$ can process this special case easily. A very small locating factor is assigned to the gene when its value is greater than 100. This situation normally happens to 3-axis machines. Repeating the aforementioned procedure for all 3-axis-based setups on every given machine yields a complete gene pool *GP*:

$$GP = \begin{bmatrix} gp_{11} & gp_{12} & .. & gp_{1I} \\ gp_{21} & gp_{22} & .. & gp_{2I} \\ : & : & gp_{li} & : \\ gp_{L1} & gp_{L2} & .. & gp_{LI} \end{bmatrix} \tag{8.25}$$

The last step of pre-processing is to create an initial GA generation by populating the $I \times L$ chromosomes to the specified size N with genes randomly selected from the corresponding gene pool gp_{li}. Figure 8.10 gives one example.

Post-processing

In the GA$^+$, chromosomes represent potential setup plans. In order to evaluate the fitness of each chromosome (setup plan) and to identify the elites for the next GA operation or to terminate the GA$^+$, *post-processing* is required. This decoding and fitness evaluation process includes the following steps, with the purpose of merging as many 3-axis-based setups as possible into a final setup:

(1) Identify a primary locating surface on each machine that can accommodate a maximum number of 3-axis-based setups.
(2) Merge the group of 3-axis-based setups as a new setup, and remove them from the set of $ST_{3\text{-}axis}$ for merging.
(3) Repeat steps 1 and 2 until all 3-axis-based setups are merged into a final setup.

(4) Form a setup plan based on the final setups, evaluate its fitness, and identify the elites for evolution. A setup plan with the best fitness is the final setup plan for the given part.

8.4 Implementation and Case Study

8.4.1 Prototype Implementation

A prototype system for adaptive setup planning is implemented in Java language and supported by MATLAB. Figure 8.11 shows the architecture of the ASP prototype. The kernel of GA-based optimisation comes from MATLAB and runs in the MCR (MATLAB Component Runtime) environment. The extended GA^+ algorithms for chromosome encoding and pre-/post-processing are developed in Java and linked to relevant MATLAB functions. MATLAB Builder for Java (also called Java Builder) is used to wrap the GA^+ and the relevant MATLAB functions into Java classes. They are further integrated with a user interface for adaptive setup planning. Since the MCR is freely available from The MathWorks, Inc. and the GA^+ is wrapped in Java classes, a MATLAB installation is no longer needed for ASP.

Figure 8.12 gives a snapshot of the integrated adaptive setup planning system, showing its user interface with both input data and optimisation results.

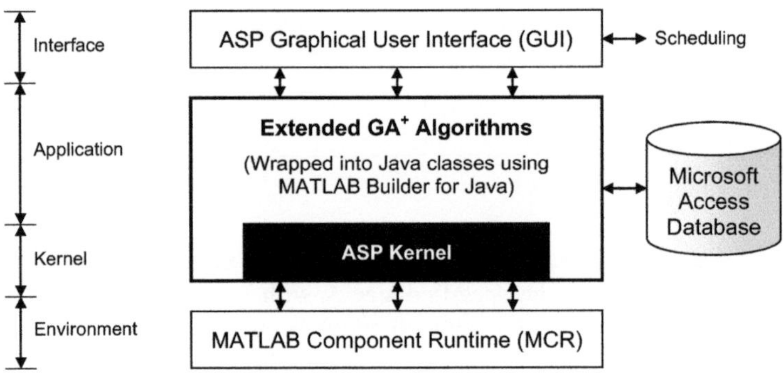

Figure 8.11. System architecture for adaptive setup planning

8.4.2 A Case Study

A slightly-revised test part shown in Figure 8.13 is chosen for the case study, particularly for algorithms validation in cross-machine ASP. The basic information of the six primary locating surfaces and 22 machining features of the test part are listed in Tables 8.1 and 8.2, respectively. Since cutting parameters and tool path optimisation is beyond the scope of setup planning, for simplification, a single unit time (UT) is assigned to every machining feature as the standard machining time. This arrangement does not affect the ASP validation. In reality, the machining time can be calculated based on cutting parameters and the length of tool path of each machining feature using our DPP system [8.28].

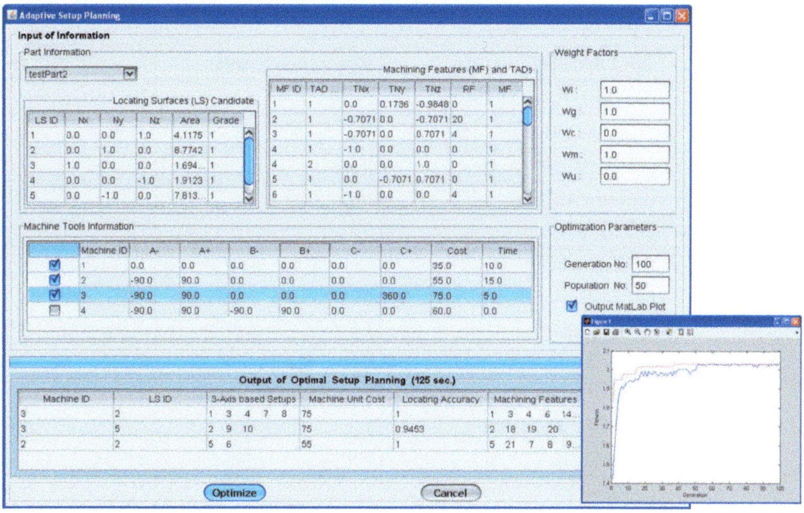

Figure 8.12. User interface of the adaptive setup planning system

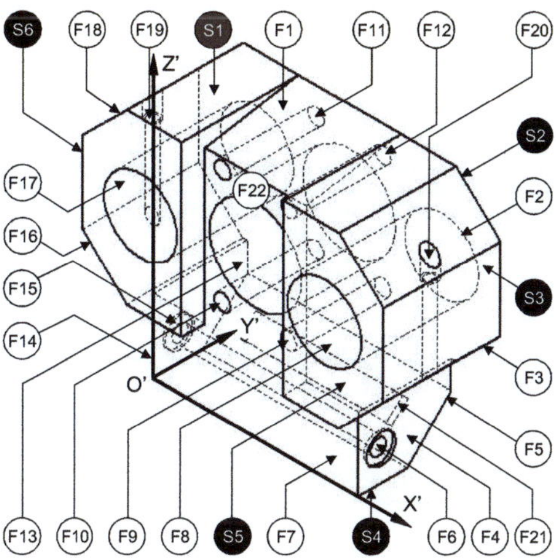

Figure 8.13. Revised test part

Table 8.1. PLS candidates of the test part

Surface ID	Surface normal	Surface area (inch^2)	Surface accuracy grade
S_1	(0, 0, 1)	4.1175	1
S_2	(0, 1, 0)	8.7742	1
S_3	(1, 0, 0)	1.69425	1
S_4	(0, 0, –1)	1.9123	1
S_5	(0, –1, 0)	7.81345	1
S_6	(–1, 0, 0)	1.69425	1

Table 8.2. Machining features of the test part

Feature ID	Tool access direction	Reference feature	Machining time (UT)
F_1	(0, sin 10°, –cos 10°)	none	1
F_2, F_3	(–sin 45°, 0, –cos 45°)	F_{20}, F_4	1, 1
F_4	(–1, 0, 0), (0, 0, 1)	none	1
F_5, F_{21}	(0, –sin 45°, cos 45°)	none, F_5	1, 1
F_6	(–1, 0, 0)	F_4	1
F_7	(0, 1, 0), (0, 0, 1)	none	1
F_8, F_{11}, F_{12}, F_{17}	(0, 1, 0), (0, –1, 0)	none	1, 1, 1, 1
F_9, F_{10}, F_{13}	(0, 1, 0), (0, –1, 0)	F_{22}	1, 1, 1
F_{14}	(1, 0, 0), (0, 0, 1)	none	1
F_{15}	(1, 0, 0)	F_{14}	1
F_{16}	(sin 45°, 0, cos 45°)	F_{14}	1
F_{18}	(sin 45°, 0, –cos 45°)	F_{19}	1
F_{19}, F_{20}	(0, 0, –1)	none	1, 1
F_{22}	(0, 1, 0)	none	1

Table 8.3. Three available machine tools for setup merging

Machine ID	Machine type	Orientation range*	Cost ($/h)	Available time (UT)
1	3-axis	(0, 0, 0, 0, 0, 0)	35	10
2	4-axis	(–90, 90, 0, 0, 0, 0)	55	15
3	5-axis	(–90, 90, 0, 0, 0, 360)	75	5

*Orientation range of *A*, *B*, and *C* axes: (A^-, A^+, B^-, B^+, C^-, C^+)

Table 8.4. 3-axis-based setups after generic setup planning

Generic setup	Tool access direction	Machining feature
ST^1_{3-axis}	(0, sin 10°, –cos 10°)	F_1
ST^2_{3-axis}	(–sin 45°, 0, –cos 45°)	F_2
ST^3_{3-axis}	(–sin 45°, 0, cos 45°)	F_3
ST^4_{3-axis}	(–1, 0, 0)	F_4, F_6
ST^5_{3-axis}	(0, –sin 45°, cos 45°)	F_5, F_{21}
ST^6_{3-axis}	(0, 1, 0)	F_7, F_8, F_9, F_{10}, F_{11}, F_{12}, F_{13}, F_{17}, F_{22}
ST^7_{3-axis}	(1, 0, 0)	F_{14}, F_{15}
ST^8_{3-axis}	(sin 45°, 0, cos 45°)	F_{16}
ST^9_{3-axis}	(sin 45°, 0, –cos 45°)	F_{18}
ST^{10}_{3-axis}	(0, 0, –1)	F_{19}, F_{20}

In order to demonstrate the cross-machine ASP concept, three typical machine tools with varying configurations are chosen for the case study. Table 8.3 summarises the machine configurations, along with the unit cost and available time of each machine. As 3-axis-based setups are generated by using the tool accessibility analysis during generic setup planning, they are treated as the known inputs in the case study and listed in Table 8.4.

In this case study, the total number of $ST_{3\text{-}axis}$ is 10 and the total number of available machines is 3. Each chromosome, therefore, contains (10×3) genes. After pre-processing, a gene pool for problem encoding is produced, and listed in Table 8.5. The gene pool provides valid values (indices of primary locating surfaces) of each 3-axis-based setup for GA^{+} operations.

Table 8.5. Gene pool for adaptive setup merging using GA^{+}

Generic setup	Machine #1	Machine #2	Machine #3
ST^{1}_{3-axis}	101	2, 5	2, 3, 4, 6
ST^{2}_{3-axis}	102	1, 3, 4, 6	2, 4, 5, 6
ST^{3}_{3-axis}	103	1, 3, 4, 6	1, 2, 5, 6
ST^{4}_{3-axis}	6	1, 4, 6	1, 2, 4, 5, 6
ST^{5}_{3-axis}	105	2, 5	2, 3, 5, 6
ST^{6}_{3-axis}	2	2	1, 2, 3, 4, 6
ST^{7}_{3-axis}	3	1, 3, 4	1, 2, 3, 4, 5
ST^{8}_{3-axis}	108	1, 3, 4	1, 2, 3, 5
ST^{9}_{3-axis}	109	1, 3, 4	2, 3, 4, 5
ST^{10}_{3-axis}	4	2, 3, 4, 5, 6	2, 3, 4, 5, 6

After encoding, it is crucial in GA^{+} to use the right GA parameters to obtain the correct results during iterative searching. The parameters to be tuned in GA^{+} are population size N and the number of generation G. Crossover operator and mutation rate, however, rely on the default values as suggested in GA toolbox of MATLAB. Figure 8.14 presents two typical tuning results with different GA parameters. By comparing the performance, it is clear that when $N = 50$ it is more efficient for fast search towards convergence, and after about 80 generations the fitness becomes quite stable. Therefore, $N = 50$ and $G = 100$ are chosen for the rest of the GA calculations during the case study.

8.4.3 Optimisation Results

According to Equation (8.23), the objective function for the cross-machine ASP considers 5 factors (locating, grouping, cost, make span, and machine utilisation). By adjusting their weights (W_L, W_G, W_C, W_M and W_U) accordingly, different scenario (or different scheduling requirements) can be handled by the ASP system. Ideally, if the ASP is integrated with a scheduling system, the values of the weights can be determined by the scheduling system according to the available resources and then passed to the ASP for setup planning to adapt to the changes on a shop floor. Hence, an adaptive setup planning to a real-world situation becomes possible.

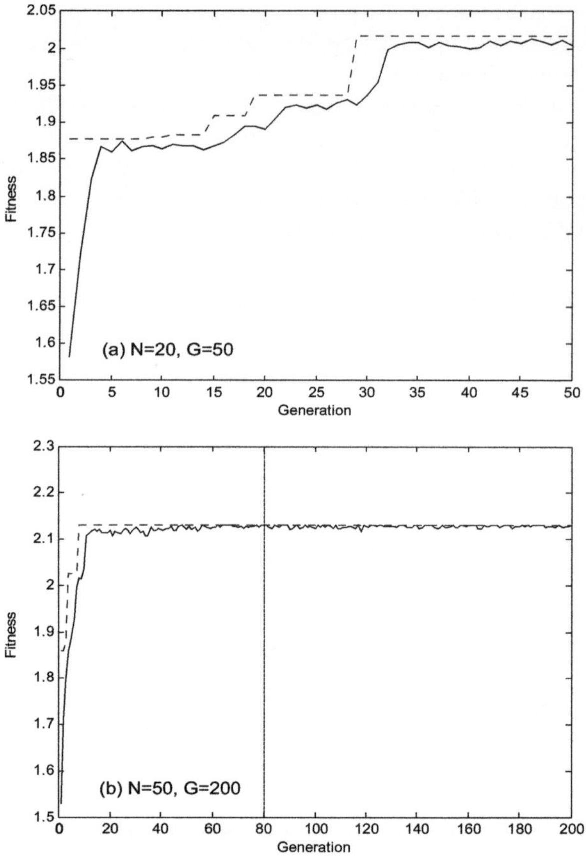

Figure 8.14. Results of parameter tuning

In the case study, a number of numerical experiments are carried out, from which six typical cases are selected and revealed in Figure 8.15:

(a) $W_L = 1, W_G = 1, W_C = 0, W_M = 0, W_U = 0$
To utilise the most capable machines first

(b) $W_L = 1, W_G = 1, W_C = 1, W_M = 0, W_U = 0$
To reduce cost while keeping minimum setups

(c) $W_L = 1, W_G = 1, W_C = 0, W_M = 1, W_U = 0$
To produce the part as soon as possible

(d) $W_L = 1, W_G = 1, W_C = 0, W_M = 0, W_U = 1$
To make use of all available machines

(e) $W_L = 1, W_G = 1, W_C = 1, W_M = 1, W_U = 1$
To trade off between all optimisation criteria

(f) $W_L = 1, W_G = 1, W_C = 2, W_M = 0, W_U = 0$
To minimise cost as much as possible

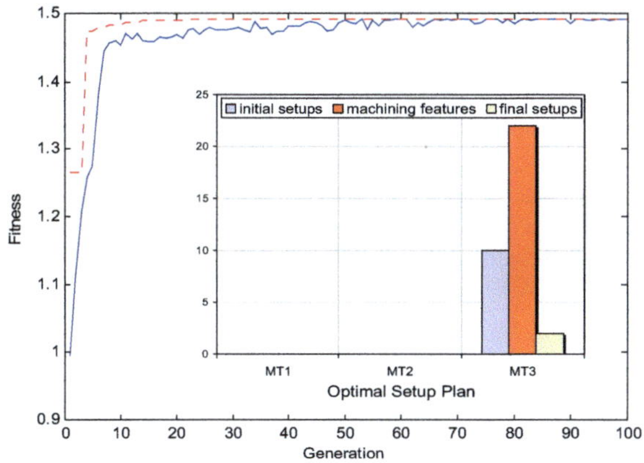

(a) $W_L = 1, W_G = 1, W_C = 0, W_M = 0, W_U = 0$

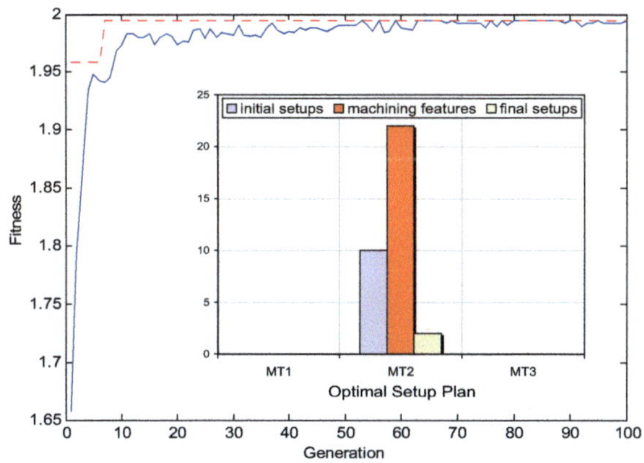

(b) $W_L = 1, W_G = 1, W_C = 1, W_M = 0, W_U = 0$

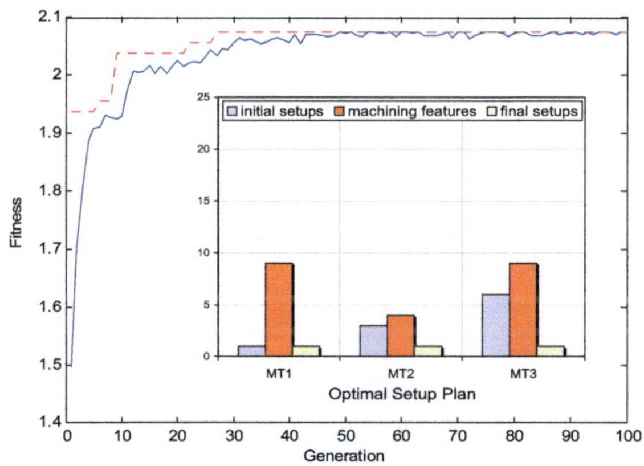

(c) $W_L = 1, W_G = 1, W_C = 0, W_M = 1, W_U = 0$

Figure 8.15. Optimisation results of six typical cases

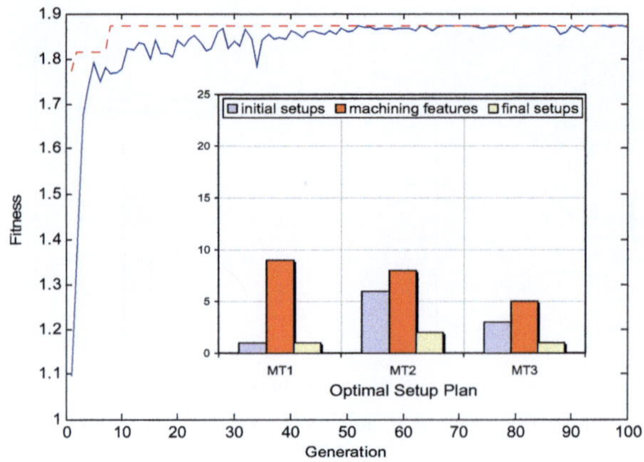

(d) $W_L = 1, W_G = 1, W_C = 0, W_M = 0, W_U = 1$

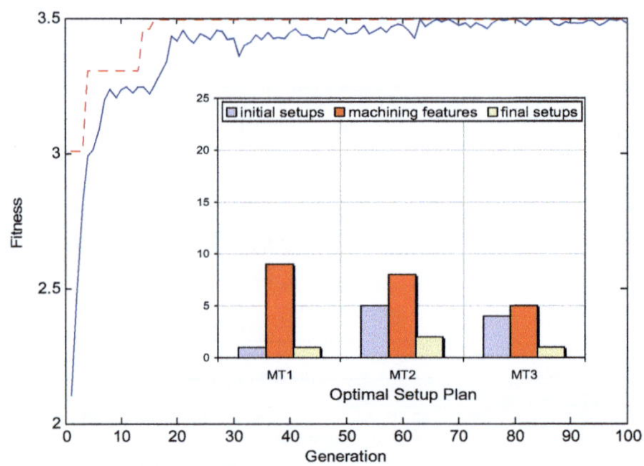

(e) $W_L = 1, W_G = 1, W_C = 1, W_M = 1, W_U = 1$

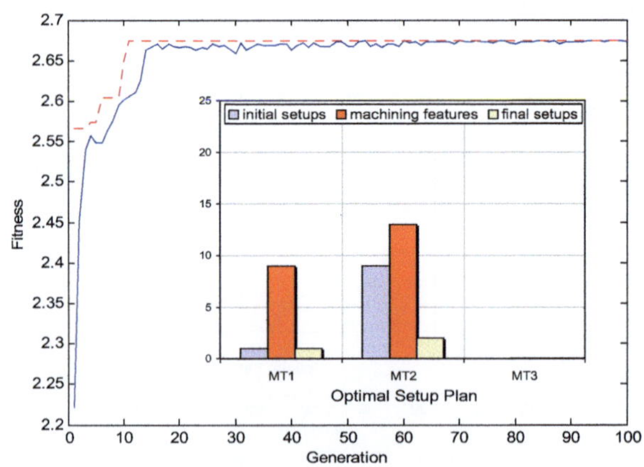

(f) $W_L = 1, W_G = 1, W_C = 2, W_M = 0, W_U = 0$

Figure 8.15. Optimisation results of six typical cases (continued)

In case (a) when no constraints are given to the cost, make span and machine utilisation, all final setups go to MT3 (the 5-axis machine). If cost reduction is required as shown in case (b), all final setups go to MT2 (the 4-axis machine). In case (c) when a minimum make span (or a quick production) is required, the final setups are distributed among all three machines. In this case, machining time is the major concern. In other words, although different numbers of setups and machining features are assigned to each machine, all machining jobs should be completed at roughly the same time, thus minimising make span. Case (d) is similar, but instead of make span, machine utilisation is considered against the available time slot of each machine. Combining cost, make span and machine utilisation together creates a trade-off situation as depicted in case (e). This all-in-one optimisation is rarely used in reality. Due to the trade-off, no single criterion is fully satisfied. Finally, when the weight of cost factor is doubled, all machining jobs of MT3 are forced to be moved to MT2, as shown in case (f). The reason for not moving them to MT1 is due to the effect of grouping factor ($W_G = 1$) that strives to reach a minimum number of final setups.

8.4.4 Discussion

8.4.4.1 Timing Issue of Adaptive Setup Planning

As mentioned in the previous sections, the fast adaptive decision making in ASP is facilitated by separating generic decisions from machine-specific ones in two steps. During the generic setup planning, machining features are grouped into 3-axis-based setups, each of which is represented by a unit vector indicating its locating direction. After the first step of decision making, a complex setup planning problem is simplified to a vector mapping and grouping problem that considers only setup merging to the available resources and is adaptive to dynamic changes. As the first step of generic setup planning is accomplished in advance, the timing issue is only concerned with the second step of adaptive setup merging.

Table 8.6 records the actual computational time (repeated three times for each case) *versus* the mean time for the six scenarios of the case study during the GA^+ optimisation. All computations are done on a personal computer (Intel Pentium® M, CPU 1.86 GHz, 1.00 GB RAM).

Table 8.6. Actual computational time *versus* mean time of GA^+ (s)

Six scenarios with different weight assignments (W_L, W_G, W_C, W_M, W_U)																	
(1,1,0,0,0)			(1,1,1,0,0)			(1,1,0,1,0)			(1,1,0,0,1)			(1,1,1,1,1)			(1,1,2,0,0)		
101	98	122	96	84	71	149	94	120	97	82	87	117	98	107	85	94	127
107.0			83.7			121.0			88.7			107.3			102.0		

From Table 8.6, it is clear that the computational time of GA^+ is within the range 1–2 minutes, and the trend of computation is relatively stable for different weight assignments. This phenomenon is compliant with GA^+ because its computational complexity has been relaxed by the two-step decision making. The efficiency of

GA^+ is only affected by the size of gene pool and the length of a chromosome. The slight variation of computational time is due to the nature of iterative search of GA and is also influenced by the weight assignments. Nevertheless, the length of GA^+ optimisation is acceptable in a machine shop, if a setup plan can be generated in a minute or two in normal operations or after a disturbance.

8.4.4.2 Accuracy Issue of GA^+

To verify the accuracy of GA^+ for adaptive setup merging, the effects of the five weighting factors are tested individually. The same scenario is considered by using exact search according to the objective function, Equation (8.23). The *exact search* means the calculation of the theoretical best solutions through simplified objectives. A comparison of the computational results of both the GA^+ and the exact search is listed in Table 8.7. It is evident that the proposed GA^+ approach can generate optimal or near-optimal solutions for adaptive setup planning problems.

Table 8.7. Fitness comparison between GA^+ and exact search

Weight assignments (W_L, W_G, W_C, W_M, W_U)	GA^+	Exact search
(1,0,0,0,0)	0.9925	0.9950 (1)
(0,1,0,0,0)	0.5	0.5 (2)
(0,0,1,0,0)	0.8462	1 (3)
(0,0,0,1,0)	0.8148	0.8148 (4)
(0,0,0,0,1)	0.8222	0.8222 (5)

Notes: (1) According to Equation (8.13) where $S_p = 2, Q_p^1 = 20, Q_p^2 = 2, lf_p^1 = 1, lf_p^2 = 0.9453$.

(2) According to Equation (8.14) where $S_p = 2$.

(3) According to Equation (8.19) where $MC_{\min} = MC_p = 35 \times 22\$$.

(4) According to Equation (8.20) where $MS_{\min} = 22/3, MS_p = 9$.

(5) According to Equation (8.22) where $MU_1 = 10, MU_2 = 7, MU_3 = 5, L = 3$.

8.5 Conclusions

In job shop machining operations, setup plans generated in advance are often subject to changes even before execution. An adaptive setup planning approach is urgently needed to deal with the uncertainty issues. This chapter presents in detail a two-step ASP approach. It first groups machining features into 3-axis-based generic setups and then generates machine-specific setups upon request by adaptive setup merging. The optimisation during setup planning and merging considers machine availability, capability and configurations, as well as scheduling requirements on cost, make span and machine utilisation. A so-generated setup plan can not only meet the scheduling requirements with respect to dynamic changes, but also can adapt to the chosen machines with optimised solutions. Due to the huge solution space, an extended GA^+ approach has been developed. The concept and algorithms are validated through a case study.

The results demonstrate that the ASP approach can provide adaptive solutions to job shop operations under uncertainty, where the availability of machines and the requirements on cost and make span change over time. The changes are brought to a separate dynamic scheduling system, which then passes the setup requirements to the ASP. Owing to the iterative search and fitness-based termination of GA, we cannot guarantee that a so-generated setup plan is truly optimal and in real time. However, it is feasible and practical in adaptive decision making within a minute or two (near real time). Being able to generate near optimal setup plans upon request makes our ASP unique to support the fluctuating job shop operations.

Acknowledgement

This ASP research is supported by the Natural Sciences and Engineering Research Council of Canada.

References

[8.1] Zhang, H.-C. and Lin, E., 1999, "A hybrid-graph approach for automated setup planning in CAPP," *Robotics and Computer-Integrated Manufacturing*, **15**, pp. 89–100.

[8.2] Tan, W. and Khoshnevis, B., 2000, "Integration of process planning and scheduling – a review," *Journal of Intelligent Manufacturing*, **11**, pp. 51–63.

[8.3] Detand, J., Kruth, J.P. and Kempenaers, J., 1992, "A computer aided process planning system that increases the flexibility of manufacturing," In *Proceedings of IPDES (Espirit Project 2590) Workshop*.

[8.4] Zhang, H.-C., Huang, S.H. and Mei, J., 1996, "Operational dimensioning and tolerancing in process planning: setup planning," *International Journal of Production Research*, **34**(7), pp. 1841–1858.

[8.5] Wu, H.-C. and Chang, T.-C., 1998, "Automated setup selection in feature-based process planning," *International Journal of Production Research*, **36**(3), pp. 695–712.

[8.6] Zhang, Y.F., Nee, A.Y.C. and Ong, S.K., 1995, "A hybrid approach for setup planning," *International Journal of Advanced Manufacturing Technology*, **10**, pp. 183–190.

[8.7] Ferreira, P.M. and Liu, C.R., 1988, "Generation of workpiece orientation for machining using a rule-based system," *Robotics and Computer-Integrated Manufacturing*, **4**(3–4), pp. 545–555.

[8.8] Contini, P. and Tolio, T., 2004, "Computer-aided setup planning for machining centres configuration," *International Journal of Production Research*, **42**(17), pp. 3473–3491.

[8.9] Öztürk, F., 1997, "The use of machining features in set-up planning and fixture design to interface CAD to CAPP," *International Journal of Vehicle Design*, **18**(5), pp. 558–573.

[8.10] Sakurai, H., 1992, "Automatic setup planning and fixture design for machining," *Journal of Manufacturing System*, **11**(1), pp. 30–37.

[8.11] Sundararajan, V. and Wright, P.K., 2002, "Feature based macroplanning including fixturing," *ASME Journal of Computing and Information Science in Engineering*, **2**, pp. 179–191.

[8.12] Joneja, A. and Chang, T.-C, 1999, "Setup and fixture planning in automated process planning systems," *IIE Transactions*, **31**, pp. 653–665.

[8.13] Lin, L., Zhang, Y.F. and Nee, A.Y.C., 1997, "An integrated setup planning and fixture design system for prismatic parts," *International Journal of Computer Applications in Technology*, **10**(3–4), pp. 198–212.

[8.14] Zhang, Y., Wu, W., Rong, Y. and Yen, D.W., 2001, "Graph-based setup planning and tolerance decomposition for computer-aided fixture design," *International Journal of Production Research*, **39**(14), pp. 3109–3126.

[8.15] Ong, S.K., Ding, J. and Nee, A.Y.C., 2002, "Hybrid GA and SA dynamic setup planning optimization," *International Journal of Production Research*, **40**(18), pp. 4697–4719.

[8.16] Huang, S.H. and Xu, N., 2003, "Automatic setup planning for metal cutting: an integrated methodology," *International Journal of Production Research*, **41**(18), pp. 4339–4356.

[8.17] Gologlu, C., 2004, "Machine capability and fixturing constraints-imposed automatic machining set-ups generation," *Journal of Materials Processing Technology*, **148**, pp. 83–92.

[8.18] Yilmaz, I.O., Grunow, M., Günther, H.O. and Yapan, C., 2007, "Development of group setup strategies for makespan minimization in PCB assembly," *International Journal of Production Research*, **45**(4), pp. 871–897.

[8.19] Yao, S., Han, X., Yang, Y., Rong, Y., Huang, S.H., Yen, D.W. and Zhang, G., 2007, "Computer aided manufacturing planning for mass customisation: part 2, automated setup planning," *International Journal of Advanced Manufacturing Technology*, **32**, pp. 205–217.

[8.20] Cai, N., Wang, L. and Feng, H.-Y., 2008, "Adaptive setup planning of prismatic parts for machine tools with varying configurations," *International Journal of Production Research*, **46**(3), pp. 571–594.

[8.21] Singh, D.K.J. and Jebaraj, C., 2005, "Feature-based design for process planning of machining process with optimisation using genetic algorithms," *International Journal of Production Research*, **43**(18), pp. 3855–3887.

[8.22] Chen, J., Zhang, Y.F. and Nee, A.Y.C., 1998, "Setup planning using Hopfield net and simulated annealing," *International Journal of Production Research*, **36**(4), pp. 981–1000.

[8.23] Amaitik, S.M. and Kilic, S.E., 2007, "An intelligent process planning system for prismatic parts using STEP features," *International Journal of Advanced Manufacturing Technology*, **31**, pp. 978–993.

[8.24] Wang, L., Feng, H.-Y. and Cai, N., 2003, "Architecture design for distributed process planning," *Journal of Manufacturing Systems*, **22**(2), pp. 99–115.

[8.25] Boerma, J.R. and Kals, H.J.J., 1989, "Fixture design with FIXES: the automated selection of positioning, clamping and support features for prismatic parts," *Annals of CIRP*, **38**, pp. 399–402.

[8.26] Rong, Y., Liu, X., Zhou, J. and Wen, A., 1997, "Computer-aided setup planning and fixture design," *International Journal of Intelligent Automation and Soft Computing*, **3**(3), pp. 191–206.

[8.27] Ma, W., Li, J. and Rong, Y., 1999, "Development of automated fixture planning systems," *International Journal of Advanced Manufacturing Technology*, **15**, pp. 171–181.

[8.28] Wang, L., Cai, N., Feng, H.-Y. and Liu, Z., 2006, "Enriched machining feature based reasoning for generic machining process sequencing," *International Journal of Production Research*, **44**(8), pp. 1479–1501.

9

Auction-based Heuristic in Digitised Manufacturing Environment for Part Type Selection and Operation Allocation

M. K. Tiwari and M. K. Pandey

Department of Industrial Engineering and Management
Indian Institute of Technology, Kharagpur, 721302, India
Email: mkt09@hotmail.com

Abstract
This chapter high lights some of the key issues involved in developing real schedule generation architecture in an e-manufacturing environment. The high cost, long cycle time of development of shop floor control systems and the lack of robust system integration capabilities are some of the major deterrents to the development of the underlying architecture. We conceptualise a robust framework, capable of providing flexibility to the system, communicating among various entities and making intelligent decisions. Owing to the fast communication, distributed control and autonomous character, agent-oriented architecture has been preferred to address the scheduling problem in e-manufacturing. An integer programming-based model with dual objectives of minimising the make span and increasing the system throughput has been formulated to determine the optimal part type sequence from the part type pool. It is very difficult to appraise all possible combinations of operation-machine allocations in order to accomplish the above objectives. A combinatorial auction-based heuristic has been proposed to minimise large search spaces and to obtain optimal or near optimal solutions of operation-machine allocations of given part types with tool slots and available machine time as constraint. The effects of exceeding the planning horizon due to urgency of part types or over time given to complete the part type processing on shop floor is also exhibited and a significant increase in system throughput is observed.

9.1 Introduction

Manufacturing industries are experiencing remarkable challenges from the consumer market due to frequent changes in product design. A common and accepted way to meet customer requirements in a manufacturing system is by connecting the industries through communication networks. With the emergence of electronic information technologies, such as the Internet and wireless communication that form the core of highly competitive and efficient manufacturing systems, the business world is entering a new era of e-manufacturing. E-manufacturing is a system methodology that enables operations to successfully integrate with the functional objectives of the enterprise through the use of the Internet, tether-free (*i.e.* wireless, web, *etc.*) and predictive technologies [9.1]. The Internet is used to monitor processes on the shop floor and other peripheral systems to assure that all are operating at optimal levels [9.2]. It reduces geographical distances and allows products to be manufactured and marketed on a global basis. E-manufacturing perceptions started originally from the idea of the factory for the future, e-business/commerce application in manufacturing and the extension of computer networking technology. Basically, e-manufacturing integrates customers, e-commerce systems, and suppliers with manufacturing process to provide an Internet-based strategic framework for the factory. In an e-manufacturing system, people, machines and organisations act as software agents connected via the Internet. Internet establishes a dynamic environment that enables the agents to move from one place to another in order to deliver services and to achieve the pre-determined goals in a similar way to people co-operating by exchanging services [9.3–9.7].

E-manufacturing bridges the gap between product development and supply chain, which exists due to the lack of lifecycle information and information about suppliers' capabilities. Figure 9.1 shows the integration of product development, supply chain and plant floor in an e-manufacturing environment [9.3, 9.4]. With advancements in the Internet and tether-free communication technologies, the philosophy of e-manufacturing, e-factory and e-supply chain has replaced traditional factory integration concepts [9.4]. The technological advancement for achieving collaborative design is based on multi-media type information-based engineering tools and a highly reliable communication system. It is also required for remote operation of manufacturing processes, and operation of distributed production systems.

In e-manufacturing, flexible and concurrent planning and scheduling can be realised using the multi-agent paradigm. Implementation of real-world agent-based system architecture, communicated through the Internet and web using Java, is growing in the manufacturing sector. This implementation provides an effective way for components to interact with each other.

E-manufacturing systems allow companies to access data in other companies, which helps in better planning and scheduling. The flow of information takes place in both directions: from the producer to the supplier as well as from the supplier to the producer. For this purpose, data is continuously put into a database to which other companies have instant access. Data collection from the plant floor needs a variety of communication functions and protocols based on a wide collection of sensors, devices, gauges, and measurement instruments, in process automation. Collected data are useful only when they are reduced and transformed into information and knowledge for responsive actions. For this, data mining tools are used for data reduction, representation and prediction adopted for shop floor data. Tools are needed to correlate data from different formats and transform them to web-deployable information systems [9.3]. These data can be gathered from traditional control input/output or through a separate wireless data acquisition system using different communication protocols. Users from different factories or locations can share this information through web tools. Shop floor-level integration occurs within the enterprise-level integration for flow of data as well as to get order status at any time (as shown in Figure 9.2). Web-centric technologies like Java

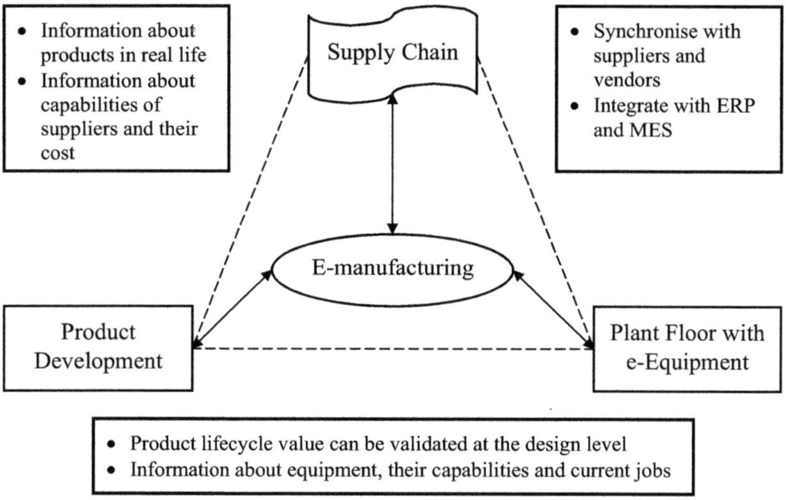

Figure 9.1. Integration of different modules in an e-manufacturing system

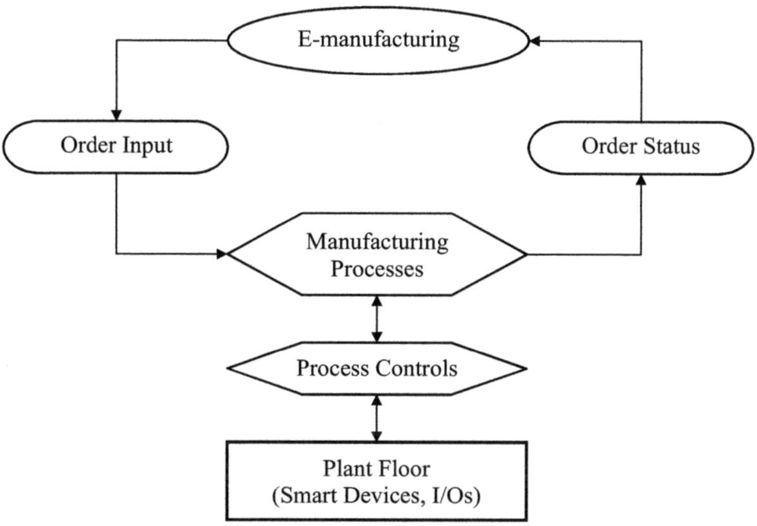

Figure 9.2. Information flow in a manufacturing system

technology, XML, and XML schema frameworks provide the bond that connects the front-end of e-business to the back-end of e-manufacturing.

Extensible Markup Language (XML) documents can be used across different platforms and applications thus is a good choice in current information technology for data exchange. XML can be used for formatting messages among all multi-agent systems. The architecture of e-manufacturing is implemented with Java technology, depending on the requirements of the manufacturing processes and their integration with the enterprise business system. Intelligent agents and portals developed with Java technology are capable of integrating the heterogeneous mix of operating systems and plant floor automations.

In the e-manufacturing environment, operation allocations are one of the greatest bottlenecks in a company's production and planning activities. Geographically distributed manufacturing plants, as well as decentralised decision levels of the company, make it a complex process to allocate the whole enterprise manufacturing capacity. It involves the selection of part types to be produced in a given planning horizon and allocation of operations on machines, tools, fixtures, pallets, *etc.*, to process the selected parts. E-manufacturing provides a wide array of product routings, allocation of resources to make the product and the scheduling of the manufacturing activities to achieve the best operational efficiency. Part selection, machine loading and tool configuration are three prominent areas interlinked with each other. Stecke [9.8] has discussed six objectives of the machine-loading problem: (1) balancing the machine processing time, (2) minimising the number of movements, (3) balancing the workload per machine for a system or group of pooled machines of equal size, (4) unbalancing the workload per machine for a system or group of pooled machines of unequal size, (5) filling the tool magazines as densely as possible, and (6) maximising the sum of operations priorities.

Buzacott and Yao [9.9] have discussed various methodologies and approaches to solve the machine-loading problem. Liang and Dutta [9.10, 9.11] have considered the part type selection and the machine-loading problem concurrently, which was treated separately by many researchers. Maturana *et al.* [9.12, 9.13] have proposed a multi-agent architecture for distributed manufacturing systems, called MetaMorph. In this architecture, two types of agents are used: resource agents for representing physical entities and mediator agents for co-ordination. Shen and Norrie [9.14] extend the MetaMorph architecture to integrate enterprise-level activities with its suppliers, partners and customers in their MetaMorph II project. In this project, hierarchical mediator and bidding mechanism used for co-operative negotiation among resource agents are employed. The interrelationships of various decisions and various hierarchies in the manufacturing system are issues that have been widely surveyed and popularly referred to by researchers like Singhal [9.15], Kusiak [9.16], Stecke [9.17], Rajagopalan [9.18], and van Looveren *et al.* [9.19]. Some of the heuristic solutions proposed by Shanker and Srinivasulu [9.20], Mukhopadhyay *et al.* [9.21], Tiwari *et al.* [9.22], Moreno and Ding [9.23] have used fixed pre-determined part sequencing rules as input to the heuristic for allocation of operations to the machines using minimisation of system unbalance and maximisation of throughput as objectives subject to constraints posed by the availability of machining time and tool slots. Turgay [9.24] presented the design of an agent-based FMS control system using Petri nets and evaluated its performance. A mathematical model is proposed that minimises the queue length during system processing. Mes *et al.* [9.25] compared the multi-agent system for scheduling transportation systems. It is proved that a multi-agent system is less sensitive to fluctuations and provides flexibility by inherently solving local problems. Wang *et al.* [9.26] presented a set of agents, each of which uses local information to generate a schedule. Filtered beam search (FBS) was used with an agent-based system to act as a scheduling engine. Meyyappan *et al.* [9.27] proposed a wasp-based control model for routing in FMS. It was shown that wasps used a non-negotiation-based model to reduce communication overhead. Hussain and Frey [9.28] applied an auction-based agent system for distributed architecture and found that a predictive environment helps in reduction of communication load.

This work attempts to develop an exact heuristic to confine the intricate details of the loading problem of a manufacturing system in an e-manufacturing environment. Here, machines are capable of performing several types of operations using several tool types. A framework has been conceptualised using a Java-based multi-agent system to provide an effective way for components to interact simultaneously by wrapping process planning and scheduling tools into the multiple agents. Tool magazine capacity and available processing time are considered as common constraints to solve the problem with dual objectives. These objectives are defined as maximisation of the throughput and minimisation of the make span (by minimising idle time) to achieve high system utilisation. An integer programming model has been used for determining the optimal part type sequence from the part type pool. An auction-based heuristic approach is then proposed using agent technologies to solve a set of random machine-loading problems in an e-manufacturing environment. The efficacy of the proposed framework and its solution methodology has been tested on different test beds. We have further shown the effects of exceeding the planning horizon due to urgency of part types or over time given to complete the part type processing on the shop floor, and observed the significant increase in system throughput.

The rest of the chapter is organised in the following manner. Section 9.2 presents a brief overview of agent technology used for developing e-manufacturing architecture. In Section 9.3, the details of auction mechanism for winner determination are described. Section 9.4 illustrates the problem definition. The proposed framework to solve the machine loading problem is elucidated in Section 9.5. It is followed by Section 9.6, where a case study is considered and numerical simulation is discussed to investigate and validate the solution methodology. Finally, this work concludes in Section 9.7, briefing some of the key findings and extension for future work.

9.2 Overview of Agent Technology

9.2.1 Definition of an Agent and its Properties

Chaib-Draa and Moulin [9.29] have categorised agent technology as a new specialisation of distributed artificial intelligence (DAI). Owing to its novelty, there is not a universally accepted definition of an agent. Singh and Tiwari [9.30] has defined agent adopting some of the definitions form Fisher [9.31], Jennings and Wooldridge [9.32], Davidson *et al.* [9.33], Nwana and Ndumu [9.34], as an object of a program, which have its own value and means to solve some subtasks independently and finally communicate its solution to a large problem-solving process to achieve the objective. For the notion of agent and autonomy used in the present context, an agent represents an object, which is either a physical object such as a worker, a machine tool, fixtures, machines, *etc.*, or a logical object such as an order, a task, *etc.* Maes [9.35] has discussed an agent-based system for more flexible and fault-tolerant systems than traditional ones. In addition to several definitions of an agent, agents may have other properties like autonomy, social ability, responsiveness, adaptability, mobility and protectiveness [9.32]. Further, agents have reasoning capabilities as discussed by Maturana [9.12]. Finally, an agent has the ability to make a plan to achieve the goal. Based on a goal-oriented plan, agents execute actions, monitor their environment to determine the effect of their actions, and accordingly pre-plan their actions to achieve their goals.

Agent technology has been applied to resolve many operations research problems and manufacturing case studies [9.36–9.38]. In an e-manufacturing environment, agents can be employed for partner selection, network design, planning and scheduling. Each agent can perform one or more functions and co-ordinates with other agents [9.39]. Agents can split the larger order into several sub-orders, and use the Internet to check the availability of capacities among the partners to achieve the optimal solution of a particular order [9.40].

9.2.2 Heterarchical Control Framework

A heterarchical control framework has no central controller, like the hierarchical control framework. With a heterarchical framework, control is distributed so that it is able to handle unplanned events such as machine breakdown. Hatvany [9.41] was one of the first to propose co-operative heterarchies as an alternative to the hierarchical control system. He recognised the requirement for designing behaviour rules, local objectives and global objectives, to prevent anarchy. Duffie and Piper [9.42, 9.43] presented the advantages of the heterarchical control architecture by comparing it with two control systems namely centralised and hierarchical. The main advantages of heterarchical include reduced complexity, high flexibility and modularity, reduced software development costs, and improved fault tolerance. Ou-Yang and Lin [9.44] have discussed the limitation of it, when a trade-off arises between local objective and the overall system performance. Duffie and Prabhu [9.45] proposed a co-operative scheduling algorithm to improve the global system performance. They also discussed some design principles for constructing a heterarchical system in a model. Lin and Solberg [9.46] proposed a heterarchical intelligent agent framework for the manufacturing system. They have considered each part and resource unit as an agent in shop floor control architecture. In real time, each agent communicates with others via a bidding mechanism to achieve individual objectives. For example, as a machine agent enters a system for auction of free machining hours, the part agent bids for machining hours depending on its processing requirements to the shop floor manager agent (SFMA). SFMA communicates with part agents to optimise an objective that can be a function of cost. SFMA evaluates the bids and selects the one that optimises its objective. An offer submitted by a part agent can be accepted or rejected by SFMA.

9.2.3 Contract-net Protocol (CNP)

A negotiation protocol is required for communication among agents. The negotiation protocol used in this chapter is derived from the CNP proposed by Smith [9.47]. Smith and Davis [9.48, 9.49] discussed the CNP and distributed sensing system to solve the allocation problem of the tasks in a decentralised system. The CNP consists of a set of nodes. Each node has the ability to take decision and to negotiate with other nodes.

The auction process with the CNP is similar to a sealed bid auction by contractors, where the winner is determined by the highest/lowest bid value. It defines a bidding mechanism that enables task allocation among multiple machine agents. The bid value depends on the agent's local criteria and assessment of its own capabilities for achieving the goal. Tilley [9.50] has discussed some protocols used in constructing bids, free machining hours announcement, *etc.*, and came up with a bidding-based heterarchical system behaviour from the communication point of view and shown time as main constraint, during announcements of free machining hours of machines by a shop floor manager. The main application of the CNP is the decomposition of complex problems, if sub-tasks are large and require intensive computation.

9.3 Overview of Auction Mechanism

Traditional auctions have some limitations and deficiencies, as they last only for a few minutes for each item sold. Here, both sellers and bidders may not get what they want. With the emergence of new technologies, auctions can be performed on the Internet to overcome the above limitation and deficiencies. Trading on the Internet has several advantages, such as:

- It is independent of geographical location.
- Business settlement will be in shorter time with lower overhead cost.

The main advantage of Internet auctions is that they allow individual bidders or a group of machines to sell their services. Auctioning on the Internet can support a greater range of potential bidders. The bidder can bid for any part type independent of geographical location and can also submit bids for more than one auction simultaneously. The Internet provides an infrastructure for executing auctions and bids more cheaply. Finally, we conclude that Internet auctions provide a new approach to solving the problems of scheduling and control of part types in an e-manufacturing environment. Several researchers have discussed the Internet auction as future business applications, including Turban [9.51] as well as Segev and Gebauer [9.52]. They have discussed auction-based manufacturing systems, in which various entities bid by themselves, accept bids, and make a selection from the available bids based on some heuristic procedure.

In this chapter, we have considered combinatorial auctions mechanism for determining the selection of part type. In combinatorial auctions, a combination of different types of resources is available for auction and the bidders bid for different combinations of these resources. It allows bidders to express their synergistic values. The determination of winners is a non-trivial problem in this class of auction. In combinatorial auctions, manufacturing capacity can be utilised for production of different part type mixes in different volumes. Bidding rules and allocation of bids to bidders are important issues in the combinatorial auction process.

Bidding Rules

A bidder bids for a combination of resources depending on his/her needs. A bid is a demand for resources (machines), and is the maximum amount of money that the bidder is willing to pay in exchange for services offered by each combination of resources. Thus, the bids determine the order in which the part types are to be processed on different machines in given time slots [9.53]. A bid is feasible up to that part type in its sequence until it satisfies the machining time and tool slot constraints. Bidders evaluate a bid value by considering the following factors before bidding:

- Machines used for processing the operations, *i.e.* the same operations on different machines can have different bid values.
- Tool type used on machines for processing the operations, as each tool type may have different tool life, different materials, *etc.*
- Required number of operations to obtain the desired features. Bid value increases when the complexity of the operations increases.
- Bid value is high during starting time slots on machines and its value decreases as time slots increase. This is due to the sudden breakdown of a machine that is unable to complete a part type in the planned time frame.

The auction protocol provides a means for this bidding communication. Various bidding languages have been discussed in the literature [9.54]. In this work, the XOR bidding mechanism is used to solve the machine loading problem.

9.4 Problem Definition

The problem considered in the e-manufacturing environment is assumed to occur in a real-time situation. In this environment, shop floor-level integration occurs at a lower level within the shop floor of the selected manufacturing partners. The shop floor of an e-manufacturing system consists of multiple machines at different locations, each machine with an automatic tool changer and a tool magazine of limited capacity. The machines can perform different operations with different tool types. The present status of scheduling of an individual machine can be determined or updated by the task assignment on that machine. A part can be produced by performing a number of operations on any machines, but the sequence of operations remains unaltered. Throughput refers to the summation of the batch size of the part types that are to be produced, and minimisation of make span can be achieved by eliminating time redundancy and reallocation of part types.

It is very difficult to appraise all possible combinations of operation-machine allocations in order to achieve the maximum throughput and minimum make span due to the large search space. Therefore, heuristics that require a wide search space to solve the machine-loading problem are avoided. A combinatorial auction-based heuristic is applied to obtain an optimal or near-optimal combination of operation-machine allocations of given part types with tool slots and available machine time as constraints. This work considers dual objectives namely: *to maximise throughput*, and *to minimize make span*. In addition, the technical constraints considered include:

(1) availability of machining hours on each machine, (2) availability of tool slots on each machine, (3) unique part type routing, and (4) non-splitting of part types.

An attempt has been made to adopt a heuristic procedure that effectively minimises large search spaces and obtains optimal or sub-optimal solutions. This can be achieved by using a combinatorial auction-based heuristic. The bids determine the order in which part types are processed on different machines. Heuristics and multi-objective functions are used in determining the winning bidders for assigning parts, machines, AGV (automated guided vehicle), and sequencing of incoming parts. A shop floor manager agent assigns the part types to the machine agent that can process the part types. The budget for processing each part type is based on the time data in the process plans and the rate of running cost (per time unit) of the partner equipment [9.55]. The following assumptions are made to minimise the complexities while analysing the problem:

- Initially, all the part types and machines are simultaneously available.
- Processing time required to complete an entire part type order is known *a priori*.
- Part type undergoing processing is to have all its operations completed before considering a new part type.
- The operation of a part type once started on a machine is continued until it is completed.
- Transportation time required to move a part type between machines is negligible.
- Sharing and duplication of tools is not allowed.

9.5 Proposed Framework

9.5.1 Agent Architecture

It is considered that machines and parts are intelligent agents having communication capabilities. The primary purpose of a part type is to accomplish all the processing as early as possible, whereas a machine tries to maximise utilisation rate. A shop floor manager agent acts as a mediator for communication between a part agent and machine agent. A communication facilitator is used for communication among all the agents through the Internet. A database agent provides required data to other agents for message exchange in order to achieve certain goals.

9.5.1.1 Part Agent (PA)

A part agent (PA) is created, when an order enters the system. The PA bids for those machines that fulfil its objective. Each part agent prevail its process plan from the database after it is created. When the PA needs to use the machines, tools, fixtures and AGV, it submits a bid for this. A PA has the following information:

1. A unique identification number.
2. Number of operations required.
3. Machining time of each operation on different machines.

4. Tool type required for each operation.
5. Locations of part types on machines.
6. NC programs related to each operation.
7. Bid calculation logic for computation of cost of each operation; summation of all operation cost gives bid the value.

When the PA enters the system, it informs the shop-floor manager agent (SFMA) about its arrival and receives feedback about the requirements of processing operations, its location, *etc.* When an assigned machine fails to process the part type, the part agent again makes contact with the SFMA.

9.5.1.2 Machine Agent (MA)

A machine agent (MA) represents a machine. Each machine may have a different set of objectives. A part type with a particular operations sequence can be processed on either one machine or different machines, depending on the cost and time involved. Machine agents determine the acceptance of a part type for the maximum utilisation of machines and revenue generated from the order. Free machining hours on each machine depends on constraints like processing time of part type, tool slots required and tool type availability. An MA has the following information:

1. A unique identification number.
2. Specifications of machines.
3. Attributes of a machine including its status, part type waiting in the buffer.
4. Free machining hours available on individual machine.
5. Capacity of each machine.
6. Locations of machines on shop floor.

An MA uses the unique identification number of a part agent to query the database agent (DA) for detailed information about the design and processing requirements of a part type. It computes processing cost based on processing time, tool materials and tool life. Processing time for each operation can be computed as

$$P_{nj} = M_{nj} + S_{nj} \tag{9.1}$$

where P_{nj}, M_{nj}, and S_{nj} are the processing time, machining time and setup time of the jth operation for the nth part type, respectively. The total processing time P_n of the nth part type yields

$$P_n = \sum_{j=1}^{m_n} \left(M_{nj} \times B_n + S_{nj} \right) \tag{9.2}$$

where B_n and m_n are the batch size and total number of operations of the nth part type, respectively.

An MA receives a signal from the SFMA, about processing a part type. The MA checks for the availability of required tool type, necessary tool slots, its capability, and buffer limit. If all constraints are satisfied, it then accepts that part type; otherwise, reject it.

9.5.1.3 Shop-floor Manager Agent (SFMA)

Through communication with MAs, an SFMA possesses knowledge about the machine capacity and free machining hours of machines. It has a bidding protocol and the capability to compute machine processing cost for each part type and can keep track of the system state, all by communicating with MAs. After communication with MAs, it broadcasts the free machining hours of different machines available in a plant, and then requests the PAs to submit bids. The SFMA interacts with bidders, and accepts a bid when the highest value of a bidder is more than the cost associated with machine schedule, machining time, setup time, tool change time and cost of tooling. After bidding, the SFMA assigns the part type to the MA that can process the part type.

9.5.1.4 Communication Facilitator (CF)

A communication facilitator (CF) is an interface, responsible for communication within a group of agents so that the selection of part types and then processing them on either the same or different machines is possible in a secured manner. CF is responsible for passing messages among agents and also able to convert an incoming message to a language that could be understood by the agents.

9.5.1.5 Database Agent (DA)

A DA acts as a database server and works through the Internet. It helps in storing/ retrieving some common data in the e-manufacturing environment, such as process plan about specific operations, resource information and shared design information, *etc.*

9.5.2 Framework with Agent Architecture

In the proposed framework, agents play a central role in co-ordination and co-operation in the e-manufacturing environment. The underlying agent architecture should be robust. The architecture facilitates users to customise agents in order to achieve their goals. In an e-manufacturing system, resources (machines, tools, fixtures, AGV, *etc.*) at machine level are represented as agents. These agents are dynamically grouped together using group technology at enterprise level.

The multi-agent system (MAS) proposed in this chapter is a network of single problem-solving agents, developed in Java based on Java Agent Template Lite (JATLite). JATLite is a prototype agent environment developed by Stanford University [9.56]. It is a set of lightweight Java packages that can be used to build an MAS (Figure 9.3). The following components are presented in each single agent:

- *Communication module*: This module is based on JATLite API, which describes communication in KQML (Knowledge Query and Manipulation Language) among agents [9.57]. Communication is the backbone of any system, as it allows agents to share information and help in determining the overall behaviour and organisation of a system [9.58]. JATLite facilitates the construction of agents, and helps agents to send and receive messages using KQML [9.59]. An agent can register, connect or disconnect to a CF by its name, password and IP address. The CF itself is connected to the Internet. After that, the agent can co-operate and co-ordinate with other agents and is able to access database remotely.

- *Problem solver*: This module is responsible for solving the problems generated by other agents. It deploys inference engine that uses accumulated knowledge (set of rules) and algorithm to solve the generated problem. An agent performs the task by parsing the incoming information. In an MAS system, agents can co-operate and co-ordinate to accomplish the task. Local/global data stored in database can be accessed remotely by agents.
- *Legacy software*: It is the application software wrapped by the single agent. In this chapter, it refers to scheduling and planning software. It permits the editing of the program by implementation of a wrapper. The legacy software is integrated over the Internet and makes communication and data sharing at peer-to-peer level with the help of agents.

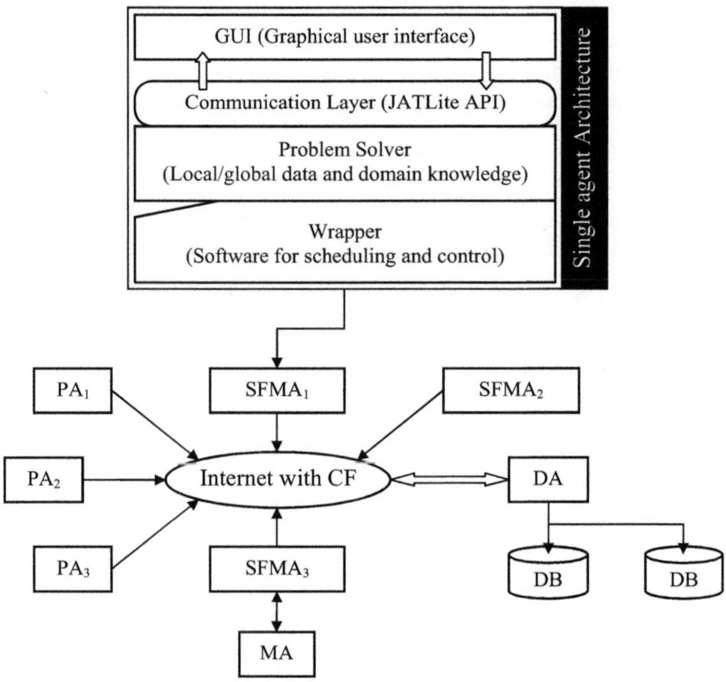

Figure 9.3. Agent architecture and communication among agents

An MAS is a software system in which program modules (individual agents) have given autonomy, intelligence and subordinate co-ordination mechanism that enables collaboration between each module (agent) to attain the systems objective.

Hence, an MAS is characterised by a number of autonomous, heterogeneous and potentially independent agents working together to solve specific problems. An MAS provides an effective means for integrating legacy software over the Internet in an e-manufacturing environment. It defines the basic rules for co-operation of the agents by limiting their autonomy.

Interlinking of Agents with the Internet

Agents are interlinked and communicate through the Internet/Intranet/LAN, and the interactions among agents are possible by messages passing. Messages are defined in terms of *request*, *reply* or *inform*. Requests are used for providing services; replies are used to answer the requests; and inform is to notify agents without expecting a response. Agents use the Internet to check availability of machine capacity for the fulfilments of a particular objective.

The main objective of the Internet is to send and receive textual information and graphics. The Internet, being platform and language-independent, is easily accessible and popular to a mass population, and is powered by information technologies such as HTTP (Hypertext Transfer Protocol), HTML (Hypertext Markup Language), XML (eXtensible Markup Language), and Java technology. With the help of these technologies, agents provide people a common 'look and feel' for information exchange.

Agent technology provides a means for e-manufacturing implementation with open system architecture, and gives some important features like flexibility, modularity, reconfigurability, scalability, and robustness. However, an Internet-based manufacturing system has to consider security and privacy of the individuals and organisations involved in the e-manufacturing environment.

9.5.3 Framework of Auction Mechanism

An auction-based heuristic approach using agent technologies has been applied to solve a set of random machine-loading problems in e-manufacturing environment and further validated on a case study described in details in the next section.

In this section, a rule-based algorithm is considered to solve machine-loading problem with dual-criteria objectives of maximising system throughput and minimising make span. We select part types by auctioning the machining hours. The number of bidders is the same as the number of part types to be processed. The number of bids that a bidder can submit is equal to the number of ways in which the part types can be processed.

Winner Determination

Given a set of bids for a subset of machines, selecting the winning set of bids is denoted by "the winner determination problem". This problem can be formulated as an integer programming problem. Let N be the set of bidders, M the set of m distinct machines, and S a subset of M. Agent j's ($j \in N$) bid for combination S is denoted by $b^j(S)$, and the winning bid is

$$b(S) = \max b^j(S), \qquad j \in N \tag{9.3}$$

Integer Programming Formulation

The integer programming formulation is used for determining the optimal part type sequence from the part type pool, and is formulated below

$$\max \sum_{j \in N} \sum_{S \subset M} b^j(S) \times y(S,j) \tag{9.4}$$

subject to

$$\sum_{S,j \in N} \sum y(S,j) = 1 \tag{9.5}$$

$$\sum_{S,j \in N} y(S,j) \leq \forall j \in N \tag{9.6}$$

$$y(S,J) = 0,1 \quad \forall S \subseteq M, j \in N$$

where $y(S, j)$ equals 1 if the subset S is allocated to bidder j, otherwise 0.

The objective of a plant manager is to maximise the revenue generated from auctioning the machines capacities. The constraint described in Equation (9.5) ensures that a particular part type can be selected only once. Equation (9.6) ascertains the assignment of a part type only once on a particular machine. The solution to the winner determination problem represents the efficient utilisation of machines in an exchange economy.

As an example of selling the resources available in the form of machining hours of different machines in a plant, let us consider a scenario in which free machining hours are available over the planning horizon as shown in Table 9.1 at the time of decision making.

Table 9.1. Auctioning machining hours over different time slots

1 day = 1shift		Machine				
1shift = 8 hours		M_1	M_2	M_3	...	M_n
Time slot	1 hour	1	1	0	1	1
	2 hours	1	0	1	1	0
	3 hours	1	1	0	1	1
	4 hours	1	0	1	0	1
	5 hours	0	0	1	1	0
	6 hours	0	1	0	0	1
	7 hours	1	1	1	0	0
	8 hours	1	1	1	0	0

The entry 1 in each cell represents the availability of a machine during the time slot, and 0 represents non-availability of the machine. A bidder bids for a combination of machining hours over different time slots, with which he can produce his product in the required quantity. Bidders compute the revenue generated from this product, and they can submit the bid value. Different bidders have different approaches to computing the bid value, so they may have different bid values for the same subset of machines. The winning bids are those that together maximise the revenue for the auctioneer (plant manager). According to the winning bids, the machining hours are allotted to different bidders.

9.5.4 Communications among Agents

Co-ordination is required in a multi-agent system to prevent chaos, satisfy global constraint and synchronise the behaviour of individual agents. Agents interact with each other and find (near) optimal results as a result of their interactions. Planning, scheduling and control through agents can be done in the following steps:

1. PA → DA: PA retrieves the information from DA by using the unique identification number of machine and gets information about machine capacity, free machining hours, and machine specification, *etc*., about the concerning plant.
2. DA → PA: DA stores all important information about a PA, *e.g*., number of operations on a part type, its location on a particular machine, processing time requirement, *etc*., of a part type.
3. MA → SFMA: MA provides its current status such as free machining hours, its location in a plant, its capacity, and its specifications, *etc*.
4. SFMA → PA: SFMA invites PAs to submit bids, and selects the bidder paying highest bid value to purchase the service.
5. PA → SFMA: PA submits a bid to SFMA for a single machine or subset of machines.
6. SFMA → MA: SFMA schedules a task to MA for processing.
7. SFMA → DA: SFMA retrieves information about a part type, *e.g*., number of operations, and its location on a particular machine, *etc*.
8. MA → DA: MA retrieves a process plan, design features, *etc*., from DA using part identification number for performing an operation.
9. DA → MA: DA stores all important information about an MA, including machine capacity, free machining hours, and its location, *etc*.

The aforementioned activities are realised through the interactions among a group of agents. All the agents are registered and connected/disconnected to the CF through the Internet/Intranet by their IDs, passwords, and IP addresses.

Figure 9.4 shows a group of agents communicating to each other, where SFMA and PA interact using the auction-based model according to communications between SFMA and MA. When a machine completes the processing on scheduled part types, an MA is instantiated and informs its arrival to SFMA. The SFMA then sends a message to all the part agents.

9.5.5 Task Decomposition/Distribution Pattern

SFMA is responsible for creating a pattern, which ensures the flow of tasks (part types) on machines by co-ordination. The techniques by which a pattern works are:

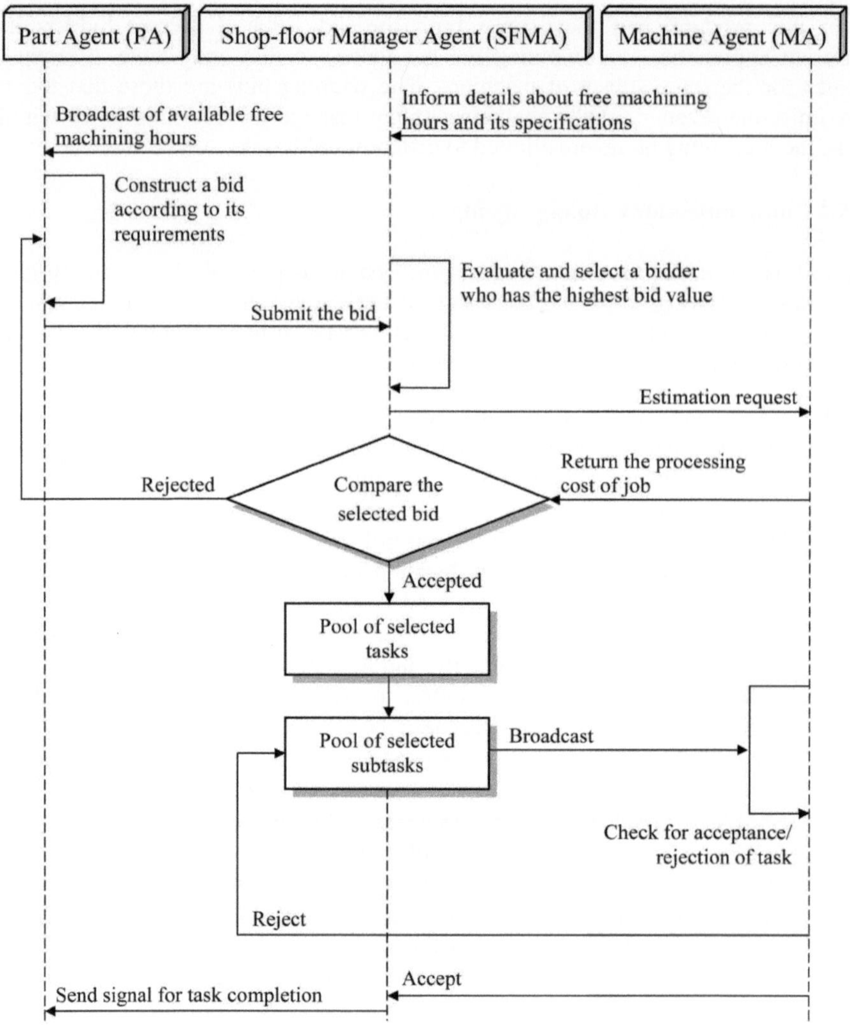

Figure 9.4. Communications among PA, SFMA and MA

1. Task breakdown into subtasks.
2. Subtask distribution among the entities of a cluster.
3. Subtask deployment among machines.

Task decomposition and processing on different machines using different tools are shown in Figure 9.5.

9.5.6 Heuristic Rules for Sequencing and Part Selection

To generate a sequence of part types from a pool of part types, we consider the following objective function f_j:

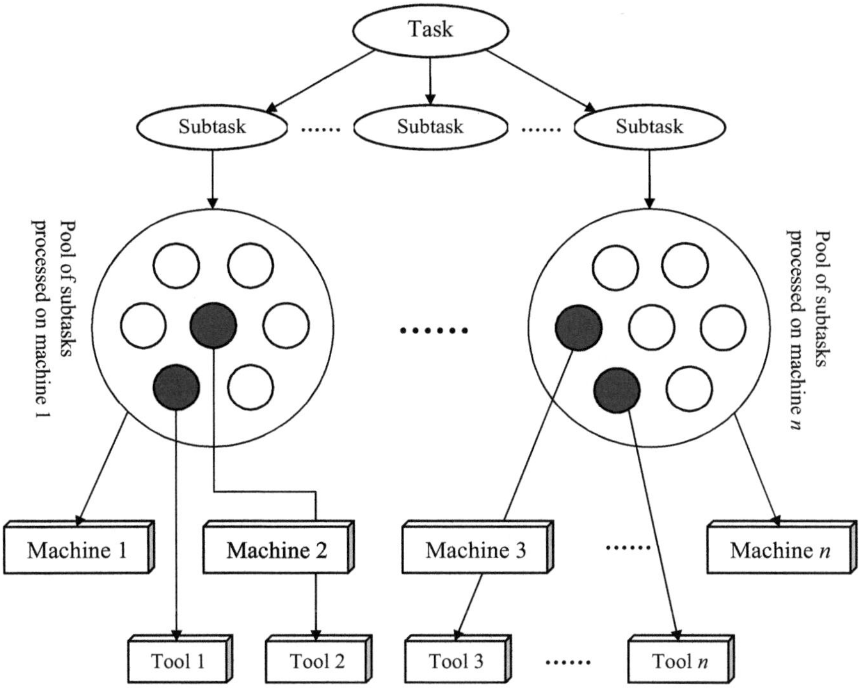

Figure 9.5. Task distribution pattern

$$\max \quad f_j = (w_1 \times t_1 + w_2 \times t_2)/(w_1 + w_2) \tag{9.7}$$

Here, two more functions are considered for minimisation of system idle time t_1 and maximisation of throughput t_2:

$$\min \quad t_1 = (S_{um} - S_{us})/(S_{um} - S_{un}) \tag{9.8}$$

$$\max \quad t_2 = (T_{hs} - T_{hm})/(T_{hm} - T_{hn}) \tag{9.9}$$

where, S_{um} is the maximum system unbalance, S_{un} the minimum system unbalance (we considered it zero), and S_{us} the system unbalance corresponding to a particular part type sequence. T_{um} is the maximum throughput, T_{hn} the minimum throughput (we considered it zero), and T_{hs} the throughput corresponding to a particular part type sequence. w_1 is the weight assigned to t_1, and w_2 the weight assigned to t_2 (in our case, $w_1 = w_2 = 1$).

Step by step implementation of the proposed heuristic is described as follows:

1. Each bidder can submit a bid as a pair (S, V) where S is a subset of machines and V is bid value that bidder is willing to pay for that subset.
2. Initially, bidder can bid for a combination of machines and suitable available time horizon in a planning horizon, after that bidder has to bid for available free machining hours.

3. Among the list of bidders, select bidders submitting the highest price value for each part type.
4. For each bid, determine the value of objective function (described in Equation (9.3)) while observing the constraints related to the available machining hours and tool slots.
5. If a tie for any bid of the part type occurs, the bid with maximum batch size and minimum SPT (shortest processing time) is selected.
6. Part types having the same bid number are grouped into one group. Group of part types are arranged in ascending order of the number of bids. In a group, part types are arranged according to the value of function f (described in Equation (9.7)) in descending order. Finally, make a sequence of part types to be processed on shop floor.
7. First operation is given the highest priority; last operation is given the lowest and all other operations are with normal priority. According to the priority level, each operation of a part type is assigned to machines. Assign low priority operation at last on each machine.
8. From the selected part types, choose the part type having low SPT for the first operation, so that the second operation if performed on a different machine can start early, which minimises make span, else select another part type having the next SPT.
9. For the next operation, if allocated machine is free at that time, assign part type to the machine; otherwise, wait for free available time.
10. If a machine is available for processing an operation, assign it to the part type having the SPT for an operation.
11. Repeat steps 8, 9, and 10 until there is no unassigned operation of a part type.

9.6 Case Study

In order to validate the proposed heuristic, an example consisting of 7 candidate part types and 4 machining centres is considered. Each machining centre is assumed to have a maximum utilisation of 100%. Although it is practically not possible, the proposed framework holds good for less utilisation (in that case, the planning horizon will be altered). The planning horizon of 8 hours is available as machining hours for each machining centre. Each machining centre has a magazine with 5 tool slots. The details of the problem are given in Table 9.2, while Table 9.3 shows the details of bids presented by part agents; from a set of bids, a winner is decided based on bid values. We have considered that each machine has a different machining cost for the same operation. Table 9.4 shows how many part types are scheduled in a given planning horizon. Finally, Gantt charts are drawn to represents free machining hours on each machine and a make span value is given.

9.6.1 Winner Determination

Initially, a bidder bids for a combination of machines and processing time per unit piece on that machine. Part types having an equal number of bids are grouped

Table 9.2. Problem description for case study

Part type	Operation #	Batch size	Unit processing time	Machine #	Tool slot needed
1	1	9	22	2	2
			22	3	2
	2		25	2	1
2	1	8	20	3	1
3	1	9	25	1	1
			25	4	1
	2		25	4	1
	3		22	2	1
4	1	7	24	3	1
	2		19	4	1
5	1	14	26	4	2
			26	1	2
	2		11	3	3
6	1	13	25	1	1
			25	2	1
			25	3	1
	2		17	2	1
			17	1	1
	3		24	1	3
7	1	10	16	4	1
	2		7	4	1
			7	2	1
			7	3	1

together and groups of part types are arranged in ascending order of the number of bids. Price of the bid can be calculated according to the time expended on individual machines by considering time slots. Here, we consider a cost-based function to calculate bid value as:

$$P_j = t_3 \times t_4 \times P_n \times t_5 + t_6 \tag{9.10}$$

where P_j is the bid value of part type j, $t_3 \in (0, 1)$ ($t_3 = 0$ for the last hour of time horizon, $t_3 = 1$ for the first hour of time horizon), $t_4 \in (1.0, 1.75)$ ($t_4 = 1.0$ for M_1, $t_4 = 1.25$ for M_2, $t_4 = 1.5$ for M_3, $t_4 = 1.75$ for M_4), t_5 is the processing cost per minute of a machine (here, $t_5 = 10$ for all four machines), and t_6 is the cost due to waiting time, setup time and also due to transportation (here $t_6 = 0$).

Table 9.3 represents the bids submitted by bidders. For example, the first bidder submitted a bid of (2, 22). Here, 2 indicates the machine number on which a part type will be processed, and 22 indicates the processing time per unit piece of the part type. Bid price is calculated on the basis of the above function, *e.g.*, calculation of price for bid offer {(2, 22) (2, 25)} = {1.25×22×10 + 1.25×25×10 = 275+315 = 590}. Finally, a winner can be decided according to the highest bid submitted by

Table 9.3. List of various bids of each part type

Bidder	Bid	Price	Winner
1	(2, 22) (2, 25)	590	
1	(3, 22) (2, 25)	645	√
2	(3, 20)	300	√
3	(1, 25) (4, 25) (2, 22)	962	
3	(4, 25) (4, 25) (2, 22)	1151	√
4	(3, 24)	360	√
5	(4, 26) (3, 11)	620	√
5	(1, 26) (3, 11)	425	
6	(1, 25) (2, 17) (1, 24)	703	
6	(2, 25) (2, 17) (1, 24)	768	
6	(3, 25) (2, 17) (1, 24)	828	√
6	(1, 25) (1, 17) (1, 24)	660	
6	(1, 25) (1, 17) (1, 24)	725	
6	(1, 25) (1, 17) (1, 24)	785	
7	(4, 16) (4, 7)	403	√
7	(4, 16) (4, 7)	368	
7	(4, 16) (4, 7)	385	

bidder 1. Here, bidder 1 submitted two bids and the offered bid values are 590 and 645. Hence, bidder 1 with bid value 645 is automatically the winner.

9.6.2 Analysis of the Best Sequence

Various simulation runs have been carried out to reach a near optimal solution. to obtain the sequence, part types are grouped according to the number of bids, and the outcome is {(2, 4), (5, 1, 3), (7), (6)}. In a group, part types are sorted in descending order of their objective function discussed in Section 9.5.6. Considering the first group (2, 4), the combined objective function $f_2 = 0.5155$ and $f_4 = 0.4715$. Hence, according to the proposed heuristic, the part type sequence is (2, 4). Similarly, the final best sequence obtained is [2, 4, 5, 1, 3, 7, 6]. Table 9.4 presents a detailed summary of the results obtained, including which part type should be accepted or rejected against the global constraints considered.

9.6.3 Results and Discussion

The efficacy of the proposed framework and its solution methodology has been tested on the example problem. The planning horizon for this case study is 8 hours (one shift). The results of experiments in the above planning horizon have been represented in Table 9.5 to show throughput, make span, and throughput after considering the Gantt chart. It can be easily elicited from Figure 9.6 and Table 9.5 that the proposed methodology can produce superior and consistent results. This can be attributed to the fact that communications among agents facilitate proper understanding of the system while bidding.

Table 9.4. Summary of part type allocations

Part type	Batch size	Machine number	Available time on machine (minute)	Processing time required on machine	Remaining time on machine	Available tool slots	Required tool slots	Remaining tool slots	Remarks
2	8	1	480		480	5		5	Part type selected for processing
		2	480		480	5		5	
		3	480	160	320	5	1	4	
		4	480		480	5		5	
4	7	1	480		480	5		5	Part type selected for processing
		2	480		480	5		5	
		3	320	168	152	4	1	3	
		4	480	133	347	5	1	4	
5	14	1	480	364	116	5		5	Part type rejected due to unavailable machining hour on machine 3
		2	480		480	5		5	
		3	152	154	–2	3		3	
		4	347		347	4		4	
1	9	1	480		480	5		5	Part type selected for processing
		2	480	198	282	5	2	3	
		3	152		152	3		3	
		4	347		347	4		4	
		1	480		480	5		5	
		2	282	225	57	3	1	2	
		3	152		152	3		3	
		4	347		347	4		4	

Table 9.4. Summary of part type allocations (continued)

Part type	Batch size	Machine number	Available time on machine (minute)	Processing time required on machine	Remaining time on machine	Available tool slots	Required tool slots	Remaining tool slots	Remarks
3	9	1	480	116	364	5		5	Part type rejected due to unavailable machining hour on machine 2
		2	57	198	–141	2		2	
		3	152		152	3		3	
		4	347		347	4		4	
7	13	1	480		480	5		5	Part type selected for processing
		2	57		57	2		2	
		3	152		152	3		3	
		4	347	160	187	4	1	3	
		1	480		480	5		5	
		2	57		57	2		2	
		3	152		152	3		3	
		4	187	70	117	3	1	2	
6	13	1	480	312	168	5		5	Part type rejected due to unavailable machining hour on machines 2 and 3
		2	57	221	–164	2		2	
		3	152	325	–173	3		3	
		4	117			2		2	

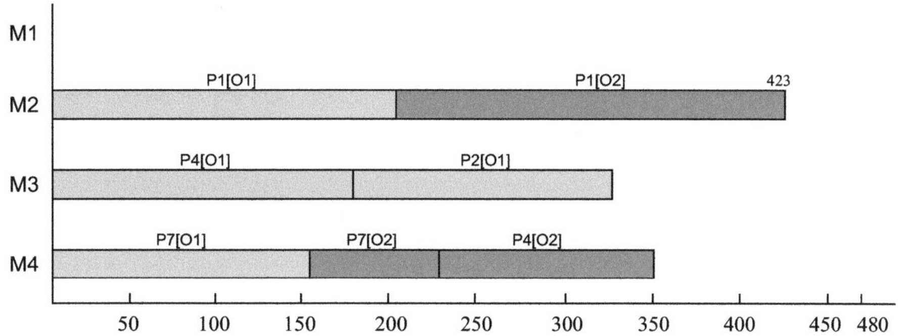

Figure 9.6. Gantt chart representing allocation of machines

Table 9.5. Results of experimentations based on heuristic

Throughput	Make span	Throughput after considering Gantt chart
34	423	34

While preparing the Gantt chart, we have considered the constraint that total processing time cannot exceed the given planning horizon of 480 minutes. If this occurs, the corresponding part type must be rejected. Figure 9.6 depicts the Gantt chart of operation allocations in a given planning horizon.

In a given example problem, if the planning horizon is exceeded due to urgency of part types or overtime given to complete part type processing on the shop floor, the throughput of the system increases. If we consider the overtime as 1 hour, then the planning horizon becomes 9 hours. The results of experiments on the new planning horizon are summarised in Table 9.6, while Figure 9.7 show the idle time of all machines in a Gantt chart. Here, a new part type 5 is selected apart from the above mentioned part types for processing on the shop floor.

We have also tested our proposed methodology on nine other example problems shown in Table 9.7. The solutions obtained are with due consideration of operation allocations and make span.

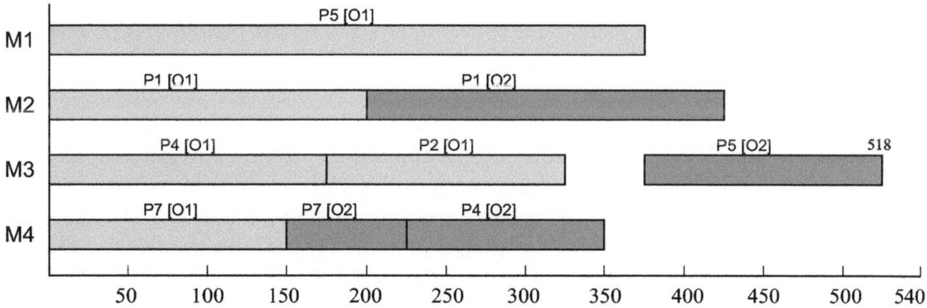

Figure 9.7. Gantt chart representing allocation of machines

Table 9.6. Results of experimentations based on heuristic

Throughput	Make span	Throughput after considering Gantt chart
48	518	48

Table 9.7. Results obtained using proposed methodology for all problems

Problem Number	Proposed auction-based heuristic		Throughput after considering Gantt chart
	Throughput	Make span	
1	39	42	42
2	51	63	63
3	63	79	69
4	51	51	51
5	62	76	61
6	51	62	63
7	54	66	48
8	79	88	88
9	44	56	55

The salient feature of the proposed methodology is to present a generic framework of real-time application in e-manufacturing environment. Application of the proposed methodology is not restricted to operation allocation and part type selection. The case study was discussed to illustrate the working of the proposed methodology. Although, some assumptions were made in order to simplify the underlying problem, the proposed methodology can be applied to similar real-time problems.

9.7 Conclusions

An agent and auction-based heuristic approach was proposed to resolve machine loading problems in an e-manufacturing environment. The objectives of the machine-loading problem are maximisation of throughput and minimisation of make span. Due to the complexity underlying the machine-loading problem, it requires a huge search space to obtain an optimal or near-optimal solution. The proposed heuristic rule is applied to achieve the objectives. The efficacy of the proposed framework and its solution methodology has been checked on different test beds and encouraging results establish that the proposed framework can cope with the complexities normally witnessed in a dynamic environment such as the one found in e-manufacturing. Real-life manufacturing operations often warrant the overtime situation where the planning horizon is exceeded due to urgency of part types or overtime given to complete part type processing on the shop floor. We have also shown the effects of exceeding the planning horizon and observed the significant increase in system throughput. The performance of the proposed auction-based heuristic depends on the bidding language and scenario. Moreover, constraints on information sharing and the conflicting nature of information pose a potential hurdle for any e-manufacturing environment. The proposed auction-based heuristic to allocate the operation on machines, supported by the agent paradigm of distributed computing, appears to be a good formalism to resolve a complex problem in e-manufacturing environment.

The proposed methodology can be extended to cover several other allocations of resources, such as pallets, fixtures, AGVs, *etc*. This work can be further extended by adding additional objective functions such as minimisation of part movement or tool changeovers, along with measures of flexibility pertaining to machines, material handling, *etc*.

Acknowledgement

This research work is part of a project 'Operation Allocation on Machines in e-Manufacturing: an Auction-based Heuristic Supported by Agent Technology' carried out by M. K. Tiwari, S. K. Jha and Raj Bardhan Anand.

References

[9.1] Koc, M., Ni, J. and Lee, J., 2002, "Introduction of e-manufacturing," In *Proceedings of International Conference on Frontiers of Design and Manufacturing.*

[9.2] Maloney, D., 2002, "E-manufacturing e'zing into factories," *Modern Materials Handling*, **57**(6), pp. 21–23.

[9.3] Koc, M., Ni, J. and Lee, J., 2003, "Introduction of e-manufacturing," *The 31st North American Manufacturing Research Conference, E-Manufacturing Panel.*

[9.4] Lee, J., 2003, "E-manufacturing – fundamental, tools, and transformation," *Robotics and Computer Integrated Manufacturing*, **19**, pp. 501–507.

[9.5] Lee, J. and Ni, J., 2002, "Infotronics agent for tether-free prognostics," In *Proceedings of the AAAI Spring Symposium on Information Refinement and Revision for Decision Making: Modelling for Diagnostics, Prognostics, and Prediction.*

[9.6] Rockwell Automation, *E-manufacturing Industry Road Map*, available on-line at http://www.rockwellautomation.com/.

[9.7] Ni, J., Lee, J. and Djurdjanovic, D., 2003, "Watchdog – information technology for proactive product maintenance and ecological product re-use," In *Proceedings on Colloquium on E-ecological Manufacturing.*

[9.8] Stecke, K.E., 1983, "Formulation and solution of non-linear integer production planning problem for flexible manufacturing system," *Management Science*, **29**, pp. 273–288.

[9.9] Buzacott, J.A. and Yao, D.D., 1980, "Flexible manufacturing system: a review of models," *Working Chapter*, No. 82-007 (Toronto: University of Toronto).

[9.10] Liang, M. and Dutta, S.P., 1992, "Combined part-selection, load-sharing and machine-loading problem in flexible manufacturing system," *International Journal of Production Research*, **30**(10), pp. 2335–2350.

[9.11] Liang, M. and Dutta, S.P, 1993, "An integrated approach to part-selection and machine-loading problem in a class of flexible manufacturing system," *European Journal of Operational Research*, **67**, pp. 387–404.

[9.12] Maturana, F., 1997, "MetaMorph: an adaptive multi-agent architecture for advanced manufacturing system," *PhD Dissertation*, University of Calgary, Canada.

[9.13] Maturana, F., Shen, W. and Norrie, D.H., 1999, "MetaMorph: an adaptive agent-based architecture for intelligent manufacturing," *International Journal of Production Research*, **37**(10), pp. 2159–2173.

[9.14] Shen, W. and Norrie, D.H., 1999, "Agent-based architecture for intelligent manufacturing: a state-of-the-art survey," *Knowledge and Information System, an International Journal*, **1**(2), pp. 129–156.

[9.15] Singhal, K., 1978, "Integrating production decisions," *International Journal of Production Research*, **16**, pp. 383–393.

[9.16] Kusiak, A., 1985, "Loading models in flexible manufacturing system," In *The Manufacturing Research and Technology* – 1, Raouf, A. and Ahmed, S.H. (eds), Elsevier, Amsterdam.

[9.17] Stecke, K.E., 1986, "A hierarchical approach to solving grouping and loading problems of flexible manufacturing system," *European Journal of Operational Research*, **24**, pp. 369–378.

[9.18] Rajagopalan, S., 1986, "Formulation and heuristic solutions for part grouping and tool loading in flexible manufacturing systems," In *Proceedings of the 2nd ORSA/TIMS Conference on FMS*, **5**, pp. 312–320.

[9.19] van Looveren, A.J., Gelders, J.L.F. and van Wassenhove, L.N., 1986, "A review of FMS planning models," In *Modelling and Design of Flexible Manufacturing Systems*, Kusiak, A. (ed.), Elsevier, Amsterdam, pp. 3–31.

[9.20] Shanker, K. and Srinivasulu, A., 1989, "Some solution methodologies for loading problems in flexible manufacturing system," *International Journal of Production Research*, **27**(6), pp. 1019–1034.

[9.21] Mukhopadhyay, S.K., Midha, S. and Krisna, V.A., 1992, "A heuristic procedure for loading problem in FMSs," *International Journal of Production Research*, **30**(9), pp. 2213–2228.

[9.22] Tiwari, M.K., Hazarika, B., Vidyarthi, N.K., Jaggi, P. and Mukhopadhyay, S.K., 1997, "A heuristic solution to machine loading problem of a FMS and its Petri net model," *International Journal of Production Research*, **35**(8), pp. 2269–2284.

[9.23] Moreno, A.A. and Ding, F.-Y., 1993, "Heuristic for the FMS loading and part type selection," *International Journal of Flexible Manufacturing System*, **5**, pp. 287–300.

[9.24] Turgay, S., 2008, "Agent based FMS control," *Robotics and Computer Integrated Manufacturing*, doi:10.1016/j.rcim.2008.02.011.

[9.25] Mes, M., Heijden, M.V.D. and Harten, A.V., 2007, "Comparison of agent-based scheduling to look-ahead heuristics for real-time transportation problems," *European Journal of Operational research*, **181**, pp. 59–75.

[9.26] Wang, S.-J., Xi, L.-F. and Zhou, B.-H., 2008, "FBS-enhanced agent based dynamic scheduling in FMS," *Engineering Applications of Artificial Intelligence*, **21**, pp. 644–657.

[9.27] Meyyappan, L., Soylemezoglu, A., Saygin, C. and Dagli, C.H., 2008, "A wasp-based control model for real-time routing of parts in a flexible manufacturing system," *International Journal of Computer Integrated Manufacturing*, **21**(3), pp. 259–268.

[9.28] Hussain, T. and Fray, G., 2006, "Auction based agent oriented process control," *Information Control Problem in Manufacturing*, pp. 425–430.

[9.29] Chaib-Draa, B. and Moulin, B., 1996, "An overview of distributed artificial intelligence," In *Foundation of Distributed Artificial Intelligence*, O' Hare, G.M.P. and Jennings, N.R. (eds), John Wiley and Sons, New York, pp. 3–55.

[9.30] Singh, S.P. and Tiwari, M.K., 2002, "Intelligent agent framework to determine the optimal conflict-free path for an automated guided vehicles system," *International Journal of Production Research*, **40**(16), pp. 4195–4223.

[9.31] Fisher, M., 1994, "Representing and executing agent based systems," In *Proceedings of the ECAI'94 Workshop on Agent Theories, Architecture and languages*, Springer, Berlin, pp. 307–323.

[9.32] Jennings, N.R. and Wooldridge, M., 1998, "Applications of intelligent agents," In *Agent Technology*, Jennings, N.R. and Wooldridge, M.J. (eds), Springer, pp. 3–28.

[9.33] Davidson, P., Astore, E. and Ekdah, L.B., 1994, "A framework for autonomous agent based on the concept of anticipatory systems," In *Proceedings of Cybernetics and Systems'94*, World Scientific, Singapore, **II**, pp. 1427–1434.

[9.34] Nwana, H.S. and Ndumu, D.T., 1997, "An introduction to agent technology," In *Software Agents and Soft Computing – towards Enhancing Machine Intelligence*, Nwana, H.S. and Azami, N. (eds), Springer-Verlag, Berlin, pp. 3-26.

[9.35] Maes, P., 1995, "Modelling adaptive autonomous agents," *Artificial Life, an Overview*, Langton, C.G. (ed.), MIT Press, Cambridge, pp. 135–162.

[9.36] Parunak, H.V.D, 1996, "Application of distributed artificial intelligence in industry," In *Foundation for Distributed Artificial Intelligence*, O'Hare, G.M.P. and Jennings, N.R. (eds), John-Wiley and Sons, New York, pp. 139–164.

[9.37] Liu, L.S. and Sycara, K.P, 1998, "Multi-agents co-ordination in tightly coupled task scheduling," In *Readings in Agents*, Huhns, M.N. and Singh, M.P. (eds), Morgan Kaufmann, San Francisco, pp. 164–171.

[9.38] Sun, J., Zhang, Y.F. and Nee, A.Y.C., 1997, "A distributed multi-agent environment for product design and manufacturing planning," *International Journal of Production Research*, **39**(4), pp. 625–645.

[9.39] Lee, J., 1997, "Overview and perspectives on Japanese manufacturing strategies and production practices in machinery industry," *International Journal of Machine Tools and Manufacture*, **37**(10), pp. 1449–1463.

[9.40] Warnecke, H.J. and Braun, J., 1999, *Vom Fraktal zum Produktionsnetzwerk – Unternehmenskooperationen Erfolgreich Gestalten*, Springer-Verlag, Berlin.

[9.41] Hatvany, J., 1985, "Intelligence and cooperation in heterarchic manufacturing systems," *Robotics and Computer Integrated Manufacturing*, **2**, pp. 101–104.

[9.42] Duffie, N.A. and Piper, R.S., 1987, "Non-hierarchical control of a flexible manufacturing cell," *Robotics and Computer Integrated Manufacturing*, **3**(2), pp. 175–179.

[9.43] Duffie, N.A. and Piper, R.S., 1986, "Nonhierarchical control of manufacturing systems," *Journal of Manufacturing Systems*, **5**(2), pp. 137–139.

[9.44] Ou-Yang, C. and Lin, J.S., 1998, "The development of a hybrid hierarchical/ heterarchical shop floor control applying bidding method in job dispatching," *Robotics and Computer Integrated Manufacturing*, **14**(3), pp. 199–217.

[9.45] Duffie, N.A. and Prabhu, V.V., 1994, "Real-time distributed scheduling of heterarchical manufacturing systems," *Journal of Manufacturing Systems*, **13**(2), pp. 94–107.

[9.46] Lin, G.Y. and Solberg, J.J., 1992, "Integrated shop-floor using autonomous agents," *IIE Transactions*, **34**(7), pp. 57–71.

[9.47] Smith, R.G., 1980, "The contract net protocol: high level communication and control in a distributed problem solver," *IEEE Transactions on Computers*, **29**(12), pp. 1104–1113.

[9.48] Smith, R.G. and Davis, R., 1981, "Frameworks for cooperation in distributed problem solving," *IEEE Transactions on Systems, Man, and Cybernetics*, **11**, pp. 61–70.

[9.49] Davis, R. and Smith, R.G., 1983, "Negotiation as a metaphor for distributed problem solving," *Artificial Intelligence*, **20**, pp. 63–109.

[9.50] Tilley, K. J., 1996, "Machining task allocation in discrete manufacturing systems," In *Market Based Control: A Paradigm for Distributed Resource Allocation*, Clearwater, S.H. (ed.), World Scientific, River Edge, NJ, pp. 225–252.

[9.51] Turban, E., 1997, "Auction and bidding on the Internet: an assessment," *International Journal of Electronic Markets*, **7**(4), pp. 7–11.

[9.52] Segev, A. and Gebauer, J., 1999, *Emerging Technologies to Support Indirect Procurement: Two Case Studies from the Petroleum Industry*, Fisher Centre for Management and Information Technology, University of California at Berkeley.

[9.53] Nisan, N., 2000, "Bidding and allocation in combinatorial auctions," In *Proceedings of the 2nd ACM Conference on Electronic Commerce*, pp. 1–12.

[9.54] Srivinas, Tiwari, M.K. and Allada, 2004, "Solving the machine loading problem in a flexible manufacturing system using a combinatorial auction-based approach," *International Journal of Production Research*, **42**(9), pp. 1879–1893.

[9.55] Kim, K., Song, J. and Wang, K., 1997, "A negotiation based scheduling for items with flexible process plans," *Computers and Industrial Engineering*, **33**, pp. 785–788.

[9.56] Stanford University, 2001, *JATLite API*, available on-line at http://java.stanford.edu/.

[9.57] DARPA, 1993, *Specification of KQML – Agent Communication Language*, available on-line at http://www.cs.umbc.edu/kqml/papers/kqmlspec.pdf.

[9.58] Bieszczad, A., Pratik, K.B., Buga, W., Malek, M. and Tan, H., 1999, "Management of heterogeneous networks with intelligent agent," *Bell Labs, Technical Journal*, pp. 109–135.

[9.59] Finin, T., Fritzon, R., McKay, D. and Mcentire, R., 1993, "KQML – a language and protocol for knowledge and information exchange," *Technical Report*, University of Maryland, Baltimore.

10

A Web-based Rapid Prototyping Manufacturing System for Rapid Product Development

Hongbo Lan

Shandong University, 73 Jingshi Road, Jinan, 250061, China
Email: hblan@sdu.edu.cn

Abstract
A web-based rapid prototyping and manufacturing (RP&M) system offers a collaborative production environment among users and RP&M providers to implement the remote service and manufacturing for rapid prototyping, to enhance the availability of RP&M facilities, and to improve the capability of rapid product development for various small and medium sized enterprises. This chapter first provides a comprehensive review of recent research on developing web-based rapid prototyping and manufacturing systems. In order to meet the increasing requirements of rapid product development, an integrated manufacturing system based on RP&M is proposed. The workflow and overall architecture of a web-based RP&M system are described in detail. Furthermore, the key technologies for developing the Web-based RP&M system, which involve deploying the running platform, determining the system model, choosing a server-side language, constructing development platform as well as designing the database and developing application, are also discussed. Finally, a case study is given to demonstrate the application of the web-based RP&M system.

10.1 Introduction

Due to the pressure of international competition and market globalisation in the 21st century, there continues to be a strong driving force in industry to compete effectively by reducing time-to-market and cost while assuring high quality products and service. Quick response to business opportunity has been considered as one of the important factors to ensure company competitiveness. The rapid prototyping and manufacturing (RP&M) technique has shown a high potential to reduce the cycle time and cost of product development, and has been considered a critical enabling tool in digital manufacturing to effectively aid rapid product development. Manufacturing industries are evolving towards digitalisation, network and globalisation. With the rapid development and applied prevalence of the Internet technologies, they have been widely employed in many developing manufacturing systems to associate with various product development activities, such as marketing, design, process planning, production, customer service, *etc.*, distributed at different locations into an integrated environment [10.1]. It has now been widely accepted that the future patterns of manufacturing organisations will be information oriented, knowledge driven and many of their daily operations will be automated around a global information network that connects everyone together [10.2]. The integration and collaboration among different partners of the product development team can largely improve the product quality, reduce the product development cost and lead-time, and extend the use capability of key equipments. Therefore, it can better satisfy the demands of different users and provide better global competitiveness of products in the marketplace [10.1, 10.3]. The Internet, together with computers and multimedia, has provided tremendous potential for remote integration and collaboration in business and manufacturing applications. RP&M techniques using the Internet can further enhance the design and manufacturing productivity, speed up the product development process, as well as share RP machines. Therefore, a web-based rapid prototyping manufacturing system provides a collaborative production platform among users and RP&M providers to implement the remote service and manufacturing for rapid prototyping, enhances the availability of RP&M facilities, and can improve the capability of rapid product development for many small and medium sized enterprises.

10.2 Web-based RP&M Systems: a Comprehensive Review

Since the mid-1990s, the research and development of web-based RP&M systems have received much attention [10.4]. Substantial investments have been made to support the research and practice of web-based RP&M systems from both the academic community and industrial bodies all over the world. A number of studies have been performed to explore the architecture, key issues and enabling tools for developing web-based RP&M systems.

10.2.1 Various Architectures for Web-based RP&M Systems

A variety of frameworks for developing web-based RP&M systems have been proposed. The Tele-Manufacturing Facility (TMF) is probably the first system that provides users with direct access to a rapid prototyping facility over the Internet. TMF allows users to easily submit jobs and have the system automatically maintain a queue. It can also automatically check many flaws in .STL (StereoLithography) files, and in many cases, fix them. A laminated object manufacturing (LOM) machine was first connected with network, and then the .STL file of a part to be built could be submitted to this machine via a command-line [10.4–10.6].

Luo and Tzou *et al.* [10.3, 10.7–10.11] presented an e-manufacturing application framework of a web-based RP system that mainly includes five parts for: (1) opening .STL file and displaying it using Open GL technology, (2) product quotation, (3) selecting a suitable RP system, (4) joint alliance, and (5) order scheduling.

Jiang and Fukuda [10.12, 10.13] described a methodology to create an Internet-based infrastructure to service and maintain RP-oriented tele-manufacturing. One of the most important applications of such an infrastructure is to support the closed-loop product development practice. The Java-enabled solution based on the Web/ Internet computing model is used to implement the infrastructure. The main functions include the remote part submission, queuing and monitoring. Under the control of different access competences, manufacturing sites and queues can be maintained, respectively, in distributed locations. A software test platform has been developed in Java to verify the methodology.

Liu *et al.* [10.14] addressed the development of a web-based tele-manufacturing service system for rapid prototyping. The system provides geographically dispersed enterprises with a platform that permits them to share RP machines. In contrast with other similar systems, the proposed system is comprised of three components: online commerce, online manufacturing services, and online data management. Three supporting software packages are also provided: for self-check and self-repair of .STL files, for real-time collaboration, and for remote monitoring and control.

Lan *et al.* [10.15, 10.16] developed a tele-service system for RP service providers to support the implementation of web-based RP manufacturing. The tele-service system consists of two components: software sub-system and hardware sub-system. The hardware involves not only the RP&M facilities of service provider itself but also the RP&M equipments from the other service providers. The software module includes eight functional components: information centre, ASP (application service provider) tool set, client management, e-commerce, manufacturing service, system navigation, and collaborative tools. The crucial issues for developing the system, which involve deploying the running platform, determining the system model, choosing a server-side language, constructing development platform as well as designing the database and application, were also discussed in detail. Finally, a case study was given to demonstrate the use of the tele-service system. This system has been developing and employing in the Northwest Productivity Promotion Centre in China.

Tay *et al.* [10.17] introduced the development of a distributed rapid prototyping system via the Internet to form a framework of Internet prototype and manufacturing for the support of effective product development.

Xu *et al.* [10.18] presented an Internet-based virtual rapid prototyping system, named VRPS-I, which was implemented by using Java and VRML (Virtual Reality Modelling Language). With the aid of this system, not only can the visual rapid prototyping process be dynamically previewed, but the forming process and some part-quality-related parameters can also be predicted and evaluated.

An RP-related tele-manufacturing investigation was also conducted at Stanford University, where both RP hardware [10.19] and software [10.20] have been developed. The hardware deals with an integrated mould SDM (mould shape deposition manufacturing) machine and the software includes an agent-based infrastructure to implement the Internet-based RP manufacturing service. It has been emphasised to use the concepts of plug-in, broker, local tool integration, *etc.*, under the Java-based agent environment named JAT (Java Application Template) [10.12].

Huang *et al.* [10.21] developed a rapid prototyping-oriented tele-service system based on the Internet and Intranet. The functional units of SL (stereolithography) oriented online pricing, online bargaining and clients' information management were implemented. Huang *et al.* [10.22] introduced an Internet/web-based rapid prototyping oriented tele-manufacturing service centre to share an RP manufacturing service over the Internet. Under the support of this virtual service centre, RP manufacturers can provide and publish their manufacturing service by configuring their RP sites, and users can submit the STL files of parts to their desired sites for manufacturing by comparing the requirements of their prototype models to the capabilities of RP sites. Fidan and Ghani [10.23] recently developed a remotely accessible laboratory for rapid prototyping.

10.2.2 Key Issues in Developing Web-based RP&M Systems

In the development of various types of web-based RP&M systems, a number of key issues were investigated. We classify these key issues into the following seven categories: (1) RP&M process selection, (2) RP price quotation, (3) .STL viewer, (4) RP data pre-processing, (5) job planning and scheduling, (6) remote control and monitoring for RP machines, and (7) security management. Details of these key issues are discussed in the following subsections.

RP&M Process Selection

There are a variety of different processes for RP and RT (rapid tooling); each of them has its characteristics and scope of application. It is especially difficult for many beginners to select the most suitable process according to the individual task requirements and actual conditions. A number of studies have been carried out into the development of methodologies, decision support techniques and software tools for assisting RP users in selecting the most suitable RP process. A web-based RP system selector has been developed by the Helsinki University of Technology [10.24]. This program is not perfect, not even close, but a friendly pointer to the right direction [10.25]. Lan *et al.* [10.26] proposed a method integrating the expert system and fuzzy synthetic evaluation to select the most appropriate RP process according to user's specific requirements. Based on the proposed approach, a web-based decision support system for RP process selection was developed. Chung *et al.* [10.27] discussed a methodology for selection of RT processes based on a number of user-specified attributes and relative cost and lead-time comparisons across a wide spectrum of available RT processes. The method has been put on the Internet for easy access. It is currently limited to only several of the most common RT processes and materials. However, the database should be further expanded to include a majority if not all of the metal casting processes.

RP Price Quotation

The current major practice in the RP industry for quoting prices for fabricating an RP part is either by experience or by comparison with a similar product. A web-based price quotation approach can provide an instant price quotation for remote customers, quickly obtain feedback from users, and show a number of advantages: easy to use and update, simple operation, high interactivity with customers, and user-friendly interface. In order to satisfy the current quotation needs of RP service providers and users, several web-based automated quotation systems have been developed. Quickparts.com (and its software QuickQuote) [10.28] and 3T RPD [10.29] are now the main commercial online price quotation systems for RP parts. Unfortunately, maybe for reasons of commercial know-how (or secret), both of them do not describe and report the implementation mechanism of quotation, application effect and development details. Two quoting approaches that include the rough quotation based on weight and the precise quotation based on build-time were proposed to determine the price quotation for SL parts by using STL models of parts to be built. Based on the proposed methods, Lan *et al.* [10.30, 10.31] developed a web-based automated quotation system that can provide instant price quotations for SL parts to support effectively a web-based RP&M system. The web-based automated quotation system can satisfy the engineering requirements of price quotation at the early stage of product development, and offers a new option for RP users to allow them to instantly quote and compare multiple rapid prototyping processes. Luo *et al.* [10.8] proposed an equation to estimate RP product cost, and developed an RP product quotation system to evaluate the cost of 3D CAD models based on the proposed equation. The quotation system was implemented by using Visual Basic 6.0 with Components Object Model (COM) object to code the interface engine. Users can open an .STL file from the client-side homepage program, and carry out the action of the product quotation.

STL Viewer

Rosen [10.32] developed a collaborative design tool, called STL Viewer, which allows remote users to view .STL models over the Internet. This software, operated through a Java Applet in the user's web browser, reads the STL file format (the *de facto* file format for rapid prototyping) and displays the model for the remote user to inspect. The user is able to rotate, pan, and zoom the model. Utilities are also provided for verifying the validity of the model. STL Viewer has a wide variety of potential applications. For instance, online 3D catalogues, infusion into an online ordering process for an SL laboratory, and sharing design concepts over the Internet, are typical potential uses of this software.

RP Data Pre-processing

Lan *et al.* [10.33] described an ASP (application service provider) tool set for RP data pre-processing based on the .STL model to aid effectively the networked service of RP&M. The architecture and functional model of the ASP tool set were established. The checking and fixing system for STL models, as a typical example, was systematically studied and developed, and was used to demonstrate the detailed development process of the ASP tool set. Liu *et al.* [10.14] developed a software package for self-check and self-repair of .STL files. Roy and Cargian [10.34] described the implementation of an object slicing algorithm that runs through the World Wide Web. In the intelligent web-based RP&M system [10.10], Luo *et al.* presented a new adaptive slicing algorithm for RP systems. According to the algorithm, the 3D CAD model can be sliced with different thicknesses automatically by comparing contour circumference or the centre of gravity of the contour with those of the adjacent layer.

Job Planning and Scheduling

Today's commercial RP systems rely on human interventions to load and unload build jobs. Hence, jobs are processed subject to both the machine's and the operator's schedules. In particular, first-in-first-out (FIFO) queuing of such systems will result in machine idle time whenever a build job has been completed and an operator is not available to unload the build job and start the next one. These machine idle times can significantly affect the system throughput and the cost-effective rate. Wu [10.35] addressed this issue by rearranging the job queue to minimise the machine idle time, subject to the schedules of the machine and the operator. Lin *et al.* [10.36] proposed a real-time scheduling approach to maximise system utilisation and minimise the average response time for scheduling non-pre-emptive aperiodic tasks so that it is suitable for the distributed web-based RP systems. The idea is to incorporate shortest-job-first (SJF) into earliest-deadline-first (EDF) scheduling algorithm. Jiang and Fukuda [10.12, 10.13] addressed the selection of a feasible RP manufacturing site and submitting and queuing a part for remote RP manufacturing. Tzou [10.7] presented a framework of an order scheduling system. In TMF [10.4], queuing a part depended only on the time when the part arrived in the queue.

Remote Control and Monitoring for RP Machines

Luo *et al.* [10.3, 10.11] profoundly discussed a tele-control and remote monitoring system for RP machines. It allows a user to directly control an RP machine via the Internet. Also the remote user can receive real-time images, *e.g.* the completed RP model parts captured by a CCD camera that is mounted on an RP machine. The remote control system has proven itself as a feasible and useful tool to share RP machines with others via the Internet. Different RP systems may not easily communicate with each other because they are built on different operation systems, using different communication protocols. CORBA (Common Object Request Broker Architecture) has been proposed to integrated heterogeneous environments so that these different RP systems can be easily integrated regardless of what language they are written in or where these applications reside with any commercial CORBA solutions [10.7, 10.9]. Instead of using a CCD camera, Wang *et al.* [10.37] used a Java 3D-based remote monitoring system for a fused deposition modelling (FDM) machine. Gao *et al.* [10.38] described the development of a tele-control system for rapid prototyping machines based on the Internet. Among different remote control methods, Winsock was adopted to implement tele-controlling of an RP machine. Kang *et al.* [10.39] investigated a remote monitoring and diagnosis of breaking-down fault for the FDM machine.

Security Management

Web-based manufacturing systems allow clients to download client application programs from a server to access the client programs using local web browsers and to execute the programs remotely at the client-side. Therefore, security management mechanisms are required to specify different levels of accessibility permissions for different users to prevent the local machines from being damaged by poor programs and viruses, and to prevent the server machines from being visited by unauthorised clients [10.1]. In the tele-RP system developed by Jiang and Fukuda [10.12, 10.13], the general security strategy is mainly concerned with (1) collection of information from clients and confirmation of the clients' certificates at different levels, (2) clients' access security considerations to web servers, directories, and files inside the servers, including individual web pages, and (3) secure encrypted data transactions over the Internet. In their research, the currently available data encryption transaction methodologies, including RSA (Rivest, Shamir, Adlernan), digital signatures of users for payment and reliability, Security Socket Layer (SSL), and Secure HTTP (SHTTP), were investigated [10.1]. In the web-based RP system presented by Lan *et al.* [10.15, 10.16], in order to prevent system hacking, two firewalls based on package filtration and proxy server, namely, Cisico2511 router and Proxy Server 2.0, were utilised.

From the literature, it is clear that most studies focused on individual functional modules and strategies as well as overall system architectures. There is still no comprehensive system applied to support the full implementation of web-based rapid prototyping manufacturing. Therefore, a practical web-based RP&M system urgently needs to be investigated and developed.

10.3 An Integrated Manufacturing System for Rapid Product Development Based on RP&M

In order to meet the increasing requirements of rapid product development, an integrated manufacturing system based on RP&M is proposed. The system is composed of four building blocks: *Digital Prototype*, *Virtual Prototype*, *Physical Prototype* and *Rapid Tooling* manufacturing system, as shown in Figure 10.1. It can aid effectively in product design, analysis, prototype, mould, and manufacturing process development by integrating closely the various advanced manufacturing technologies that involve CAD, CAE, reverse engineering (RE), rapid prototyping and rapid tooling. The main function of the Digital Prototype is to perform 3D CAD modelling. The CAD model is regarded as a central component of the entire system or project information base, meaning that the same data is utilised in all design, analysis and manufacturing activities. The product and its components are directly designed using a 3D CAD system (*e.g.* Pro/Engineer, Unigraphics, CATIA, IDEAS, *etc.*) during the creative design process. If a physical part is ready, the model can be reconstructed by the RE technique. In order to reduce the iterations of design-prototype-test cycles and increase the product process and manufacturing reliability, it is necessary to guide the optimisation of product design and manufacturing process through the Virtual Prototype (VP). VP is a process using the 3D CAD model, in lieu of a physical prototype, for testing and evaluation of specific characteristics of a product or a manufacturing process. It is often carried out by CAE and virtual manufacturing systems. Although the VP is intended to ensure that unsuitable designs are rejected or modified, in many cases, a visual and physical evaluation of the real component is needed. This often requires physical prototypes to be produced. Hence, once the VP is finished, the model may often be sent directly to physical fabrication. The CAD model can be directly converted to a physical prototype using an RP technique or high-speed machining (HSM) process. Once the design has been accepted, the realisation of its production line represents a major task with a long lead time before any product can be put to the market. In particular, the preparation of complex tooling is usually in the critical path of a project and has

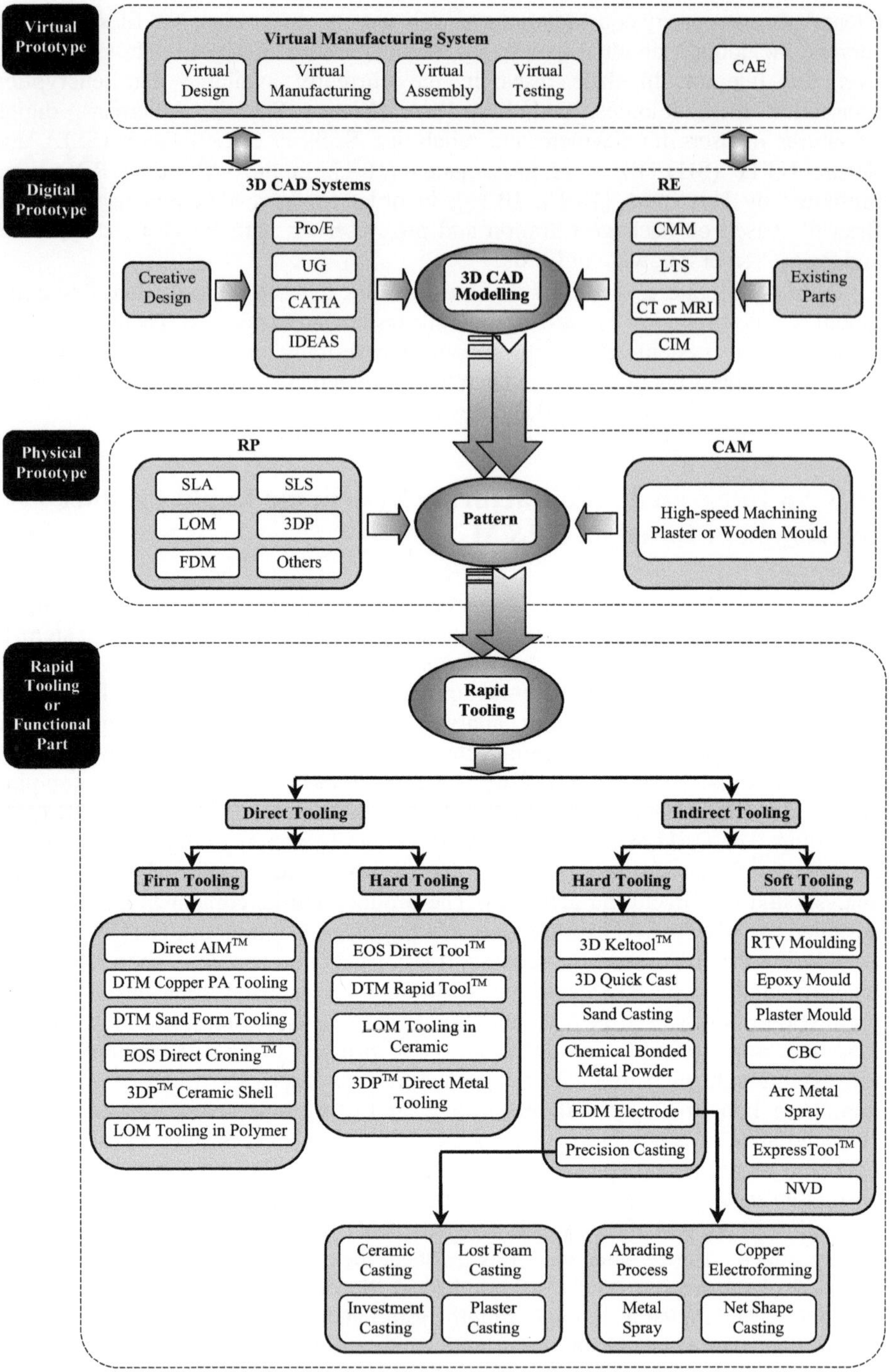

Figure 10.1. Architecture of an integrated system for rapid product development

therefore a direct and strong influence on time-to-market. In order to reduce the product development time and cost, the new technique of RT has been developed. RT is a technique that transforms the RP patterns into functional parts, especially metal parts. It offers a fast and low-cost method to produce moulds and functional parts. Furthermore, the integration of both RP and RT into a development strategy promotes the implementation of concurrent engineering in companies.

10.4 Workflow of a Web-based RP&M System

The workflow of the web-based RP&M system is shown in Figure 10.2 and described as follows. The first step is to login to a website of an RP&M service provider (SP). Users have to key in their names and passwords. Those without registration or authorisation can also login to the system, but they are limited to view information that is open to the public such as "typical cases" in this system. The password entered by the user is verified by the system. After logging in to the SP website successfully, the system will check the security level of users, and determine which modules they can access or employ. According to authentication for the system, all users are divided into four categories: general users (without registration), potential clients, real clients, and system administrator.

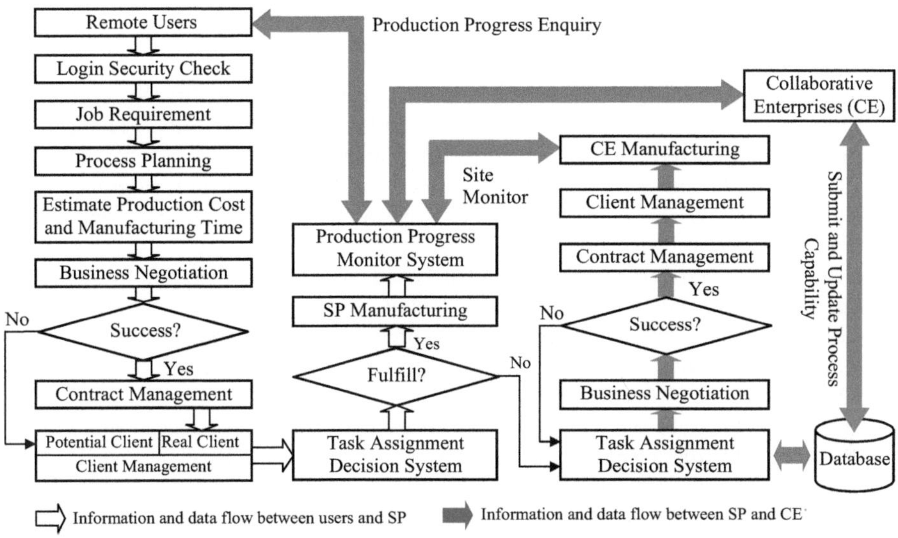

Figure 10.2. The workflow of the Web-based RP&M system

Having received job requirements from clients, the system will first perform process planning that completes the task decomposition and selects the most suitable process method. It is necessary for users to get the preliminary product quote and production cycle time from the SP before the subsequent process continues. If the results may be accepted initially, the SP may further negotiate with the user via a videoconferencing system. Once having come to terms with each other, a contract is to be confirmed, and the user becomes a formal client. The jobs submitted by formal clients are better performed by the service provider itself. However, if the SP has no such process capabilities or cannot accomplish the jobs in time, an effective solution is for the service provider to take full advantage of external sources from other service providers to carry out the remaining jobs. The following step is to determine the appropriate collaborative service providers to form a virtual enterprise to complete the rest of the tasks using the job assignment decision system. In addition, in order to monitor the schedule to ensure smooth production, both collaborative service providers and this service provider itself must provide as quickly as possible the essential information related to the production progress and schedule for the production monitor module. Any firms falling behind schedule or failing to meet quality standards will be closely examined by the SP and the user to ensure that precautionary or remedial measures are taken ahead of time or any damaging effects are predicted.

10.5 Architecture of a Web-based RP&M System

The framework of a Web-based RP&M system is proposed as shown in Figure 10.3. The tele-service system consists of two parts: software module and hardware module. The hardware section involves not only the RP&M facilities of the SP itself but also the RP&M equipments from the other SPs. Referring to the aforementioned workflow of the Web-based RP&M system and the functional requirements of the digital tele-service system, this system includes eight functional components: information centre, ASP tool set, client management, electronic commerce, manufacturing service, system navigation, and collaborative tools. The specific components for each module are illustrated in Figure 10.4.

These eight components work together seamlessly to aid effectively the implementation of the web-based RP&M system. One of the purposes of the information centre is to enable users to increase the awareness of the relevant knowledge of rapid product development based on RP&M. In order to help users better understand and apply these new techniques, the system illustrates a large number of real-life cases. Depending on the predominance of specialty and expertise in RP&M, the system can reply to the enquiries of clients and communicate with users to solve their problems.

The ASP tool set provides five useful components: the process planning of RE/RP/RT, STL model checking and fixing, the optimisation of built orientation, support structure generation, and the optimisation of part slicing. There are various processes for RE, RP and RT; each of them has its characteristics and scope of application. It is especially difficult for many users to select the most suitable process according to the individual task requirements and actual conditions. Three selectors based on ASP mode, namely, "RE selector", "RP selector", and "RT selector" have been developed to perform process planning automatically in the web server side. Development of a decision support system for RP process selection has been discussed in detail in [10.21], which states that .STL files created from solid models have anomalies about 10% of the time and those created from surface models have problems about 90% of the time. Error rates in this range make it clear that automated error checking is important for all RP operations. Based on

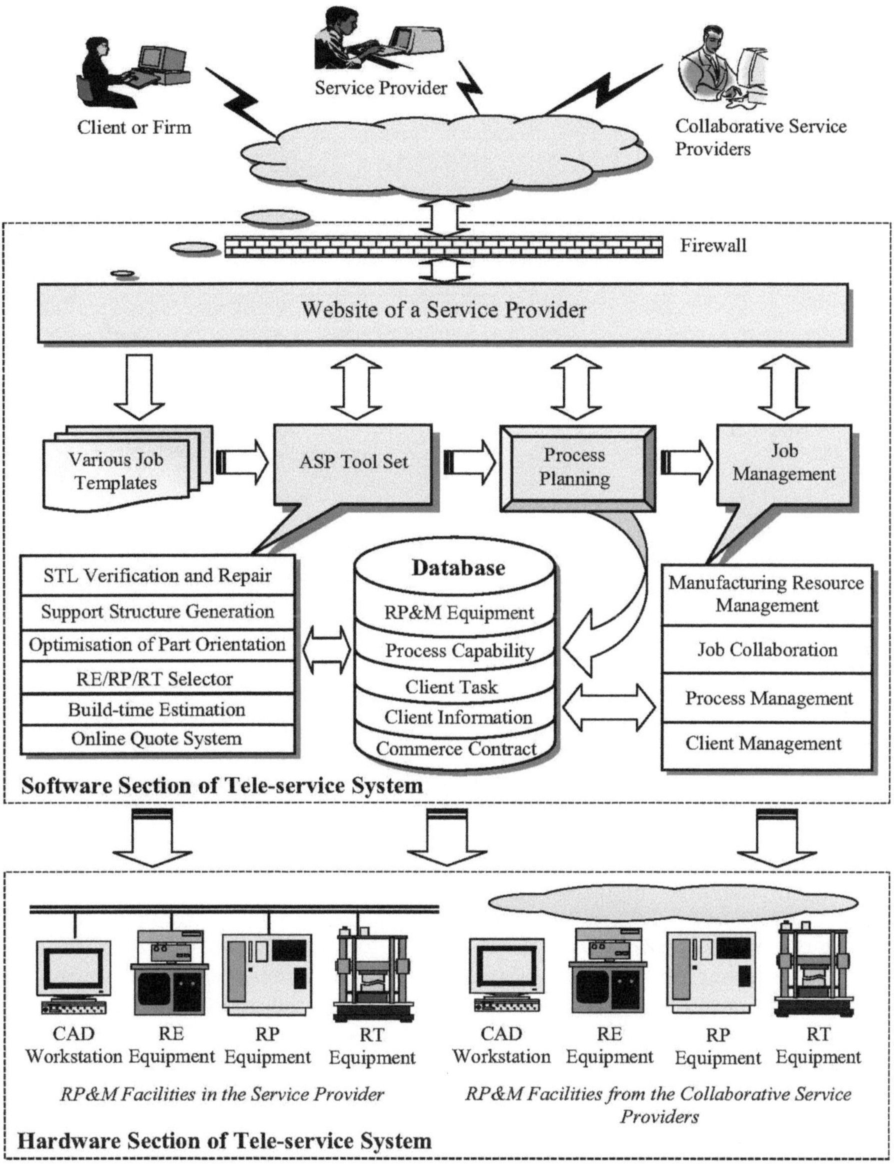

Figure 10.3. Architecture of Web-based RP&M system

our experiences supplying web-based resources, we know that it is especially crucial to conduct automatic error checking for .STL models when the RP operation is not at the designer's site. We have developed various algorithms to detect, and in some cases, automatically fix geometric and topological flaws. There are two "firewalls" to detect those flaws: one is integrated with the online pricing engine that will be operated by the user before the .STL file is submitted to the system, while

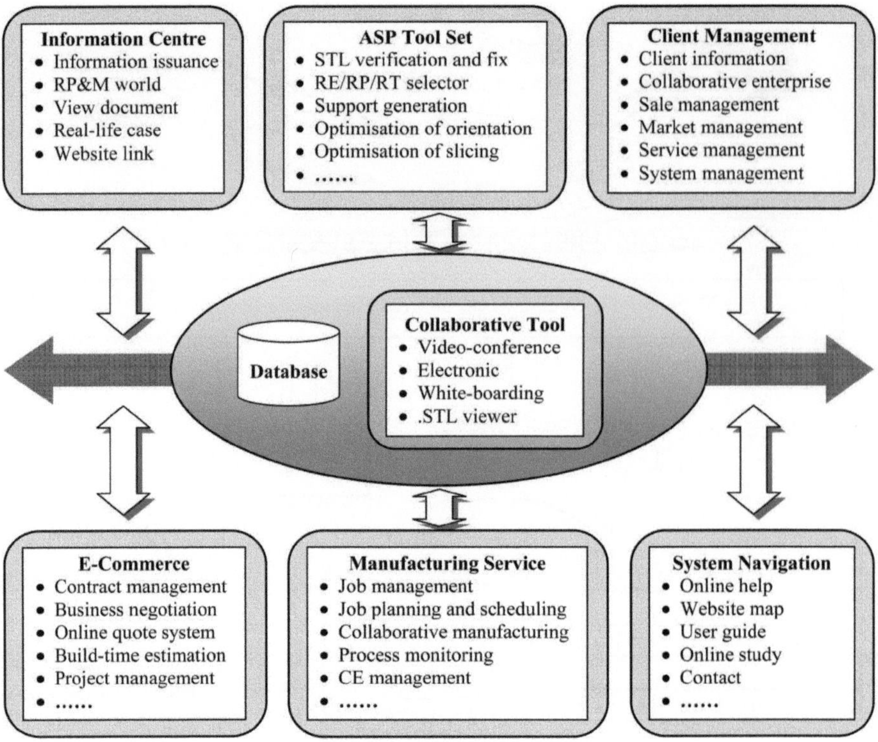

Figure 10.4. Functional modules of the tele-service system

the other is to run on the system's server-side after the .STL file is submitted. Because the fixing function for .STL model is quite limited, if an .STL file has fatal flaws or loses some data during transferring, it would have to be uploaded again from the client-side. Parts built using RP&M technique can vary significantly in quality depending on the capabilities of manufacturing process planning. The process planning of RP is conducted to generate the tool paths and process parameters for a part that is to be built by a specific RP machine. The required steps are determining the built orientation, support structure generation, slicing, path planning, and process parameter selection. Therefore, it is also important for remote users that the networked system can provide these capabilities of process planning. Three sub-modules, including the optimisation of part built orientation, support structure generation, and adaptive slicing, have been developed to aid users in setting RP process variables in order to achieve specific build goals and desirable part characteristic. Both potential clients and real users can employ freely the ASP tool set.

Electronic commerce module is composed of four sections: the online quote, the build-time estimation, the online business negotiation, and the electronic contract management. Conventionally, the RP&M providers may quote according to the client's offerings (*e.g.* CAD models, 2D drawings or physical prototypes) utilising their experiences or just obtain payment after the RP model has been built. But for the tele-service system, it is necessary for the remote users to inquire about the service expense of making RP before the follow-up process continues. Hence, an online pricing engine (OPE) has been developed. The details of the OPE have been discussed in [10.30]. Accurate prediction of the build-time required is also critical for various activities such as job quoting, job scheduling, selection of build parameters (*e.g.* layer thickness and orientation), benchmarking, *etc.* Two build-time estimators based on the sliced process and the STL model respectively have been exploited. The paper [10.30] presented the principle of a build-time estimation algorithm for stereolithography based on model geometrical features. After clients accept the quote, they may negotiate with RP&M providers on the business and technological details. The Microsoft NetMeeting, together with .STL Viewer, which can set up a collaborative environment to implement information sharing, file transferring, video and audio communication, *etc.*, provides an ideal tool for the online negotiation. As a result of negotiation, an electronic contract is signed. To manage and operate these electronic contracts, the system also provides a contract management component. It is especially convenient and prompt for clients to submit, inquire, and search contracts through this module.

The manufacturing service module that covers job management, job planning and scheduling, collaborative manufacturing, process monitoring, and collaborative enterprises management, *etc.*, is regarded as one of the most important functional modules in the web-based RP&M system. When a contract is confirmed, clients will formally submit their job requirements (*e.g.* RE, 3D CAD modelling, CAE, RP prototype, or rapid tooling) and initial source materials (*e.g.* object parts, digitised data cloud, 2D models, 3D models, or .STL files).

In order to help many end users submit quickly and easily the manufacturing tasks, various job and source templates have been established, while the client can search, modify, and even delete the manufacturing tasks itself if the occasion arises. The job planning and scheduling optimisation plays a particularly important role for the web-based manufacturing systems. Lin *et al.* [10.36] and Wu [10.35] have performed investigations on this issue. The system utilised a real-time scheduling approach proposed by Lin *et al.*, which can maximise the system utilisation and minimise the average response time for scheduling non-pre-emptive aperiodic tasks so that it is suitable for the distributed web-based RP&M systems.

The collaborative manufacturing system (CMS) is responsible for the selection of collaborative enterprises (CE) to form a virtual alliance. Many research results related to partner selection have been reported. In addition, it is important and necessary to monitor the manufacturing schedule and control product quality to ensure smooth production. In the past, an RP&M provider had to spend much time dealing with enquires from the clients via phone calls or faxes. Now, the process monitor system (PMS) provides various facilities that can guarantee the tasks to be completed timely. For example, during and after the building process, users receive live images, via the Internet, of the physical model in the RP machine from a CCD camera located in the RP machine. With the PMS and real-time job scheduling, any partners falling behind schedule or failing to meet quality standards will be closely examined by RP&M providers and users to ensure that precautionary measures are made ahead of time. Therefore, new requests can be accepted dynamically and all accepted requests can be finished before deadlines.

All information involved in the service process are managed and maintained by a special database. These data provide strong support for both online business and manufacturing service. To create a collaborative environment among RP&M providers, users and collaborative enterprises, the system relies on multi-media, collaborative tools (*e.g.* .STL Viewer) and the Internet. Therefore, the service system offers three enabling tools: videoconferencing system, collaborative tools, and FTP. In order to make use of the system as quickly as possible, users can get help from the system navigation module. The eight components form a fully-integrated system that is able to carry out tasks in an efficient and effective way. Figure 10.5 illustrates the network structure of the entire system.

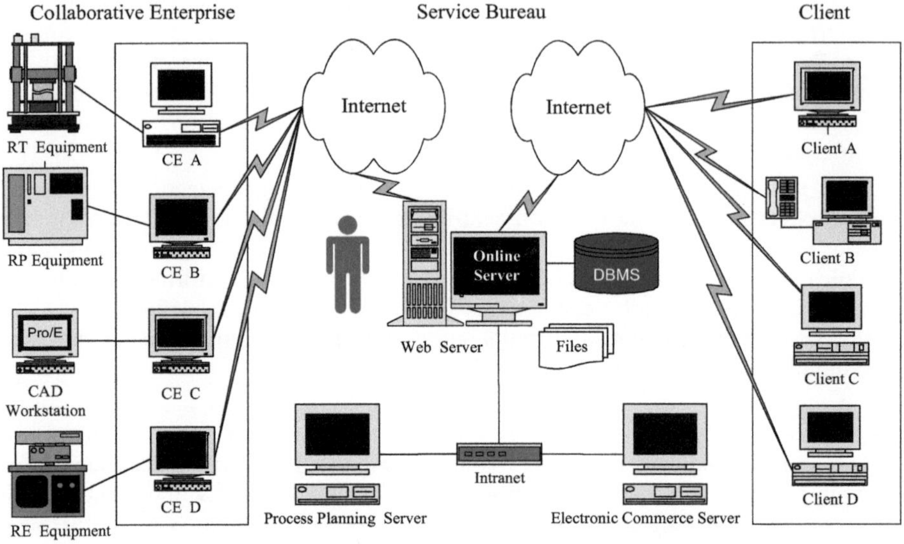

Figure 10.5. Network structure of the entire system

10.6 Development of a Web-based RP&M System

In order to run this web-based system effectively, constructing a suitable running platform is necessary and crucial. An operating system is a basic software for running this system. The operating systems commonly found on web servers are UNIX, OS, Linux, and Windows, *etc*. Windows 2000 Advanced Server is a better platform for running business applications. It is, therefore, selected as the operating system, whereas Internet Information Services (IIS) 5.0 is chosen as the web server of the running platform. Normally IIS cannot execute Servlet and Java Server Pages (JSP), configuring IIS to use Tomcat redirector plug-in lets IIS send Servlet, and JSP requests to Tomcat (and this way, serve them to clients). Tomcat 3.2 of Apache Software Foundation is selected to be the engine for JSP and Servlets. As for a database system, SQL Server 2000 Relational Database is chosen instead of other databases due to its seamless integration with Windows 2000 Advanced Server and the ease of use. Exchange Server 5.5 is used for mail service function. In order to prevent system hacking, two firewalls, Cisico2511 router and Proxy Server 2.0, have been established. The overall configuration of the running platform for the tele-service system is shown in Table 10.1.

Table 10.1. Configuration of running platform

Software component	Software product
Operating system	Windows 2000 Advanced Server
Web server	IIS 5.0 + Tomcat 3.2
Database server	SQL Server 2000
Mail server	Exchange Server 5.5
Proxy server	Proxy Server 2.0

Due to the distributed and heterogeneous trait of users and collaborative service providers, this service system adopts the popular B/S (browser/server) structure that satisfies the requirements of distributed and heterogeneous networked environment. Creating dynamic web pages and showing customised information is the pith of web applications for B/S model. Four server-side scripting languages, including Common Gateway Interface, Active Server Pages, Person Home Pages, and JavaServer Pages (JSP), are frequently used now. JSP is a better solution for generating dynamic web pages in contrast to the others. Together, JSP/Servlets provide an attractive alternative to the other types of dynamic web pages scripting/ programming that offers platform independence, enhanced performance, separation of logic from display, ease of administration, extensibility into the enterprise, and most importantly, ease of use. Hence, it is a better decision to develop dynamic web pages using JSP and Servlets.

Currently, the development platform of distributed applications includes Microsoft .NET and J2EE of Sun Microsystems. According to the requirements of the web-based manufacturing system and server-side scripting language, J2EE is selected to be the development platform of distributed applications for the web-based system. The J2EE-based architecture of the development platform is shown in Figure 10.6. The development environment and development kits of this application are presented in Tables 10.2 and 10.3, respectively. In order to ensure that the web-based RP&M system runs well and effectively, it is crucial to establish a good database system to organise and manage the vast amount of data in this system. The construction of tables in the database is important, as it affects future amendments to the database. A total of 52 tables have been constructed.

10.7 Case Study

A typical example from the prototype development of a TV frame is used to demonstrate the application of the web-based RP&M system. Let us assume that a company was developing a new TV frame for which a physical model would be required. The mock-up was purely for the purpose of design visualisation and would be used as a means of communication with other functional departments in the firm. The job requirements and source materials were submitted to an RP&M service

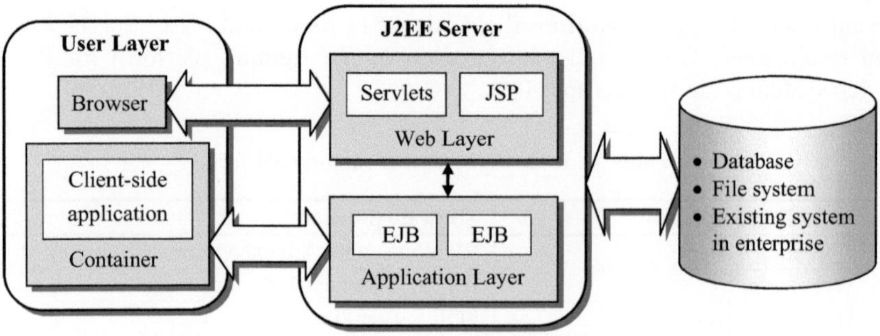

Figure 10.6. Architecture of J2EE-based development platform

Table 10.2. Development environment of applications

Software component	Software product
Operating system	Windows 2000 Advanced Server
Web server	Internet Information Server 5.0
Database server	SQL Server 2000
Web browser	Internet Explorer 5.5
Java 2 development kit	J2EE SDK 1.4.0
JSP/Servlets engine	Tomcat 3.2

Table 10.3. Application development kits

Software component	Software product
JSP	JRun Studio 3.0
JavaBeans	JBuilder 6.0
Web pages generation	FrontPage 2002
Image processing	Adobe Photoshop 5.0 CS
Animation	Flash 5.0

provider using the job submitting module in the remote tele-service system. After receiving the job requirements, the system first performs the process planning by which the job is decomposed into 3D CAD modelling and prototype making, and SL was selected as the most suitable process for building the mock-up through the RP Selector, while the system offered the preliminary product quote and production cycle time to the user. After accepting the initial results, the user continuously negotiates with the RP&M service provider via a videoconferencing system. Once come to terms with each other, a contract is confirmed. Subsequently, the 3D CAD modelling and prototype making are assigned to the collaborative enterprise A (an RP&M service provider) and collaborative enterprise B, respectively, by the job planning and scheduling component as well as the CMS module. During the building process, the remote user can receive real-time images, *e.g.* the completed RP model parts, captured by a CCD camera that is mounted on a RP machine. Finally, the green part is inspected online and delivered to the end user by DHL or

EMS. The detailed process is illustrated in Figure 10.7. The results of the process planning for the case study are reported in Table 10.4. The 3D CAD model submitted by the collaborative enterprise A is illustrated in Figure 10.8. The mock-up fabricated by the collaborative enterprise B is shown in Figure 10.9.

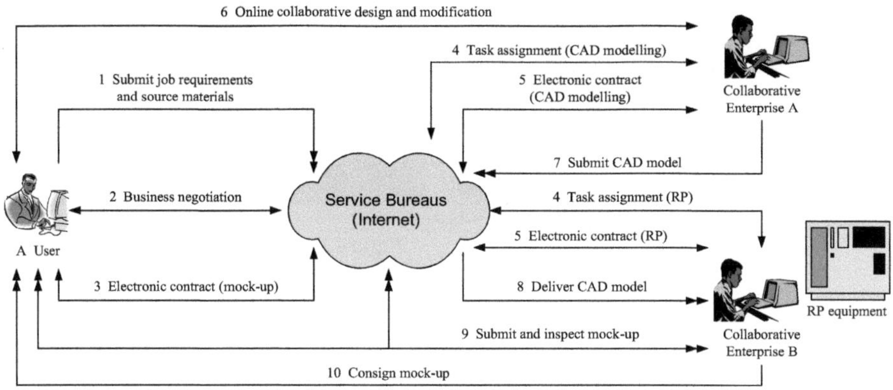

Figure 10.7. Detailed processes of the case study

Table 10.4. Results of the process planning

Process stage	Collaborative enterprise	Finish time (day)	Price (RMB)	Remark (extra day)
CAD modelling	A	4	2,200	0.5 (Deliver CAD model)
Mock-up build	B	5	7,600	2 (Consign mock-up)

10.8 Conclusions

Collaborative digital manufacturing is a new manufacturing paradigm in the 21st century. The Internet, together with computers and multi-media, has provided tremendous potentials for the remote integration and collaboration in business and manufacturing applications. In order to meet the increasing requirements of rapid product development, this chapter presents a web-based RP&M system that offers a collaborative production environment among users and RP&M providers to implement the remote services and manufacturing for rapid prototyping. Such a collaborative product development platform enables remote RP manufacturing, enhances the availability of RP facilities, and improves the capability of rapid product development for various small and medium sized enterprises. The implementation of such a system represents a fundamental shift of enterprise strategy and manufacturing paradigms in organisations. The web-based RP&M system is a new manufacturing mode in terms of mission, structure, infrastructure, capabilities, collaboration, and design process, which need more in-depth research. Further research will focus on collaborative product commerce (CPC), collaborative service support, and the detailed structure and formulation of the central-monitoring mechanism of such a partnership system.

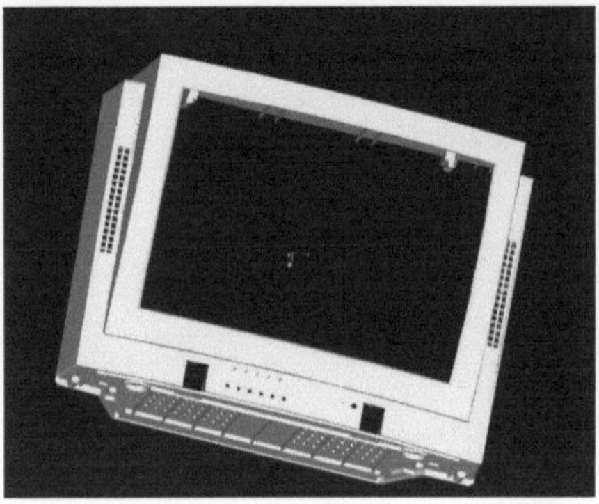

Figure 10.8. 3D CAD model

Figure 10.9. Mock-up of the TV frame

Acknowledgement

This study was partially supported by The National High Technology Research and Development Program (863 Program) under the title "RP&M Networked Service System" (No. 2002AA414110).

References

[10.1] Yang, H. and Xue, D., 2003, "Recent research on developing Web-based manufacturing systems: a review," *International Journal of Production Research*, **41**(15), pp. 3601–3629.

[10.2] Rahman, S.M., Sarker, R. and Bignall, B., 1999, "Application of multimedia technology in manufacturing: a review," *Computers in Industry*, **38**(1), pp. 43–52.

[10.3] Luo, C.R., Tzou, J.H. and Chang, Y.C., 2001, "Desktop rapid prototyping system with supervisory control and monitoring through Internet," *IEEEE/ASME Transactions on Mechatronics*, **6**(4), pp. 399–409.

[10.4] Bailey, M.L., 1995, "Tele-manufacturing: rapid prototyping on the Internet," *IEEE Computer Graphic and Applications*, **15**(6), pp. 20–26.

[10.5] San Diego Supercomputer Centre, 2007, http://www.sdsc.edu/tmf.

[10.6] Bailey, M.L., 1995, *Tele-manufacturing: Rapid Prototyping on the Internet with Automatic Consistency-checking*, White Paper, University of California at San Diego.

[10.7] Tzou, J.H., 2004, "Distributed Web-based desktop e-manufacturing system," In *The 2nd International Conference on Autonomous Robots and Agents*, pp. 356–361.

[10.8] Luo, C.R., Lan, C.C., Tzou, J.H. and Chen, C.C., 2004, "The development of Web based e-commerce platform for rapid prototyping system," In *Proceedings of the 2004 IEEE International Conference on Networking, Sensing & Control*, pp. 122–127.

[10.9] Luo, C.R. and Tzou, J.H., 2003, "The development of distributed Web-based rapid prototyping manufacturing system," In *Proceedings of the 2003 IEEE International Conference on Robotics & Automation*, pp. 1717–1722.

[10.10] Luo, C.R. and Tzou, J.H., 2004, "The development of an intelligent web-based rapid prototyping manufacturing system," *IEEE Transactions on Automation Science and Engineering*, **1**(1), pp. 4–13.

[10.11] Luo, C.R., Lee, W.Z., Chou, J.H. and Leong, H.T., 1999, "Tele-control of rapid prototyping machine via Internet for automated tele-manufacturing," In *Proceedings of the 1999 IEEE International Conference on Robotics & Automation*, pp. 2203–2208.

[10.12] Jiang, P. and Fukuda, S., 2001, "TeleRP – an Internet web-based solution for remote rapid prototyping service and maintenance," *International Journal of Computer Integrated Manufacturing*, **14**(1), pp. 83–94.

[10.13] Jiang, P. and Fukuda, S., 1999, "Internet service and maintenance for RP-oriented tele-manufacturing," *Concurrent Engineering: Research and Applications*, **7**(3), pp. 179–189.

[10.14] Liu, X.G., Jin, Y. and Xi, J.T., 2006, "Development of a web-based tele-manufacturing service system for rapid prototyping," *Journal of Manufacturing Technology Management*, **17**(3), pp. 303–314.

[10.15] Lan, H.B., Chin, K.S. and Hong J., 2005, "Development of a tele-service system for RP service bureaus," *Rapid Prototyping Journal*, **11**(2), pp. 98–105.

[10.16] Lan, H.B., Ding, Y.C., Hong, J., Huang, H.L. and Lu, B.H., 2004, "A web-based manufacturing service system for rapid product development," *Computers in Industry*, **54**(1), pp. 51–67.

[10.17] Tay, F.E.H., Khanal, Y.P., Kwong, K.K. and Tan, K.C., 2001, "Distributed rapid prototyping – a framework for Internet prototyping and manufacturing," *Integrated Manufacturing Systems*, **12**(6), pp. 409–415.

[10.18] Xu, A.P., Hou, H.Y., Qu, Y.X. and Gao, Y.P., 2005, "VRPS-I: an Internet-based virtual rapid prototyping system," *Journal of Integrated Design and Process Science*, **9**(3), pp. 15–27.

[10.19] Cooper, A.G., Kang, S., Kietzman, J.W., Prinz, F.B., Lombardi, J.L. and Weiss, L.E., 1999, "Automated fabrication of complex moulded parts using Mould Shape Deposition Manufacturing," *Materials and Design*, **20**(2), pp. 83–89.

[10.20] Rajagopalan, S., Pinilla, J.M. and Losleben, P., 1998, "Integrated design and manufacturing over the Internet," In *ASME Design Engineering Technical Conferences*, Atlanta, GA, pp. 13–16.

[10.21] Huang, H.L., Ding, Y.C. and Lu, B.H., 2000, "Research of the rapid prototype-oriented tele-service system based on Internet and Intranet," *Journal of Xi'an Jiaotong University*, **34**(7), pp. 52–57.

[10.22] Huang, J., Jiang, P.Y., Yan, J.Q., Ma, D.Z. and Jin, Y., 2000, "Implementing Internet/Web-based rapid prototyping tele-manufacturing service," *Journal of Shanghai Jiaotong University*, **34**(3), pp. 433–436.

[10.23] Fidan, I. and Ghani, N., 2007, "Remotely accessible laboratory for rapid prototyping," In *Proceedings of ASEE Annual Conference and Exposition*, pp. 7.

[10.24] Helsinki University of Technology, 2007, http://ltk.hut.fi/RP-Selector.

[10.25] Gibson, I., 2002, *Software Solutions for Rapid Prototyping*, Professional Engineering Publishers, London, UK.

[10.26] Lan, H., Ding Y. and Hong, J., 2005, "Decision support system for rapid prototyping process selection through integration of fuzzy synthetic evaluation and expert system," *International Journal of Production Research*, **43**(1), pp. 169–194.

[10.27] Chung, C.W., Hong, J., Lee, A., Ramani, K. and Tomovic, M.M., 2003, "Methodology for selection of rapid tooling process for manufacturing application," In *2003 ASME International Mechanical Engineering Congress*, pp. 23–30.

[10.28] Quickparts.com, 2007, http://www.quickparts.com.

[10.29] 3T RPD, 2007, http://www.3trpd.co.uk/request-a-quoye.htm.

[10.30] Lan, H.B. and Ding, Y.C., 2007, "Price quotation methodology for stereolithography parts based on STL model," *Computers & Industrial Engineering*, **52**(2), pp. 241–256.

[10.31] Lan, H.B., Ding, Y.C., Hong, J., Huang, H.L. and Lu, B.H., 2008, "Web-based quotation system for stereolithography parts," *Computers in Industry*, doi: 10.1016/ j.compind.2008.03.006.
[10.32] Rosen, D., 2001, "STL viewer," http://srl.gatech.edu/Members/cwilliams/classes/ CBW.ME6104.Report.pdf.
[10.33] Lan, H.B., Ding, Y.C., Hong, J. and Lu, B.H., 2007, "Research on ASP tool set for RP data pre-processing based on STL model," *China Mechanical Engineering*, **18**(5), pp. 559–563.
[10.34] Roy, U. and Cargian, M., 1997, "Tele-manufacturing: object slicing for rapid prototyping on the Internet," In *Proceedings of 4th ISPE International Conference on Concurrent Engineering: Research and Applications*, pp. 301–307.
[10.35] Wu, Y.X., 2001, "Rapid prototyping job scheduling optimisation," *Master Thesis*, Virginia Polytechnic Institute and State University.
[10.36] Lin, H.H., Hsueh, C.W. and Chen, C.H., 2003, "A real-time scheduling approach for a Web-based rapid prototyping manufacturing platform," In *Proceedings of the 23rd International Conference on Distributed Computing System Workshops,* pp. 42–47.
[10.37] Wang, Q., Ren N.F. and Chen, X.L., 2006, "Java 3D-based remote monitoring system for FDM rapid prototype machine," *Computer Integrated Manufacturing Systems*, **12**(5), pp. 737–741.
[10.38] Gao, C.Y., Wang, X., Yu, X.J. and Xu, Q.S., 2005, "Research and development of tele-control system for rapid prototyping machine based on Internet," *Journal of Jiangsu University (Natural Science Edition)*, **26**(1), pp. 16–19.
[10.39] Kang, Y.Y., Xi, J.T. and Yan, J.O., 2003, "Remote monitoring and diagnosis of breaking-down fault for FDM rapid prototyping machine," *Computer Integrated Manufacturing Systems*, **9**(9), pp. 771–775.

11

Agent-based Control for Desktop Assembly Factories

José L. Martinez Lastra[1], Carlos Insaurralde[1] and Armando Colombo[2]

[1] Tampere University of Technology, FIN-33101 Tampere, P.O. Box 589, Finland
Emails: jose.lastra@tut.fi, carlos.insaurralde@tut.fi

[2] Schneider Electric, Seligenstadt, P&T HUB, 63500 Steinheimer Street 117, Germany
Email: armando.colombo@de.schneider-electric.com

Abstract
New generations of manufacturing systems have been strongly influenced by the miniaturisation revolution in the design and development of new short-lifecycle products. Multi-agent systems (MAS) and holonic manufacturing systems (HMS) are enabling the vision of the plug & play factory and paving the way for future autonomous production systems that address rightly the above trends. This chapter reviews the implementations of agent-based manufacturing systems and identifies the lack of engineering tools as a technological gap for widespread industrial adoption of the paradigm. One of the current challenges for the design and implementation of intelligent agents is the simulation and visualisation of the agent societies. This issue is significant as long as the software agent is embedded into a mechatronic device or machine resulting in a physical intelligent agent with 3D-mechanical restrictions. These mechanical restrictions must be considered in the negotiations among agents in order to co-ordinate the execution of physical operations. This chapter presents an engineering framework that contributes towards overcoming the identified technology gap. This framework consists of a comprehensive set of software tools that facilitate the creation, simulation and visualisation of agent societies. The documented research describes the methodology for the 3D representation of individual physical agents, the related identified objects present in the interaction protocols and the assembly features and clustering algorithms.

11.1 Introduction

Actual manufacturing systems that follow global trends such as shorter product life cycles and mass customisation have been strongly influenced by the miniaturisation revolution in the design and development of new market products. This tendency requires smaller facilities and equipments since big production systems imply high manufacturing costs when fabricating small products. According to the results reported in 2000 [11.1], a concordance between sizes of the production equipments and facilities and manufactured products, decreases costs and is one of the keys that lead toward future mini- and micro-factories. Small manufacturing systems or "micro-factories" [11.2] are a true approach to addressing the minimisation of production systems that match the size of the manufactured products. They also save resources such as energy, material utilisation, floor space, operational costs and load on operators.

Nowadays, the environmental impact is not a minor issue. In Europe, it is legislated by a Waste Electrical and Electronic Equipment (WEEE) directive of the European Commission (EC) and implemented in the Community Member States by focusing on material and energy preservation. Moreover, micro-factories increase the manufacturing equipment speed and positioning precision by facilitating product design modifications. This new industrial challenge causes radical changes in manufacturing practices where smaller production systems are placed next to the customers, producing tailored products for them by utilising very fast reconfigurable production systems that can change according to market demands.

Following this micro-factory direction, the Institute of Production Engineering of the Tampere University of Technology contributes permanently with scientific proposals for this research field and, in particular, has been involved in a project named "Towards Mini and Micro Assembly Factories" (TOMI) [11.3], which was reported in [11.4]. The project is a Finnish Research and Development Project that is organised under the Tekes (National Technology Agency of Finland) PRESTO (Future Products – Added Value with Micro- and Precision Technologies, 1999–2002) programme.

At present, many technical issues related to micro-factory systems are still unresolved. In particular, a modular approach for manufacturing systems that involve assembling micro-parts and micro-manipulations (*e.g.* assembly micro-robots) of materials is very suitable for the agile reconfiguration required. Thus, the basic technology and devices employed in the manufacturing modules are being replaced by more sophisticate information technologies (intelligence-based control systems) and more powerful control devices that allow reconfiguring systems for product lifecycles in short time. Predictions from the Micromachine Centre (MMC) in Japan say that there are two different markets for the micro-factories. One market replaces existing machine with new ones such as micro-manufacturing machines (industrial robots, machine tools, *etc*.). The other one is a completely new market (on-site manufacturing equipments, micro-chemical plants, *etc*.). Therefore, micro-factory is a promising technology that can easily be introduced into the market, if the investment costs are the point for saving [11.5].

Addressing installations of micro-factories for the above customised production systems (whether planning new plants or retrofitting existing ones) demands highly flexible and highly reconfigurable working environments [11.6]. However, flexibility and reconfigurability of the working environment often conflict with the requirement for high productivity. The solution is to migrate from conventional factory floor control strategies to flexible and collaborative micro-factory automation systems. The hierarchical management and control philosophy needs to be broken down into intelligent, collaborative and autonomous production units [11.7], which are intelligent physical agents suitable for micro-factory automations.

During recent years, a number of research projects have studied the application requirements for various types of collaborative automation systems in a number of domains, and have resulted in the development of a series of architectures for intelligent agent-based control systems and successful prototype implementations.

However, as yet, none of these generic approaches has resulted in large-scale industrial application trials. This reticence is probably due principally to both commercial risks involved and a series of remaining technological gaps [11.8].

One of the identified technological gaps in the field of agent-based control is the lack of powerful and well-integrated engineering tools to efficiently support the design, implementation and lifecycle support of automation applications. The lack of tools limits the implementation of agent-based manufacturing systems within reach of only a handful of domain experts, who need to be trained in disciplines such as control, mechanical engineering, communications, and artificial intelligence. A systematic approach is needed for designing, emulating, implementing, validating, testing and debugging, deploying, commissioning, monitoring, reconfiguring and recycling agent-based manufacturing systems.

This research work presents an engineering framework that contributes towards overcoming the identified technology gap in micro-factory. The framework consists of a comprehensive set of software tools that facilitate established engineering practices for the different lifecycle stages of agent-based manufacturing systems, from design through operation to recycling. Supported engineering practices include, among others, computer-aided design, simulation and emulation, commissioning, reconfiguration, and system visualisation.

The types of agent-based manufacturing systems that are developed using the framework employ the actor-based assembly systems (ABAS) reference architecture [11.9]. While agents are considered to be pure software entities, the term *holon* has been used extensively to describe an agent that is integrated with physical manufacturing hardware. However, holonic manufacturing systems (HMS) have gained prominence as a specific type of social organisation of holons, which have specific behaviour and goals as defined by Van Brussel *et al.* [11.10]. ABAS define a new type of autonomous mechatronic unit called *actor* that is differentiated from the existing characterisation of holons by adopting a different social organisation. ABAS are reconfigurable systems built by autonomous mechatronic devices, called actors that deploy auction- and negotiation-based multi-agent control in order to collaborate towards a common goal: the accomplishment of assembly tasks. These assembly tasks are complex functions generated as a composition of simpler activities called assembly operations, which are the individual goals of the actors.

The rest of this chapter is structured as follows. Section 11.2 reviews the fundamentals of agent-based control and the state of the art in agent architectures, pilots and tools. Section 11.3 presents the ABAS architecture. Section 11.4 details the ABAS tools that compose the engineering framework and Section 11.5 describes the experiments performed to illustrate the proposed methodology. Finally, Section 11.6 considers lessons learned and presents some concluding remarks.

11.2 Agent-based Manufacturing Control

This section reviews the role of agent-based control in the context of collaborative industrial automation. Subsequently, it looks into selected projects that have resulted in successful prototype implementations, and that represent the state of the art in agent-based manufacturing control. Considering the state of the art, the availability of engineering tools is also assessed.

11.2.1 Collaborative Industrial Automation

Addressing the need for more agile and reconfigurable production systems has led to growing interest in new automation paradigms that model and implement production systems as sets of production units/agents/actors interacting/collaborating in a complex manner in order to achieve a common goal [11.5]. Traditional sequential engineering methods, while appropriate for largely monolithic production systems, are inappropriate in the context of these new distributed unit/agent/actor-based approaches to system implementation. New engineering environments are needed, which are capable of supporting inherently multi-disciplinary, parallel system engineering tasks. The realisation of appropriate engineering tools requires not only a broad appreciation of mechatronics, manufacturing strategies, planning and operation but also a deep understanding of the required integration of communication, information and advanced control functionality.

One promising approach, that has the potential to overcome the technical, organisational and financial limitations inherent in most current approaches, is to consider the set of production units/agents/actors as a conglomerate of distributed, autonomous, intelligent, fault-tolerant, and reusable units, which operate as a set of co-operating entities. Each entity is typically constituted from hardware, control software and embedded intelligence. Due to this internal structure, these production entities (intelligent automation unit/physical agent/holon/actor) are capable of dynamically interacting with each other to achieve both local and global production objectives, from the physical/machine control level on the shop floor to the higher levels of the factory management systems.

The umbrella paradigm, encompassing this general form of automation system, is recognised in this research work as "collaborative automation". As depicted in Figure 11.1, it is the result of the integration of three main emerging technologies/

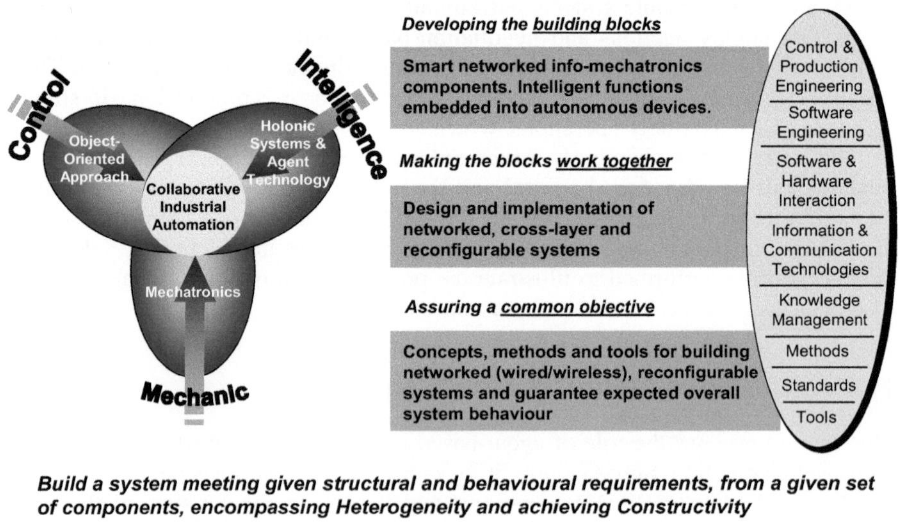

Figure 11.1. Convergent technologies in the collaborative automation paradigm

paradigms: holonic control systems utilising agent-based technology, object-oriented approaches to software, and mechatronics. The aim is to effectively utilise these technologies and methods to achieve flexible, network-enabled collaboration between decentralised and distributed intelligent production competencies. Autonomous automation units with embedded local supervisory functionality, installed in each production site, are able to collaborate to achieve production objectives at the shop floor level, and interact/co-operate to meet global (network-wide) supervisory needs (for example, related to control, monitoring, diagnosis, human-machine interface, and maintenance).

An innovative aspect of this approach is that the control of production sequences is achieved by means of negotiation and autonomous decision-making inherent in the co-ordinated operation of the functional production automation entities (intelligent, collaborative automation units), for example, system devices, machines and manual workstations. This collective functionality distributed across many mechatronic system devices and machine controls, replaces the logical programming of manufacturing sequences and supervisory functions in traditional production systems.

11.2.2 Agent-based Control: the State of the Art

Recently, a number of strands of both industrial and academic research have been undertaken that have contributed in different ways to the state of the art in the field of agent-based manufacturing control. The following subsections review selected agent-based technologies and their achievements. The assessment is not exhaustive, but instead attempts to portray the different stages of maturity achieved in representative domains. For an exhaustive survey of agent-based technologies in manufacturing, refer to [11.8].

11.2.2.1 Architectures and Demonstrators

Multi-agent systems (MAS) are applied to manufacturing control at several levels. When agents are integrated with the physical manufacturing hardware, referred to as holons or holonic agents, they enable distributed and collaborative real-time control at the factory floor. Other types of intelligent agents, usually implemented as stand-alone software units, are also used at the production management level. Moreover, agent technology is also applied to the virtual enterprise, an abstraction used to represent conglomerate of enterprises that complement each other with a particular business objective [11.11]. According to [11.12], three different approaches have been adopted in agent-based control: (1) auction- and negotiation-based; (2) artificial markets; and (3) stigmergy and ant colony co-ordination. These approaches differ in that the first two agents explicitly interact and negotiate with each other, while the latter one belongs to the category of agents that indirectly interact and coordinate by changing their environment.

The implementation of agent technology at the paint shop of the General Motors Corp. was a milestone in the use of artificial intelligence for the control of manufacturing activities. This implementation was significant in that it defines agents within the architecture not only as traditional software elements but also as individual physical devices, such as humidifiers, burners, chillers, and steam. Even though agent-based control is a new approach, the potential advantages that can be achieved make it possible to implement other applications such as those at Daewoo Motors in which the task, resource and service agents interact in a market-driven approach, building communities via hierarchical aggregation [11.13]. The field of planning and scheduling has been very receptive to this new technology. However, implementations in the field of real-time control are also common in the literature. Another example of an agent-oriented application is AMROSE in the application domain of shipbuilding; this application is especially interesting since it not only belongs to the field of robotic control but also builds the robot by assigning an agent to each of the links, with each agent deriving its positioning goal from the next agent closer to the end effector [11.14].

In 1997 the former vice president of Allen-Bradley, Dr Odo Struger, initiated the "HMS" project within the international Intelligent Manufacturing Systems (IMS) program. The approach adopted, took its inspiration for a solution to problems in modern manufacturing from Arthur Koestler's book *The Ghost in the Machine* [11.15]. Koestler describes a very particular perspective on the principle, design and function of biological and social systems. Following these design patterns enabled the creation of systems with behavioural characteristics well matched to meeting the requirements of advanced manufacturing. The technical basis for the HMS was subsequently identified as agent technology emerging from the IT sector [11.16]. In the activities of the holonic research community, two well-established approaches are reported in the literature, PROSA [11.10] and MetaMorph [11.17].

In parallel with the HMS initiative, and mutually inspired by the work of Stefan Bussmann [11.18], the first industrial agent controlled manufacturing line was developed by Schneider Electric Automation and successfully set in operation in a car production facility. This line is still in operation and proves the concept of reconfigurable systems in the control of manufacturing systems [11.19].

Modular build for distributed systems (MBODY) is the current phase of a research initiative in the Department of Mechanical and Manufacturing Engineering at Loughborough University, which began in 1999. A major goal of this work has been to achieve more efficient machine reconfigurability via a functionally modular, component-based approach to automation. A new application is created by selecting machine modules from a library and then configuring them graphically in a 3D engineering environment, which supports the lifecycle of the machine. To date, two industrial demonstrator machines have been implemented in the automotive sector, and applications in both supermarket warehousing and electronics manufacturing have also been evaluated [11.20, 11.21].

The ABAS® claims to not only attain but also exceed the objectives of mass, lean, agile and flexible manufacturing. A highly dynamic, reconfigurable assembly solution was demonstrated in a pilot installation located in Tampere, Finland. The ABAS® concept is extensively reviewed throughout the rest of this chapter.

The ADAptive holonic COntrol aRchitecture for distributed manufacturing systems (ADACOR) is a control architecture, developed and implemented at the Polytechnic Institute of Bragança, Portugal. ADACOR is built upon a set of autonomous and co-operative holons, each one being a representation of a manufacturing component that can be either a physical resource (numerical control machines, robots, pallets, *etc.*) or a logic entity (products, orders, *etc.*) [11.22]. The

ADACOR adaptive production control approach is neither completely decentralised nor purely hierarchical, and supports other intermediate forms of control implemented through holonic autonomy factors and propagation mechanisms inspired by ant-based techniques.

In another research, two production sections of a holonic enterprise, each controlled by its own multi-agent-based control system [11.23] are presented as collaborative components within intra-enterprise architecture. This work, which culminated in one of the first implementations of the integration of the two multi-agent-platforms, was presented to the scientific and industrial community at the Hanover Fair in 2002. The interoperability demonstrated, was achieved based on the use of: (1) the Agent Communication Language (ACL) released by the Foundation for Intelligent Physical Agents (FIPA) as the communication tool, and (2) socket technology for the technical implementation.

11.2.2.2 Simulation Tools

Simulation is a widely employed engineering practice for the design stages of manufacturing systems. Simulation enables recreating the environment in which a system is deployed, and foreseeing behaviour. By analysing simulation results, designs can be adjusted to meet requirements. Simulation enables tests to be carried out independently of the mechanical system to be deployed, offering potential cost savings if designs are changed, as well as time savings from concurrent design and mechanical system development.

The HMS consortium delivered an interactive, virtual simulation of a holonic material flow simulation tool [11.24]. The tool targets an automated workshop production, where the material flow is carried out by automated guided vehicles (AGVs). As a result of the research effort under the IMS framework, Rockwell Automation in co-operation with different partners has designed and developed Manufacturing Agent Simulation Tool (MAST), a graphical visualisation tool for multi-agent systems reported in different forums. The main target is the material-handling domain and it is built on the JADE standard FIPA platform. In MAST, a user is provided with the agents for basic material-handling components as for instance manufacturing-cell, conveyor belt, diverter and AGVs. The agents co-operate via message passing using common knowledge ontology developed for material-handling domain. MAST represents the state of the art in graphical simulation tools for modelling and simulation of multi-agent systems in manufacturing control. However due to the fact that only material-handling systems are targeted, the tool does not cover complex application from a 3D geometric viewpoint such as robotic manipulation.

11.2.3 Further Work Required

The aforementioned projects have studied the application requirements for various prototype forms of collaborative automation systems in a number of domains and have resulted in the development of a series of architectures for intelligent control systems and successful prototype implementations. The results have indicated that the approach may have the potential to reduce the total amount of time for production system engineering (for example, from design, via configuration, to operation) from years towards a scale of months. However, none of these generic approaches has resulted in large-scale industrial application trials. This reticence is probably due to the commercial risks involved and remaining technological gaps.

One of the major industrial requirements emerging from these projects is the need for powerful and well-integrated engineering tools for design, implementation and lifecycle support of automation applications. The effort required to develop a commercially viable engineering platform of this type is considerable. However, an even more important task is to raise the level of user awareness and to educate the industrial community (both end-users and machine builders) of the characteristics and potential benefits of adopting the collaborative automation paradigm. The role of humans in the collaborative physical networks where hybrid automatic/manual workstations are deployed has not yet been adequately researched. New ways of working need to be adopted but, due to the lack of industrial applications, little practical experience has been gathered. If collaborative automation is adopted, it is likely to have major implications on the role of humans not only on the end-user's shop floor but also at the machine builders and control system vendors sites.

The authors' vision is the creation of a new approach to automation, based on the *collaborative automation* paradigm, which in the next five to ten years will have as profound an impact as the appearance of the programmable logic controller (PLC) in the 1970s. This practical realisation of collaborative automation will only be achieved through the development and industrial exploitation of new enabling technologies in the fields of intelligent control. The success of the PLC paradigm would not have been possible without the availability of supporting engineering tools such as ladder diagram (LD) editors and others. An equivalent development framework is required in order for collaborative automation to become more than an academic experiment and enter the industrial environment.

11.3 Actor-based Assembly Systems Architecture

ABAS presents a collaborative electronics assembly automation architecture that defines a set of intelligent mechatronic devices/modules that map their functionality to basic assembly activities, which are named *assembly operations*. More complex activities are referred to as *assembly processes* that are formed by aggregating these basic operations. By rearranging these mechatronic modules or by populating the system with different modules, the system is able to accomplish different assembly processes. The mechatronic modules are assembled manually by placing them (via configuration or reconfiguration) according to the process stages requested by the products. A highly dynamic, reconfigurable test bed located in Tampere, Finland, is illustrated in Figure 11.2 together with its 3D model. This micro-assembly station (conveyers and transfer units) is configured to serve a Cartesian geometry robot arm. There is no any particular product manufactured, rather it is configured to build trial scenarios (*e.g.* inserting and screwing processes) for potential marketable solutions.

This section provides a comprehensive view of the ABAS reference architecture. Emphasis is placed not only on the agent-based aspects of the system but also on the physical aspects, which are equally critical when configuring the intelligent mechatronic devices or physical agents that are building blocks of the architecture. For complete architecture documentation, refer to [11.9].

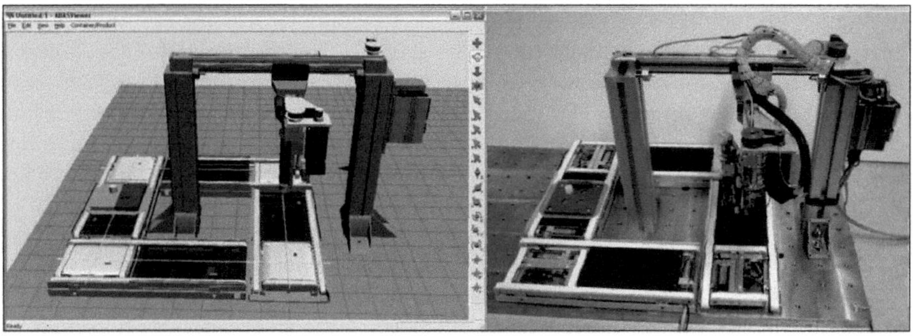

Figure 11.2. A 3D view of a system built by intelligent physical agents (left) and the correspondent physical implementation (right)

11.3.1 Architecture Overview

In order to deploy ABAS systems, a set of architectural elements has been defined. These are the actor, register, recruiter, and actor cluster. The functionality of these is reviewed in the following paragraphs.

Assembly actors are mechatronic entities built on a well-defined set of resources for accomplishing typically one assembly operation. In order to participate in an ABAS society, actors must report themselves to the register. Actors then perform assembly operations as requested by recruiters. Actors contemplate the collaboration with other actors, and have appropriate behaviour to cope with that kind of situation.

The register is a software entity that resides in the ABAS platform or may be distributed among different actors of the ABAS society. Its overall goal is to regulate the admission of actors into the society and to keep a record containing the different members accommodated in the given society and their addresses in a *white-page-list*. In addition, the register provides the ABAS with a *yellow-page-list* for each registered actor.

Recruiters are software entities that seek to enrol and secure the accomplishment of assembly tasks by recruiting assembly actors with the necessary skills for executing the requested assembly operations. Three recruiters work sequentially for the completion of a composing task. The first recruiter deals with the operations required for the part that is composed together with an assembly. The second recruiter deals with the operations that deal with the assembly where the part is composed. The third recruiter is in charge of the required operations once the part and assembly become a single entity.

ABAS identifies through the use of recruiter agents the actors capable of solving the required assembly needs by grouping the individuals into a set of actors called cluster. Clusters are software entities that keep record of the potential groups having the necessary skills for solving a given problem. There may be more than one actor cluster for a given job. The selection of a particular cluster requires the implementation of optimisation algorithms, which can contemplate one or more optimisation goals (*e.g.* least number of actors, shortest time, lowest cost, *etc.*). The lifecycle of the cluster ends once the requested job has been performed [11.25].

Figure 11.3 shows a sequence diagram of an example of actor-recruiter-register interaction at the system start-up time. The start-up scenario illustrates the situation for which a society is organised in order to cope with the requirements of a mission requested by a new product. The request protocol is executed by the initiator, which can be any member of the actor society, including the product information travelling in a bar code label or RF devices. A new software component is dynamically created called the recruiter. The mission of the recruiter is to secure the services provided by actors and actor clusters. The recruiter will poll the request for services to all those members of the society that are potentially capable of providing such services. The recruiter takes a collaborative approach to maintain those actors involved in a particular process. The cluster is a dynamic component that is subsequently destroyed once that process is no longer required by the society.

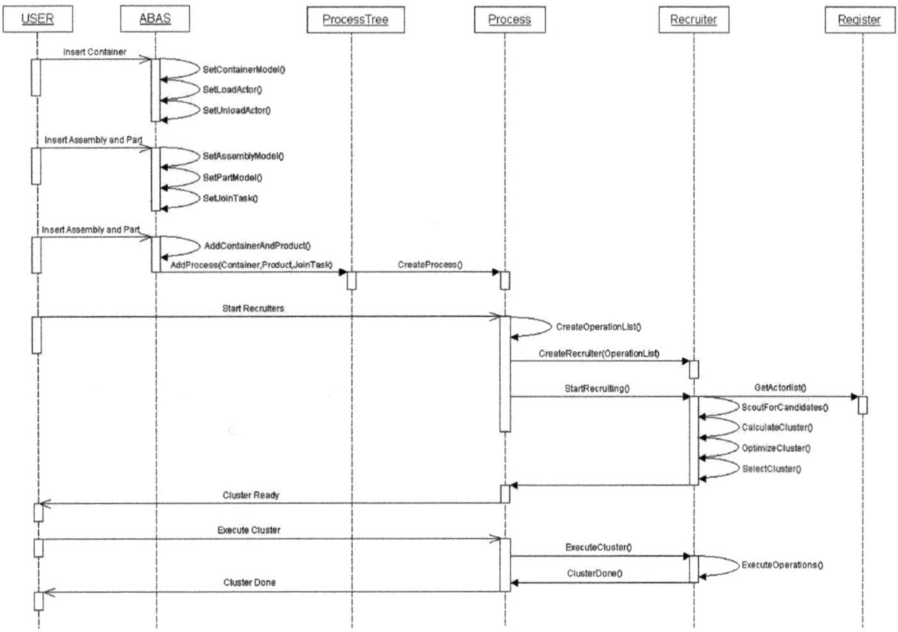

Figure 11.3. Actor, agent, object and mechatronic relationship

11.3.2 Intelligent Physical Agents: Actors

Actors can be considered a specific type of intelligent physical agent; being intelligent physical agents defined as mechatronic devices augmented with agent behaviour (Figure 11.4). Each actor is assigned, as functional objective, one of the atomic assembly operations of Figure 11.5. This figure also illustrates the atomic assembly operations needed to perform complex assembly tasks. These tasks can therefore be accomplished by forming actor clusters, where each composing actor provides one of the required operations.

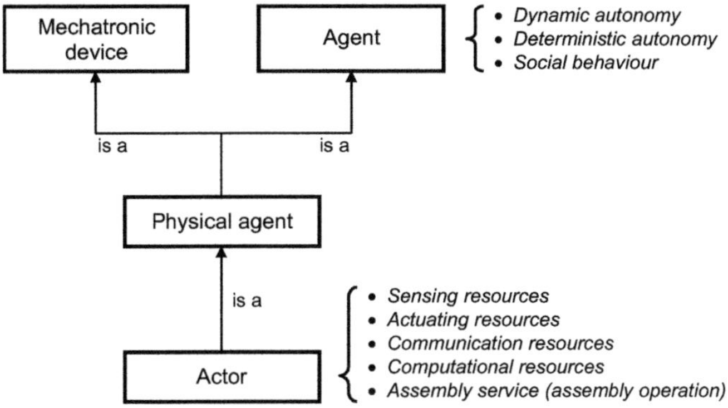

Figure 11.4. Actor, agent, object and mechatronic relationship

(a)

	Translate	Rotate	Apply force	Apply torque	Material properties change	Add material	Remove material	Preserve position	Preserve orientation
Retrieve	X	X							
Grasp			X						
Move	X	X							
Join	X	X	X	X	X	X	X	X	X
Fixture			X	X				X	X
Release			X	X					

(b)

	Translate	Rotate	Apply force	Apply torque	Material properties change	Add material	Remove material	Preserve position	Preserve orientation
Retrieve	X	X							
Store								X	X
De-fixture			X	X					

Figure 11.5. Operations generally presented in each assembly task related to: (a) parts, and (b) assemblies/products

In order to achieve the desired individual functionality and the necessary skills for combining this functionality with the ones exhibited by others, each actor has a set of resources. These resources are computation, communication, actuating, and sensing resources and are used for providing the necessary internal services. Thus, the actor is able to model the physical environment where it is placed and its current status by making use of the sensing resources, and to control that environment by executing different control algorithms that have been encapsulated into software modules using the computational actor resources. As a consequence of those control laws, the actor uses its actuating resources in order to change its environment. At any time, the actor must ensure the necessary communication links in order to interact within the society that is a member; this requirement is guaranteed by the external communication resources that the actor has.

11.3.3 Agent Societies: ABAS Systems

One of the key features of ABAS systems is that when a complex (composite) assembly process is introduced to the system, a set of recruiter entities have the responsibility of identifying the basic assembly operations that compose the process. The recruiters enroll actors existing in the ABAS society that are capable of providing those basic operations, and thus the complex process is fulfilled. When a group of actors collaborates in this way to perform a complex process, they are grouped into a cluster. In ABAS, matching atomic assembly operations to actors is a trivial process because each actor is assigned a single atomic operation as functional objective. These atomic operations were listed in Figure 11.5. However, in order to decompose complex (composite) processes to operations, a taxonomy for the domain of light assembly has been created and documented [11.9]. Current research is investigating the specification of this taxonomy using formal ontology, in order to enable reasoning agents to infer the atomic operations corresponding to a composite process, instead of having all combinations of the taxonomy hard coded.

In order to construct clusters, recruiters need to distinguish the relations between actors in order to know how they can work together to accomplish a certain output. In other words, recruiters must understand how an actor can affect its surrounding environment (other actors) in order to accomplish the required goal. For recognising the relationships between actors, the use of assembly features is proposed. Features were originally used to model geometric and non-geometric (functional) properties of the relationships between parts in an assembled product. This work proposes that an actor cluster is, like an assembled product, an assembly of parts (the actors). Therefore, assembly features are used to model the geometric and functional properties of the relationships between actors in a cluster.

Combining actors in order to form societies aligns well with the definition of assembly actions, putting together components to form a more meaningful entity. Analogies can therefore be constructed by viewing actors as parts or societies, and clusters as assemblies of those parts. The assembly features of actors are: location, physical dimension (PD), working dimension (WD) and actor interfaces (AIFC). These together with the service type, which defines the basic assembly operation that an actor is able to perform, form the common attributes of any actor. Table 11.1 illustrates those attributes and their format, the last four elements of the table can be considered as assembly features since they enclose both functional and geometry information. Figure 11.6 illustrates the features in a prototype actor. The presented features are used to create relationships between actors. Those relationships define the physical effect that an actor can have on another actor, and how they as a composed entity can be requested to perform certain assembly task. The procedure by which those features are learned by different actors is by using the FIPA query interaction protocol. The following subsections further define those features. Figure 11.6 illustrates all the attributes of an actor prototype including: (a) actor location; (b) actor dimension; (c) actor interface; and (d) working dimension.

Table 11.1. Actor attributes, including assembly features

Assembly feature	Format	Abbreviations
Service type	Assembly operation	ST
Location	$P_1(x_1, y_1, z_1)$ $\theta_1, \theta_2, \theta_3$	LC
Physical dimension	$P_{1p}(a_{1p}, b_{1p}, c_{1p})$ $P_{2p}(a_{2p}, b_{2p}, c_{2p})$	PD
Working dimension	$P_{1w}(a_{1w}, b_{1w}, c_{1w})$ $P_{2w}(a_{2w}, b_{2w}, c_{2w})$	WD
Actor interface	$P_r(x_r, y_r, z_r)$ $\{R\}\theta_1, \theta_2, \theta_3$	AIFC

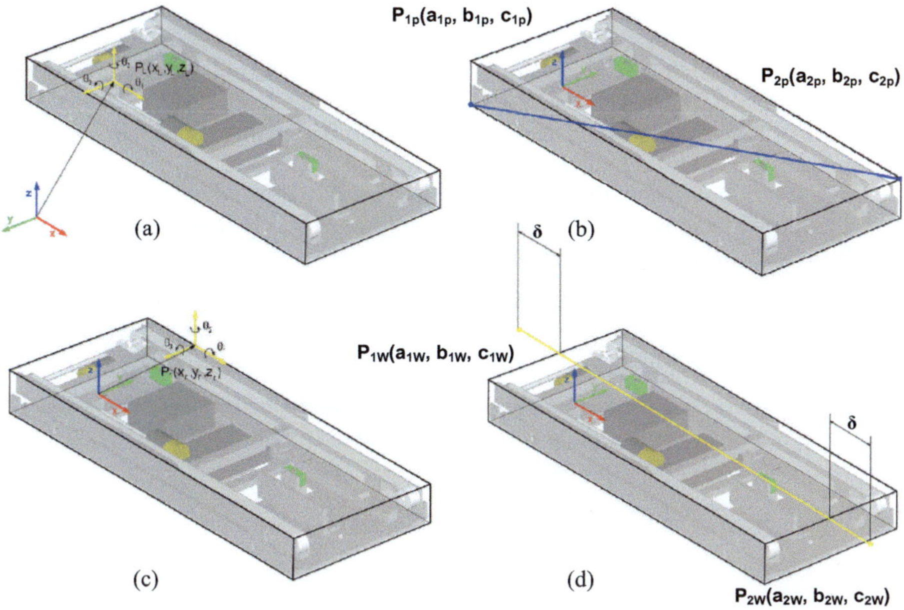

Figure 11.6. Example of the actor features identified in an actor prototype: (a) actor location; (b) actor dimension; (c) actor interface; and (d) working dimension. The service type for this prototype is the translation of assemblies, parts and products.

11.3.3.1 Location

The location of an actor in space involves three co-ordinates for the position and three angles for the orientation. ABAS systems, as in any robotics manipulation system, consider parts (products) and tools (actors) that are moved in space. This leads to the need for representing both the position and orientation of those entities, *i.e.* their location. The location is needed for recruiters given that the rest of the features are related to the actor's coordinate system, with origin in the actor location. In most cases, it is needed to refer (transform) those features to the world co-ordinate system (WCS) so that they can be compared with the features of other actors. The information provided by the location represents the actor co-ordinate system related to the WCS, and can be used to create homogeneous transformation matrices that transform those points related to the actor co-ordinate system to the WCS. The homogeneous matrices are useful when the calculations are performed by a computer, since it is possible to express transformations of vectors with matrix products, instead of combination of additions and multiplication of matrices.

Physical dimension. The physical dimension (PD) is represented by six scalars that provide the co-ordinates for two points. These two points define the diagonal of an orthogonal hexahedron containing the actor's body. The PD can be composed of more than one hexahedron (or bounding volume).

Working dimension. Using the same format as the PD, the working dimension (WD) represents the potential dimension of actuation under the service type of operation that the actor can perform.

Actor interface. The actor interface (AIFC) provides information in order to determine the current relative location between different actors. This information is calculated and defined when an actor is designed, and represents reference point actors, *e.g.* home positions in the servo actuators flag-points at the transportation units, such as loading and unloading points, waiting points, *etc.*

The AIFC goal is to reduce the location uncertainties when motion between actors and products is started. In this research, the AIFC is represented as 3D co-ordinate frame, which is justified below:

1. To be used for PD vertex mating. As the PD is represented by hexahedrons (bounding boxes), it is possible to fit one of the eight vertexes of a hexahedron with the AIFC of another actor (Figure 11.7(a)).

2. To be used for frame mating between products and actors where the co-ordinate system of parts matches with the AIFC. In this case, the AIFC is used as a frame as illustrated in Figure 11.7(b).

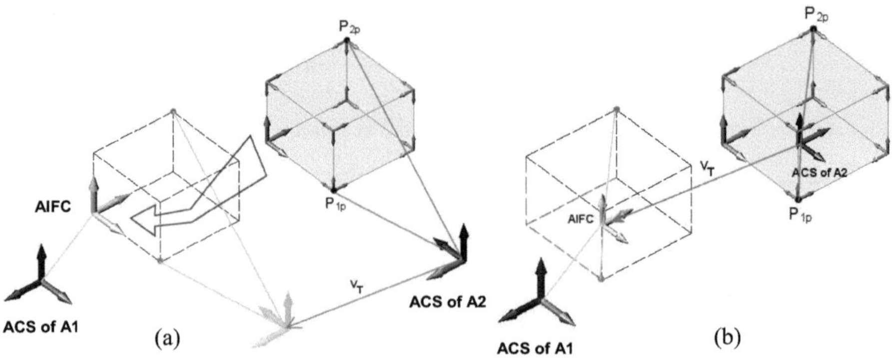

Figure 11.7. (a) Mating PD vertex of actor A2 with the AIFC of actor A1, (b) mating frame of actor A2 with AIFC of actor A1

11.3.4 Actor Contact Features

Actor contacts occur when two actors are physically attached, juxtaposed, *etc.* Figure 11.8 shows the possible contact types (CT) between two actors. Like previous attributes, contacts can be defined as assembly features since they represent functional and geometric information. Contacts have information not only about what actor dimensions (PD and WD) are in contact but also the geometry formed by that contact. The contact geometry or contact dimension (CD) is also represented with hexahedrons. By knowing the CT, it is possible to know the effect of certain actor(s) over other actor(s). Table 11.2 presents the functional information of actor contact features.

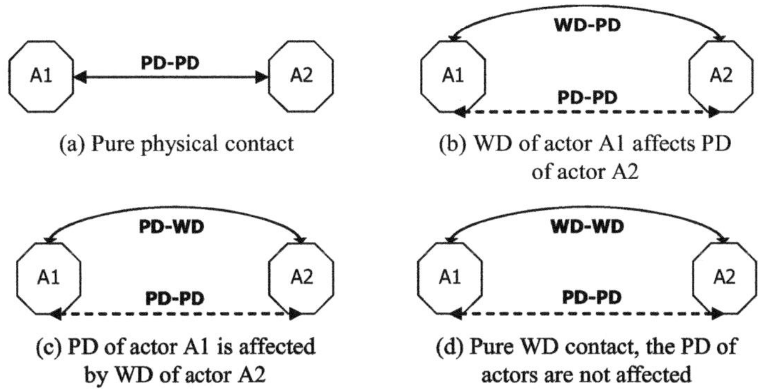

Figure 11.8. Representation of all possible contacts between two actors A1 and A2

Table 11.2. Contact assembly features

Contract type	Functionality	Detection priority
PD-PD	A relocation in actor A1 may relocate with the same magnitude the location of actor A2	Low (rigid attachment)
WD-PD	The WD of actor A1 can affect the PD of actor A2	Normal (non-rigid attachment)
PD-WD	The body of actor A1 can be affected by the assembly operation of actor A2	
WD-WD	Actor A1 and actor A2 are able to co-operate without affecting their PDs	High

When two actors have more than one CT and there is a physical contact type (PD-PD) between them, it is necessary to discern if they are either rigidly attached or not. Two actors can be in physical contact because one of them is sliding over the other, *e.g.* a container actor slides along a transporter actor (non-rigidly attached). On the other hand, it is necessary to consider the situation where an actor body is attached to another actor body without any WD intersections (rigidly attached). Therefore, it is necessary to prioritise the contact detection between actors since, in the case of multiple CTs, only one defines the real behaviour between actors. Table

11.2 shows the priority level defined for each CT. The functional information of actor contact features in Table 11.2 is needed by recruiters for two reasons:

- The recruiter knows the effects of certain actor(s) over other actor(s), and therefore it can calculate the necessary requests to them. By using the CD geometry it is possible to calculate the distances related to each actor.
- It is possible to use the CD in order to calculate WD amplifications (if they exist), and therefore to know if the cluster is capable to perform certain assembly task.

The priority helps the recruiter to choose one interpretation out of the detected contact types between certain actors.

11.3.4.1 Contact Detection Algorithm

The method used for contact detection is similar to the collision detection methods used in computer simulations. When having objects of different shapes moving in a simulation environment, it is sometimes necessary to determine not only if those objects are colliding but also the resulting collision dimension. Figure 11.9 shows the class diagram for the specification of the collision dimension.

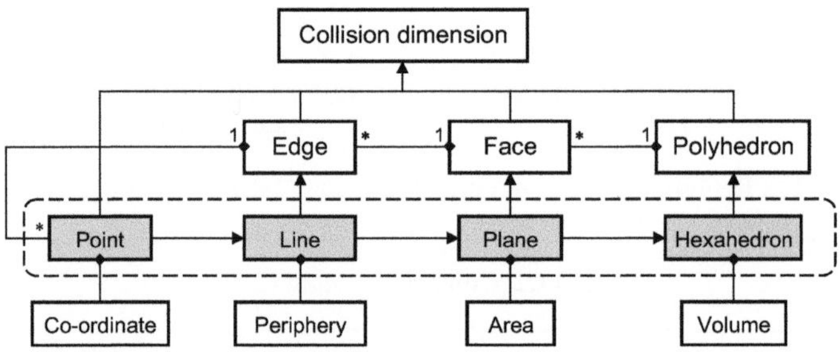

Figure 11.9. Class diagram for collision dimension

Contact detection algorithms depend on the representation of the objects in the computer environment. For two-dimensional shapes, for example, a collection of points that define lines is typically used, *e.g.* a hexagon is defined by six points. However, for three-dimensional shapes the representation of those bodies is more complex, and a collection of points is not enough. A widely used technique to represent complex three-dimensional shapes in computer environment is based on triangles. Similar to storing points for two-dimensional shapes, regarding three-dimensional shapes it is possible to store triangles; each of these triangles forms a plane, and by combining all those triangles together they can form complex shapes. By increasing the number of triangles used to form complex shapes it is possible to create more detailed shapes that describe an object with more accuracy. This approach requires significant computational power for operations (*e.g.* translation, rotation, screen rendering of those objects), and for collisions detection between them. This increase in computational power also includes the case of calculating the resulting collision area or contact of those kinds of bodies.

Considering the shortcomings of triangle-based representations, an optimised model for representing the body of an actor in a virtual environment leads to a kind of structure that can be used not only for the simplification of algorithms but also for use as a generic structure. According to Figure 11.9, a collision dimension can be specified as a body, surface, curve or point. A body is composed of surfaces; those surfaces are composed of curves; and curves are composed of points. Using a more specialised structure, a solid can be represented by a polyhedron that is composed of faces, those faces are composed of edges, and finally those edges are composed of points. A hexahedron is considered as a special case of a polyhedron that can be defined by two 3D points representing the opposite corners. Therefore, hexahedrons can be equally used to represent planes, lines and points as special cases, *e.g.* a plane can be seen as a hexahedron with zero volume but certain area. This generic structure enables representation of different shapes in which the principle of substitutability can be used, since the hexahedron structure would represent a body, surface, line or point. Therefore, the representation of PD and WD as hexahedrons facilitates the design of algorithms for collision detection and it reduces the computational power required for executing such algorithms. Moreover, the resulting collision body (or CD) as a result of collisions between WD and PD in any combination would result in another hexahedron (or any of its special cases). The disadvantage in using bounding volume would be the inaccuracy.

11.3.4.2 Working Dimension Amplification

When two or more actors are assembled together, the combination of their skill sets may result in WD amplification. This situation is illustrated in Figure 11.10, in which two servo-translator actors are attached. The combination of the WD of the two actors, which are two orthogonal lines, results in a composed WD that is an area.

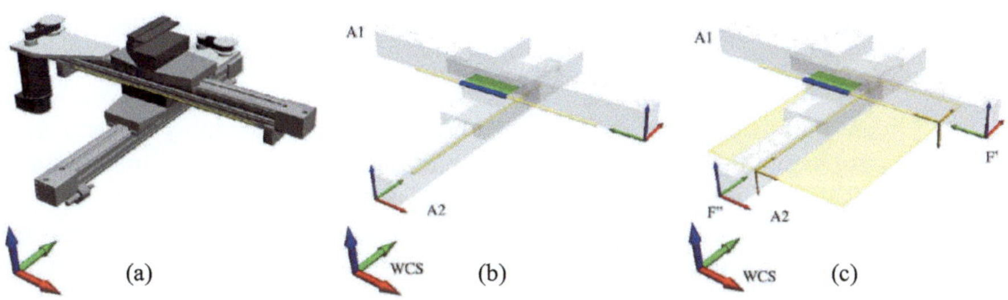

Figure 11.10. Working dimension amplification example for (a) two attached servo-translator actors; (b) with orthogonal linear WD; and (c) creating an area WD

In order to calculate WD amplifications, a special purpose algorithm was created. The contact detection algorithm is used to detect the contact features for an actor assembly. The contact type becomes critical for the WD amplification, as different contact types cause different combination actor skills. In particular, PD-PD contact types cause no amplification as none of the two actors performs an action on the other. Likewise, WD-WD contact types cause no amplification as none of the

PDs of the actors involved is affected. PD-WD and WD-PD contact types may cause amplification, depending on the type of assembly operation that is performed on the contact WD.

11.4 ABAS Engineering Framework

In order to build a comprehensive framework for ABAS, a set of engineering tools were developed. In developing these tools, the following goals were considered:

- To offer a 3D computer-aided design (CAD) environment to develop and configure actor prototypes.
- To create a blueprint for developing the agent behaviour of actors. This blueprint should serve both for developing new actors and for enhancing legacy mechatronic devices with actor features.
- To augment the CAD environment with the possibility to organise actors into actor societies.
- To provide a platform for emulating the behaviour of individual actors as well as actor societies. This platform should be accompanied by a simulation environment to replicate the physical aspects of ABAS systems.
- To provide a runtime platform that includes all necessary support services for deploying ABAS systems, both emulated and real systems. Among the support services, there should be a register, providing the white and yellow pages services for all actors in the society.
- To implement an agent-based execution control system. This system should provide the functions of the ABAS recruiters, which dynamically create and destroy actor clusters to perform complex assembly processes on demand. The system should only require the configuration of process goals, with no need for hard-coding of recruiter behaviour.
- To provide a visualisation platform in order to monitor actor societies, whether physically deployed or emulated.
- Given the mechanical and behavioural nature of assembly actors, a 3D environment should be provided to assist in concurrently manipulating physical attributes such as working and physical dimensions and software attributes such as agent behaviour. Established graphical user interface (GUI) input mechanisms should also be supported to facilitate user interactions.

Existing CAD/CAM and 3D simulation tools did not suffice to meet the aforementioned goals, as they do not facilitate the development and deployment of fully emulated agents. For example, it is not possible to deploy two agents that exchange messages within existing simulated 3D environments, and behave in exactly the same way as if they are deployed in the physical environment. Therefore, two stand-alone software applications and one reusable software component were developed. They are ABAS WorkBench, ABAS Viewer, and Actor Blueprint. The ABAS WorkBench encapsulates all functions needed for designing, emulating and configuring actors and actor societies. The ABAS Viewer encapsulates all functions required for deploying, executing and visualising actors and actor societies. When the ABAS WorkBench is used for emulation purposes, the emulated actors are deployed in a real runtime platform, typically the ABAS Viewer, which can equally have deployed real actors. The Actor Blueprint serves as a software component that facilitates the implementation of actor behaviour for mechatronic devices. These tools are discussed in detail in the following sections.

11.4.1 ABAS WorkBench

ABAS WorkBench is a tool used for designing and emulating ABAS systems, from the atomic design (actor) to the system level design (actor society). The goal of this software is to provide designers with the ability to produce actor prototypes and experiment with them before the real implementation.

As shown in Figure 11.11(a), the ABAS designer is able to create actor prototypes, which can be tested later in an emulated society; if an actor needs to be refined, it is possible to edit it until the desired behaviour is achieved. When an actor prototype is modelled, the resulting information is stored in flat files (text files) that can also be used in the implementation of real actors. The actor prototype design can be started using a CAD model of the actor, commonly implemented in commercial CAD tools like AutoCAD®, 3D Studio® or any software that could produce 3ds file format that can then be exported to Xfile format.

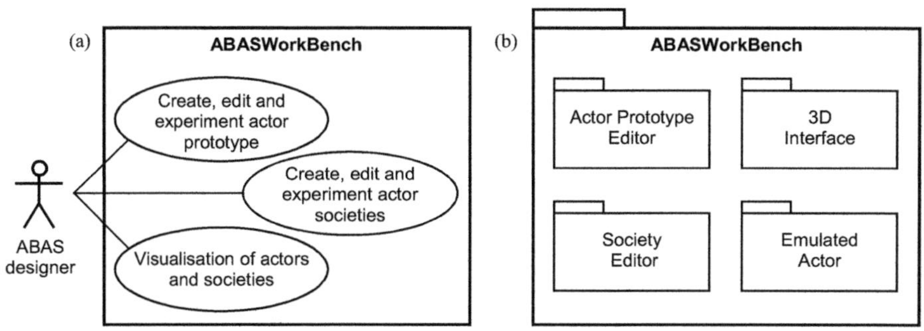

Figure 11.11. ABAS WorkBench: (a) use case diagram, and (b) main software packages

The visualisation and edition of societies is one of the salient features of ABAS WorkBench, since it enables time saving in ABAS systems design. This is possible because designed actors can be emulated, arranged and configured as it would be in the real implementation of the assembly system. Emulated actors and societies, in the same way of real actors and societies, need to be deployed in a platform where they can interact with each other. This platform has been implemented in the ABAS Viewer, presented in the next section. Even though the ABAS WorkBench needs a platform, be it ABAS Viewer or any other that can be developed by a third party, the designed societies can be saved and loaded as projects to enable running multiple design sessions for one system, even if no platform (ABAS Viewer) is connected.

The implementation of ABAS WorkBench is divided into four packages, as shown in Figure 11.11(b). The Emulated-Actor package contains all classes used to represent an actor in the ABAS WorkBench emulation environment. In general, an emulated actor is an instance of the class AWBActorEmu inherited by definition from Actor Blueprint class (CActor), which is detailed further in this research work. Emulated actors exist in a simulated space, typically provided by ABAS Viewer, but implement the full behavioural characteristics of real actors that later exists in the physical space.

The Actor-Prototype-Editor package is composed of classes that are used for modelling the physical attributes of actor and product prototypes. These classes provide graphic user interfaces for introducing, erasing and modifying actor attributes, such as PD, WD, *etc.* This package makes use of the 3D-Interface package for representing the actor attributes in tree-dimensional view on the screen in order to have a better understanding of the actor model.

The Society-Editor package contains the classes used to provide the necessary tools for editing and visualising actor societies. This package makes use of the 3D-Interface package in order to show in three-dimensional view the emulated actor societies. The society editor assists the ABAS designer to have a better understanding and control of the functionality of the system such as type of connections between actors, their position, and orientation related to other actors, among other properties.

The 3D-Interface package, as already mentioned, is used by the other packages. This package has classes that belong to the subset of classes of Direct3D software development kit from Microsoft; these classes access the hardware resources in order to create three-dimensional views.

11.4.2 ABAS Viewer

The ABAS Viewer incorporates several features required for deploying ABAS. The main goal of this tool is to provide a runtime platform for actors and other architecture elements, as well as to visualise and monitor the deployed ABAS. A key implemented element is the register, which is utilised by all architecture elements including actors and recruiters. In order to facilitate deployment, required algorithms, such as recruiting, clustering, broadcasting and loading assembly processes, have been incorporated, even though these functions could also be implemented externally if desired. ABAS Viewer provides all application logic needed for actor societies to operate and perform assembly processes.

Figure 11.12(a) shows the use case diagram of ABAS Viewer where not only the user interacts with the system but also the actors, whether emulated (created from ABAS WorkBench) or real. The designer/operator is able to visualise ABAS using a similar Direct3D-based interface, which gathers for visualisation all actor attributes that are received by the register when certain actor joins the society. The designer/operator can also edit and load assembly processes, as well as start the execution of them by commanding the internal recruiters. The actor, on the other hand, uses the ABAS Viewer to register and to request for communication channels (message transporters) according to FIPA standards. The ABAS Viewer register corresponds to the FIPA agent-directory-service.

The ABAS Viewer is divided into four packages as shown in Figure 11.12(b). The Society-Visualisation package contains the classes needed to visualise societies, whether emulated (created by ABAS WorkBench) or real. Among other properties, the Society-Visualisation package has classes for the register creation and GUIs. The Actor-Mirror package has the necessary classes used to store the attributes received from an actor when being registered, mostly used for visualisation. The Assembly-Process-Editor package has the GUIs needed to define assembly processes in the system; this package also has the classes for internal recruiter creation that implement the algorithms for cluster creation and execution.

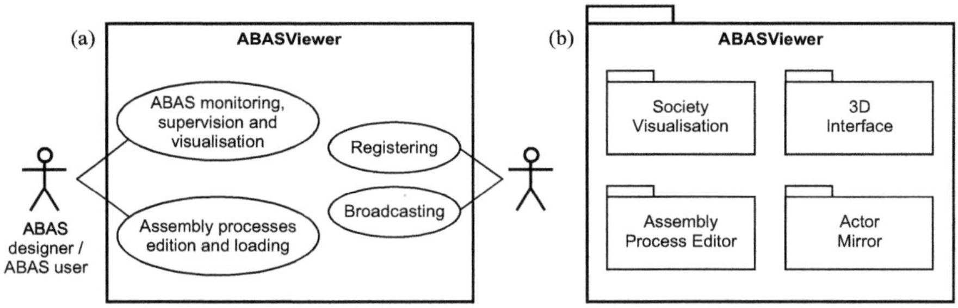

Figure 11.12. ABAS Viewer: (a) use case diagram; and (b) the main software packages

11.4.3 Actor Blueprint

Given that certain elements of behaviour are common to all actors, a reusable software component that encapsulates this common behaviour was developed. This component can be used as the foundation to build actors, be they physical actors or emulated actors. Using architecture terms, this component is named Actor Blueprint.

The programming language that was chosen for the development of the Actor Blueprint was C++. This language provides a model of memory and computation that closely matches that of most computers. In addition, it provides powerful and flexible mechanism for abstractions, that is, language construct that allow the programmer to introduce and use new types of objects that match the concepts of an application. Programmers can leverage the Actor Blueprint simply by creating a class that extends on it.

The common aspects of all actors are the previously-described features (WD, PD, *etc.*) as well as the agent interaction protocols. The Actor Blueprint provides a placeholder for these attributes, and an implementation of the interaction protocols. These protocols can be used to query particular values of the common attributes, as well as to perform commands. The Actor Blueprint can also be used for entities that do not perform any assembly operation but still need to interact with actors, such as recruiters.

Different actors perform different assembly operations, thus it is expected that their resources (sensing, actuating and communication resources) are different. It is the responsibility of the actor designer to implement this part of the behaviour of an actor and to define its attributes. Even when considering actors that perform the same basic operation, the underlying technology used to achieve that functionality can be different. Thus, the implementation of actuating, sensing and computational resources can vary greatly. Still, the reuse of the Actor Blueprint enables the rapid prototyping of actors.

11.5 Case Studies

Two experiments were carried out for proof of concept. Insertion and screwing were selected as the joining processes for the experiments, as they are representative operations in the application domain. In order to carry out these experiments, it was necessary to design and develop six different actor prototypes. Two of them handle the assemblies and products; the remaining four actors are dedicated to executing operations related to handling parts and to executing the joining processes.

11.5.1 Experimental Development of Actor Prototypes

The designed test bed consists of eight actors building four ABAS transporter systems, allowing translational motion of the container in a closed circuit. Two actor prototypes of servo-translator type dedicated to the translation of the part are mounted on an *X*-*Y* composition providing 2 DOF (degrees of freedom). Another translational prototype, this time of pneumatic double-action type, provides motion along the normal of the plane defined by the two servo-translator actors, thus providing one DOF in *Z*. In order to execute the insertion process, the equipment for grasping is the actor prototype that applies force required for insertion and is attached to the actor commanding motion in *Z*. In order to execute the screwing process, the application of a torque by means of a rotational motion along the screwing axis is required. Therefore, the society is populated by adding the torque and force applicator prototype and removing the force applicator. The physical implementation of the prototype is illustrated in Figure 11.13.

Initially, the actors were prototyped in ABAS WorkBench and emulated in order to validate the design of the actors, and the suitability of the society to perform the insertion and screwing processes. The actors were designed using one of the editor

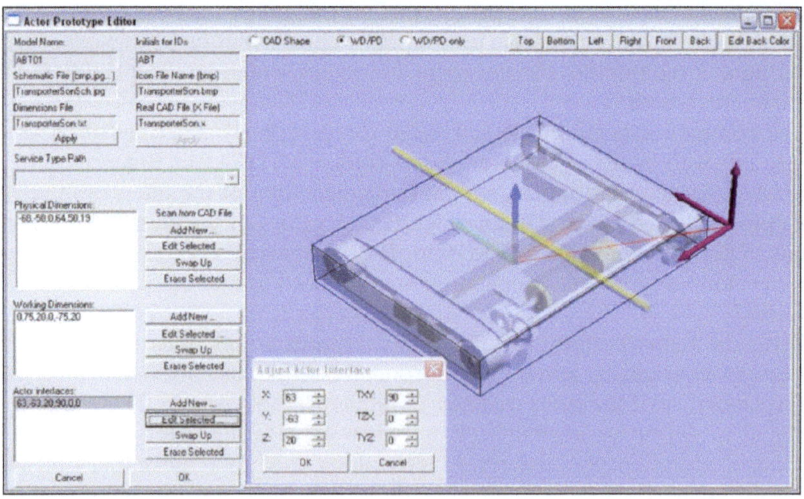

Figure 11.13. Actor prototype editor GUI, illustrating among other attributes the physical dimensions, working dimensions and actor interface

GUIs provided by ABAS WorkBench, as illustrated in Figure 11.13. Within this GUI, attributes such as WD and PD are edited and visualised. In order to enhance visualisation, a detailed visual design can be imported from a CAD file. The emulated agent behaviour is inherited from the Actor Blueprint.

After having defined all emulated actor prototypes, the actor society is constructed. In order to accomplish this objective, the prototypes are dragged into the main view of the ABAS WorkBench, and subsequently are snapped together. The snapping feature is implemented by utilising the Actor Interface attributes. A partial construction is illustrated in Figure 11.14, where servo-translation actors are attached in order to create a Cartesian geometry robot arm.

The emulated society was deployed in the ABAS Viewer platform. The emulated actors autonomously reported to the register, and were visualised according to the registered data. Subsequently, the assembly tasks required for the insertion and screwing processes were introduced.

The internal recruiters then decomposed the tasks into the required assembly operations, and calculated the actor clusters that could perform the processes. The operator can manually command the recruiters to execute the process in system validation mode, or can be executed autonomously, as illustrated in Figure 11.15.

After the actor society was validated through system emulation, the physical prototype implementation was deployed. The behaviour of the physical prototype was exactly as anticipated during society emulation. However, the authors recognise that the emulation tool provides significant flexibility and the ease to experiment, correct and validate both individual actors and societies, especially if compared with working directly with physical actors and with no customised support tools.

11.5.2 Experimental Results and Future Directions

During the experiments, an important target was to observe the reaction of the system to the introduction of new actors or removal of current members (system population) during the execution of the experiments. Thus an important feature of the experiments was the recalculation of the clusters dedicated to retrieve or store the assembly (container paths). The experiments may be completed with a time analysis for the exchange of messages in case that performance issues need to be considered. The assembly operation identification process has been executed manually. Nevertheless, a methodology for automatically generating a list of assembly operations should be developed. Since the software prototypes implement a link with the database containing the CAD information for the pallet, assembly and part, it should not be difficult to accommodate this methodology. The planning of assembly activities in a scenario with multiple products and/or with multiple containers of the same product should be addressed.

However, the required modifications for handling these scenarios do not affect the architecture definition, and allow the optimisation rules to be modified during cluster selection processes. The software tools implement the described architecture in terms of generating an assembly activity for each of the required assembly processes, and providing an error message in the case that the current members of the society lack the required skills for performing that assembly activity. A remarkable new feature is the ability of the society to propose which actors could

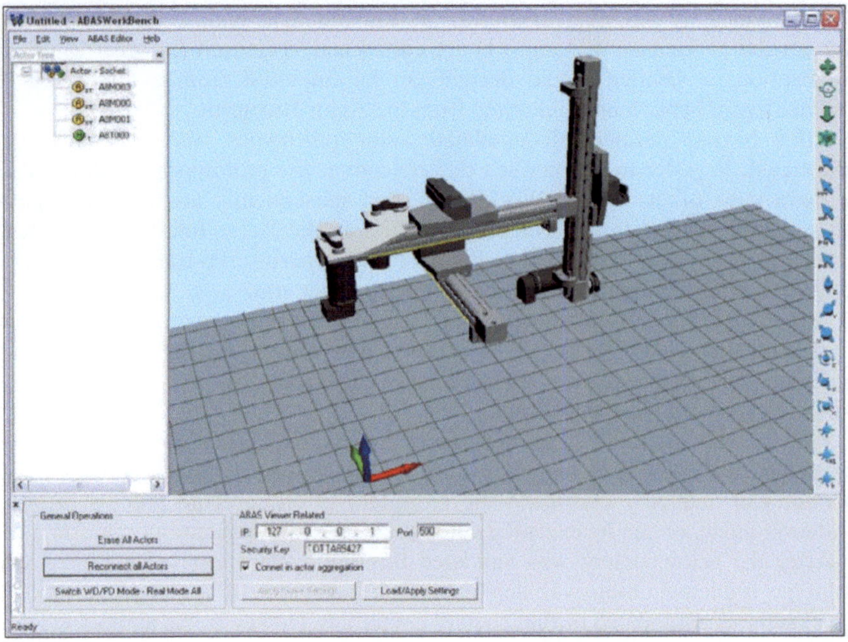

Figure 11.14. Actor society creation GUI, in this case illustrating a Cartesian geometry arm

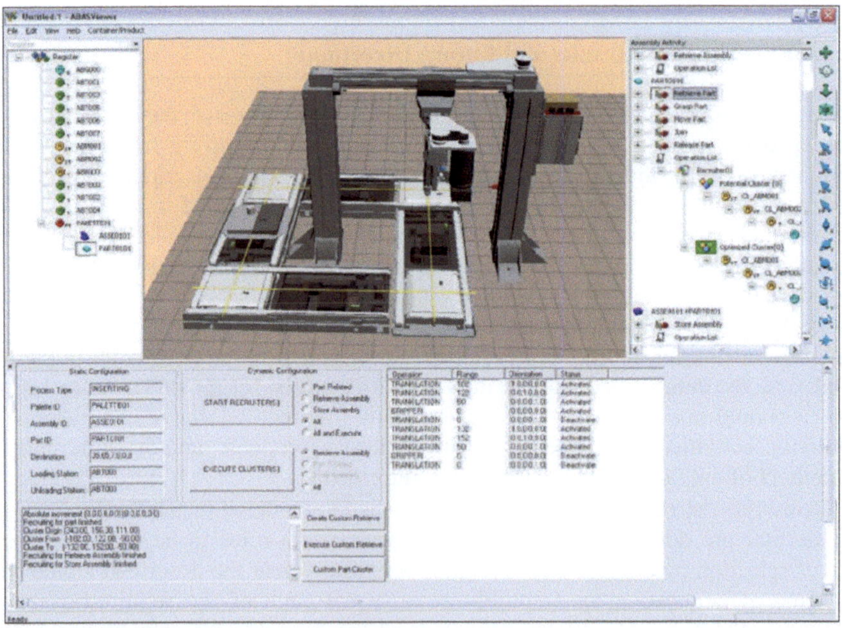

Figure 11.15. ABAS Viewer main GUI, illustrating the register, registered actor society, and assembly task input

perform the unfulfilled assembly activity. Towards this new feature, the upcoming version of the software prototypes implements algorithms for managing this situation in actor-based material-handling systems.

An important issue faced during the experiments was the accommodation of a commercial device that was not designed with the ABAS approach in mind, the pneumatic screwdriver dedicated to applying force and torque. The accommodation was successful and the prototype was integrated seamlessly. However, the methodology for dealing with legacy devices has not been discussed in this research. Therefore, the *agentification* of equipment that was not designed from an ABAS perspective seems to be an important research topic. Furthermore, the integration of humans as units in this type of assembly environments was not addressed.

11.6 Conclusions

Current manufacturing system trends for assembling small parts of changing products are leading to the micro-factory concept that mainly allow saving energy, space, material utilisation and other costs. Addressing this tendency, fundamental features of an agent-based approach and their implementation by utilising ABAS tools for developing micro-factory automation systems were proposed.

The use of ABAS architecture has been fundamental in guiding the design of the systems. This experience is highly encouraging for following architecture-based approaches in future work. Moreover, the use of technologies from other domains has helped enrich the usefulness of the tools. In particular, established techniques, *e.g.* 3D visualisation and GUI-based data entry widely used in many other domains, facilitate the use of the domain-specific tools. Thus, the use of the tools is accessible to a wide range of potential users, rather than a reduced set of tool experts.

Within the specific domain of micro-assembly system design, it has recognised the value of modelling and emulating the behaviour of intelligent physical agents in the design stages of new micro-assembly systems. Utilising this approach, the designer can focus on the configuration and control aspects of the design without making use of a mechanical setup. The configuration of the mechatronic devices can be done concurrently or even after the design has been validated.

Likewise, two real scale experiments were carried out to illustrate the ABAS concept. Insertion and screwing were selected as the joining processes for these experiments since they are representative scenarios within the application domain.

ABAS techniques have also been applied to bigger scale assembly systems with successful results, as for micro-assembly systems.

References

[11.1] Berguet, J., Schmitt, C. and Clavel, R., 2000, "Micro/Nanofactory: concept and state of the art," In *Proceedings of SPIE*, Boston, USA, November 5–6, pp. 1–11.

[11.2] Okazaki, Y., Mishima, N. and Ashida, K., 2004, "Microfactory – concept, history, and developments," *ASME Journal of Manufacturing Science and Engineering*, **126**, pp. 837–844.

[11.3] Tuokko, R., Lastra, J.L.M. and Kallio, P., 2000, "TOMI – a Finnish joint project towards mini and micro assembly factories," In *Proceedings of the Second*

International Workshop on Microfactories, Fribough, Switzerland, October 9–10, pp. 183–186.

[11.4] Pillay, R., Österholm, J. and Tuokko, R., 2002, *Sustainability Aspects and Scenarios for Mini Micro Assembly Factories*, Report 60, Institute of Production Engineering, Tampere University of Technology.

[11.5] Hirano, T., Nemoto, K. and Furuta, K, 2000, "Industrial impact of the microfactory," In *Proceedings of the Second International Workshop on Microfactories*, Fribough, Switzerland, October 9–10, pp. 35–38.

[11.6] National Research Council, 1998, *Visionary Manufacturing Challenges for 2020*, National Academy Press, Washington, DC.

[11.7] Colombo, A.W. and Martinez Lastra, J.L., 2004, "An approach to develop flexible & collaborative factory automation systems (FLEXCA)," In *Proceedings of the 4th CIRP International Seminar on Intelligent Computation in Manufacturing Engineering*, pp. 549–554.

[11.8] Harrison, R. and Colombo, A.W., 2005, "Collaborative automation from rigid coupling towards dynamic reconfigurable production systems," In *Proceedings of the 16th IFAC World Congress*.

[11.9] Martinez Lastra, J.L., 2004, *Reference Mechatronic Architecture for Actor-based Assembly Systems*, Doctoral Dissertation No. 484. Tampere University of Technology, ISBN 952-15-1210-5.

[11.10] Van Brussel, H., Wyns, J., Valckenaers, P., Bongaerts, L. and Peeters, P., 1998, "Reference architecture for holonic manufacturing systems: PROSA," *Computers in Industry*, **37**, pp. 255–274.

[11.11] Mařík, V., 2004, "Industrial application of the agent-based technology," In *Proceedings of 11th IFAC Symposium on Information Control Problems in Manufacturing*, Salvador do Bahia, Brazil.

[11.12] Valckenaers, P., 2003, "Tutorial on multi-agent manufacturing control," In *Proceedings of the First IEEE International Conference on Industrial Informatics*, Banff, Canada.

[11.13] Chung, K. and Wu, C., 1997, "Dynamic scheduling with intelligent agents: an application note," *Metra Application Note 105*, Metra, Palo Alto, CA.

[11.14] Overgaard, L., Petersen, H. and Perram, J., 1994, "Motion planning for an articulated robot: a multi-agent approach," In *Proceedings of Modelling Autonomous Agent in a Multi-Agent World*, Odense University, pp. 171–182.

[11.15] Koestler, A., 1968, *The Ghost in the Machine*, MacMillan.

[11.16] Holonic Manufacturing Systems, http://hms.ifw.uni-hannover.de/.

[11.17] An Adaptive Multi-agent Architecture for Advanced Manufacturing Systems, http://isg.enme.ucalgary.ca/research.htm.

[11.18] http://www.esinsa.unice.fr/etfa2001/Etfa-MFA/index.html.

[11.19] Colombo, A.W., Schoop, R. and Neubert, R., 2004, "Collaborative (agent-based) factory automation," In *The Industrial Information Technology Handbook*, Zurawski, R. (Ed.), CRC Press, Boca Raton.

[11.20] Harrison, R., Lee, S.M. and West, A.A., 2003, "Component-based distributed control systems for automotive manufacturing machines under the foresight vehicle program," *Transactions of Society of Automotive Engineers, Journal of Materials and Manufacturing*, **111**(5), pp. 218–226.

[11.21] Harrison, R., Lee, S.M. and West, A.A., 2004, "Lifecycle engineering of modular automated machines," In *Proceedings of the Second IEEE International Conference on Industrial Informatics*, Berlin, Germany, pp. 501–506.

[11.22] Leitão, P., Colombo, A.W. and Restivo, F., 2005, "ADACOR, a collaborative production automation and control architecture," *IEEE Intelligent Systems*, **20**(1), pp. 58–66.

[11.23] Neubert, R., Colombo, A.W. and Schoop, R., 2001, "An approach to integrate a multiagent-based production controller into a holonic enterprise platform," In *Proceedings of the First IEEE International Conference on Information Technology in Mechatronics (ITM001)*, Istanbul, Turkey.

[11.24] Simulation, Holonic Control System Driving AGVs, Holonic Manufacturing Systems website, http://hms.ifw.uni-hannover.de.

[11.25] Lopez Torres, E., 2004, "Multi agent-based configuration and visualization tools for ABAS," *Master of Science Thesis*, Tampere University of Technology.

[11.26] Deutsches Institute für Normung e.V. DIN 32561, 2000, *Production Equipment for Microsystems –Tray –Dimensions and Tolerances*, (in German).

12

Information Sharing in Digital Manufacturing Based on STEP and XML

Xiaoli Qiu[1] and Xun Xu[2]

[1] Department of Mechanical Engineering, Southeast University
Nanjing, Jiangsu, China 210096
Email: qiuxiaoli@seu.edu.cn

[2] Department of Mechanical Engineering, School of Engineering, University of Auckland
Private Bag 92019, New Zealand
Email: x.xu@auckland.ac.nz

Abstract
Information sharing and information management over the Internet is the key to success in today's digital manufacturing world, where different design and manufacturing applications with heterogeneous data formats often make up a common working environment. The additional requirement imposed on any ''neutral data format'' such as STEP (Standard for the Exchange of Product data), has been a Web-enabled data representation. STEP allows dynamic sharing of data between different systems with the standard data accessing interfaces. This chapter describes a method for STEP and XML to be combined in presenting product information. EXPRESS language (*i.e.* SCHEMA) is used for defining the data structure and DTD is used for XML transactions and presentations. Technologies on integrating STEP with XML have been discussed. A prototype system has been developed making use of STEP and XML as the data modelling and presentation tools.

12.1 Introduction

Today, companies often have operations distributed around the world, and production facilities and designers are often in different locations. The increased use of outsourcing and supply chains further complicates the manufacturing world. Globalisation of manufacturing business means that companies should be able to design anywhere, build anywhere and maintain anywhere at any time.

Manufacturing engineers can also employ collaborative tools during planning to help improve production processes, plant designs and tooling, and to allow earlier impact on product designs. Collaboration can be used for a number of activities such as (a) reviewing designs and changing orders with the design team; (b) interfacing with tooling designers; (c) verifying tooling assembly and operation; (d) reviewing manufacturing process plans and factory layouts; (e) discussing manufacturing problems with suppliers; and (f) co-ordinating tooling among dispersed sites. In larger companies, collaboration is becoming increasingly important in design and manufacturing. Everyone knows something, but no one knows everything. There is an evolution from individuals working independently to functioning in workgroups, as well as enterprise collaboration and collaboration throughout a supply chain. Within a supply chain, sharing knowledge becomes paramount.

This chapter describes a prototype system; its major functionality is to support digital manufacturing. The main goal is to provide a team environment enabling a group of designers and engineers to collaboratively develop a product in real time. STEP [12.1] and XML [12.2] are used to represent product data for heterogeneous application systems and data formats. In a nutshell, STEP is used to define a neutral data format across the entire product development process, and this neutral data is made available to the users over the Internet as well as in an Intranet (Figure 12.1) with the help of the XML file format.

Figure 12.1. Neutral translators

12.2 STEP as a Neutral Product Data Format

STEP is intended to support data exchange, data sharing and data archiving. For data exchange, STEP defines the form of the product data that is to be transferred between applications [12.3–12.5]. Each application holds its own copy of the product data in its own preferred form. The data conforming to STEP is transitory and defined only for the purposes of exchange. STEP supports data sharing by providing access to and operation on a single copy of the same product data by more than one application, potentially simultaneously. STEP is also suitable to support the interface to an archive. As in product data sharing, the architectural elements of STEP may be used to support the development of the archived product data itself. Archiving requires that the data conforming to STEP for exchange purposes is kept for use at some other time. This subsequent use may be through either product data exchange or product data sharing [12.6].

Another primary concept contributing to the STEP architecture is that the content of the standard is to be completely driven by industrial requirements. This, in combination with the concept that the re-use of data specifications is the basis for standards, led to developing two distinct types of data specifications. The first type entails a set of reusable, context-independent specifications. They are the building blocks of the standard. The second type contains application-context-dependent specifications and exists in form of application protocols (APs). This combination enables avoiding unnecessary duplication of data specifications between application protocols.

12.2.1 Components of STEP

The architectural components of STEP are reflected in the decomposition of the standard into several series of parts. Each part series contains one or more types of ISO 10303 parts. Figure 12.2 provides an overview of the structure of the STEP documentation.

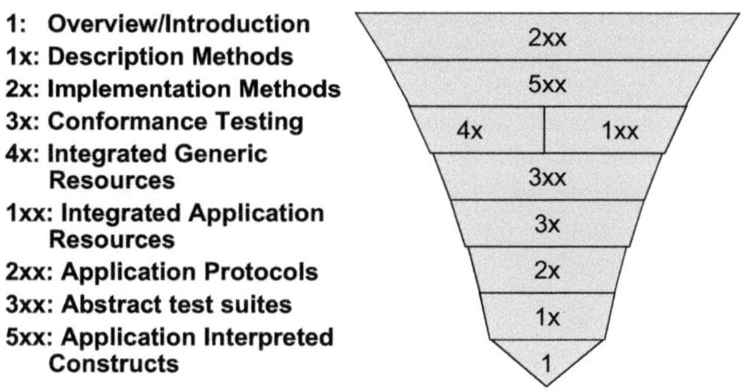

Figure 12.2. STEP document architecture

Description Methods

The first major architectural component is the description method series. Description methods are common mechanisms for specifying the data constructs of STEP. They include the formal data specification language developed for STEP, known as EXPRESS [12.7]. Other description methods include a graphical form of EXPRESS (EXPRESS-G) [12.8], a form for instantiating EXPRESS models, and a mapping language for EXPRESS. Description methods are standardised in the ISO 10303-10 series of parts (Figure 12.2).

Implementation Methods

The second major architectural component of STEP is the implementation method series. Implementation methods are standard implementation techniques for the information structures specified by the only STEP data specifications intended for implementation, application protocols. Each STEP implementation method defines the way in which the data constructs, specified using STEP description methods, are mapped to that implementation method. There are several implementation technologies available:

- A product model specific file format called Part 21 physical file [12.9].
- A variety of programming language bindings that allow an application programmer to open a data set and access values in its entity instances. Bindings have been developed for C, C++ and Java [12.10–12.12].
- The three methods for mapping EXPRESS defined data into XML described by Part 28 [12.13, 12.14].

STEP Part 21 is the first implementation method, which defines the basic rules of storing EXPRESS/STEP data in a character-based physical file. Its aim is to provide a method so that it is possible to write EXPRESS/STEP entities and transmit those entities using normal networking and communication protocols (*i.e.* FTP (File Transfer Protocol), e-mail and HTTP (Hyper Text Transfer Protocol)). A Part 21 file does not have any EXPRESS schemas included. It only defines the relationships between entities that are defined by external EXPRESS schemas. The Part 21 file format uses the minimalist style that was popular before the advent of XML. In this style the same information is never written twice so that there is no possibility of any contradictions in the data. The style assumes that normally the data will only be processed by software that people will only look at the data to create test examples or find bugs, and that making the data more easily readable by these people is less important than eliminating redundancies.

STEP data access interface (SDAI) reduces the costs of managing integrated product data by making complex engineering applications portable across data implementations. Currently, four international standards have been established for SDAI:

- Standard data access interface.
- C++ language binding to the standard data access interface.
- C language binding of standard data access interface.
- Java™ programming language binding to the standard data access interface with Internet/Intranet extensions.

Each standard defines a specific way of binding the EXPRESS data with a particular computer programming language. Binding is a terminology given to an algorithm for mapping constructs from the source language to the counterparts of another. Generally speaking, the binding defined in SDAI can be classified into early and late binding. The difference between them is whether the EXRESS data dictionary is available to the software applications. There is no data dictionary in an early binding, whereas in a late binding, the EXPRESS schema definition is needed by late binding applications at run-time.

The early binding approach generates specific data structure according to the EXPRESS schemas and the programming language definitions. The entities defined in EXPRESS schemas are converted to C++ or Java classes. The inheritance properties in the EXPRESS schemas are also preserved in those classes. The advantage of an early binding is that the compiler of the programming language can perform additional type checking. But because of the complexities of EXPRESS schemas, the initial preparation, compiling and link of an early binding approach can be time-consuming. The late binding approach, on the other hand, does not map EXPRESS entities into classes. It uses EXPRESS entity dictionaries for accessing data. Data values are found by querying those EXPRESS entity dictionaries. Only a few simple functions need to be defined in the late binding approach to get or set values. A late binding is simpler than an early binding approach because there is no need to generate the corresponding classes. However, the lack of type checking means that the late binding approach is not suitable for large systems.

XML consists of different rules for defining semantic tags that breaks a document into parts and identifies the different parts of the document. Furthermore, it is a meta-markup language that defines a syntax in which other field-specific markup languages can be written [12.2, 12.10]. Essentially, XML defines a character-based document format. XML is flexible because there is no restriction to those tag names. Hence, it is possible to assign more human-understandable tag names in an XML document, while computers just interpret an XML document according to a pre-defined formula. It is obvious that the use of meaningful tags can make an XML document human-understandable as well as computer-interpretable.

When representing EXPRESS schemas, Part 28 [12.13, 12.14] specifies an XML markup declaration set based on the syntax of the EXPRESS language. EXPRESS text representation of schemas is also supported. The markup declaration sets are intended as formal specifications for the appearance of markup in conforming XML documents. These declarations may appear as part of document type definitions (DTDs) for such documents.

Like the method used in SDAI, STEP Part 28 defined two broad approaches for representation of data corresponding to an EXPRESS schema. One approach is to specify a single markup declaration set that is independent of the EXPRESS schema and can represent data of any schema. This approach is called XML late binding. The second approach is to specify the results of the generation of a markup declaration set that is dependent on the EXPRESS schema. This approach is called XML early binding. STEP Part 28 defines one late binding approach and two early binding approaches.

Conformance Testing

The conformance testing methodology and framework series provide an explicit framework for conformance and other types of testing as an integral part of the standard. This methodology describes how testing of implementations of various STEP parts is accomplished. The fact that the framework and methodology for conformance testing is standardised reflects the importance of testing and testability within STEP. Conformance testing methods are standardised in the ISO 10303-30 series of parts.

An abstract test suite contains the set of abstract test cases necessary for conformance testing of an implementation of a STEP application protocol. Each abstract test case specifies input data to be provided to the implementation under test, along with information on how to assess the capabilities of the implementation. Abstract test suites enable the development of good processors and encourage expectations of trouble-free exchange.

Data Specifications

The final major component of the STEP architecture is the data specifications. There are four part series of data specifications in the STEP documentation structure, though conceptually there are three primary types of data specifications: integrated resources, application protocols, and application interpreted constructs. All of the data specifications are documented using the description methods. Most application-relevant of all is perhaps the APs, which are the implementable data specifications of STEP. APs include an EXPRESS information model that satisfies the specific product data needs of a given application context. APs may be implemented using one or more of the implementation methods. They are the central component of the STEP architecture, and the STEP architecture is designed primarily to support and facilitate developing APs. The first implemented and also most widely used AP is AP203 file [12.15].

12.3 XML as the "Information Carrier"

XML is the universal format for data on the web. It is developed by the World-Wide Web Consortium (W3C) from 1996 and has been a W3C standard since February 1998. XML is a set of rules, guidelines, or conventions for designing text formats for data such that the files are easy to generate and read, they are unambiguous, and they are extensible.

XML is a flexible solution to publish documents on the web (serves many other purposes as well). It does not define elements but lets the developer define the structure needed. XML editors can accept any document structure. There are very little opportunities to optimise the XML editors because, by definition, they must be as generic as XML is.

There is a potential conflict between flexibility and ease of use. As a rule, more flexible solutions are more difficult. Specific solutions might also be optimised for certain tasks. The DTD is an attempt to bridge that gap. DTD is a formal description of the document. Software tools can read it and learn about the document structure. Consequently, the tools can adapt themselves to better support the document structure.

DTD is a mechanism to describe the structure of documents. It defines the constraints on the structure of an XML document, and declares all of the document's element types, children element types, and the order and number of each element type. It also declares any attributes, entities, notations, processing instructions, and comments in the document. A document can use both internal and external DTD subsets. That is, a DTD can be declared inline in an XML document, or as an external reference.

The purpose of a DTD is to define the legal building blocks of an XML document. It defines the document structure with a list of legal elements. With DTD, the XML files can carry a description of its own format with it. With a DTD, independent groups of people can agree to use a common DTD for interchanging data. Applications can use a standard DTD to verify that the data received from the outside world is valid.

12.3.1 Development and Application Domain of XML

The four main characteristics of XML, fine data storage format, extensibility, structural, easily transfer on web, give it remarkable performance. There are also numerous XML applications.

XML can support interactive data exchange between different sources. Data may have different complex formats, for it may come from different databases. However, the characteristics of XML, *i.e.* self-definition and extension, make it possible to express almost all kinds of data. With the received data, customers can process or transfer the data between the different databases. In this kind of application, XML gives data a unified interface. Unlike other data transmission standards, XML does not define any specific standard for the data in data files. Instead, XML adds tag in data to express the logical structure and implication of data. In doing so, XML becomes a standard that can be understood by the program itself.

XML can help utilise dispersed resources. This is the situation where a source of dispersed computing power at different client site is to be utilised to carry out various calculations at the request of a task. This task can be coded in XML and sent out from the server to the clients with ease.

With the traditional client-server mode, the server responds to customer's request. Thus, the server's load is often great, and the web managers have to investigate each customer's requirement to make a corresponding program. When the customer's requirements become diverse and changeable, the programmers at the server end may not have time to meet the numerous requests in good time. XML turns the initiative of processing data to the clients, so that the server only needs to package the data as complete and accurate as possible into an XML file. XML's characteristic of self-interpretation helps the client understand the logical structure and the meaning of data upon receipt.

XML is flexible also in that it allows for the same data to be presented in different style often over the Internet. XML data can easily be tailored down to only provide a sub-set of the data. This is useful when the client-end has a need to customise the user interface for different internal users.

12.3.2 EXPRESS-XML DTD Binding Methods

Using XML DTD, the following contents can be coded,

- One or more EXPRESS schema
- One or more data group, each corresponding to an EXPRESS schema
- Combination of EXPRESS schema and its corresponding data

For each data group, it is necessary to determine its corresponding EXPRESS schema. EXPRESS schema, however, does not need to be converted to XML. In this system, late binding DTD is designed to give a basic structure for defining an early binding DTD, so that the early binding DTD can be accessed according to the late binding DTD. The late binding DTD makes use of some of the specification items of the early binding DTD. The basic principle of the late binding format is that one DTD can correspond to more than one EXPRESS schema. Different data blocks can be included in one single file while each data block conforms to its own specific schema.

Late Binding XML DTD Element of EXPRESS

Based on the syntax of the EXPRESS language (ISO 10303-11) [12.7], the following XML element specification is defined. The XML element "?" defines the root of its structure.

```
<?xml version="1.0" encoding="UTF-8" ?>
<!DOCTYPE ISO-10303-data [<!ELEMENT ISO_10303_data
  (documentation?, (schema_decl | data)*)>]>
```

Every element here must be used to represent the EXPRESS structure, which is determined by some EXPRESS syntax. The name of the applied syntax is the name of the element, unless the representation method of the XML attribute has previously named others. In that case, the element must represent the structure of the syntax, and must be named by the corresponding attribute value of the representation method. The details of element are presented in ISO 10303 Part 28 [12.13]. The data group corresponding to the EXPRESS schema is coded by these elements. Moreover, it is necessary to identify the DTD element for the late binding XML represented by EXPRESS-driven data.

Early Binding DTD

The relationship of the tag element is established by XML through its inner structure. In the case of early binding DTD, it corresponds to the EXPRESS schema. Its structure, however, conforms to the structure of the late binding DTD. In order to employ some of the main operating steps to process EXPRESS schema in developing a series of XML specifications, the following prerequisites are to be observed,

- EXPRESS schema must have a correct syntax;
- The schema should not involve EXPRESS with "xml" as its identification mark.

12.4 A Digital Manufacturing Support System

There is no shortage of research work combining STEP with XML in various industries. Zhu *et al.* [12.16] developed a unified BOM method based on STEP and

XML standards. It is used to secure the product data's uniformity in a collaboration environment. Balakrishna *et al.* [12.17] developed an interface program to communicate CAD/CAM/CAE data in a client-server environment by converting STEP files into XML files. In order to semantically and schematically integrate distributed product information, Kim *et al.* [12.18] presented product metadata represented by an XML schema, which is compatible with the ISO STEP PDM (Product Data Management) schema standard. Chan *et al.* [12.19] proposed mediators as a middle layer that provides added value by converting the basic data into information required by the clients. They presented the ways of automated conversion from STEP into XML, and the exchange of STEP data through the XML-based mediator architecture. At the STEP-NC front, researchers have also been using XML as a viable data formats. Amaitik and Kilic [12.20] developed an intelligent process planning system using STEP features (STEP AP224) for prismatic parts. The system outputs a STEP-NC process plan in the XML format. Lee *et al.* [12.21] presented a five-axis STEP-NC milling machine that is run on STEP-NC in XML format.

The system described in this chapter supports digital manufacturing where diverse computer-aided applications are used. These applications, such as CATIA, Pro/Engineer, SolidWorks and ACIS, all have the capability of saving as, or loading, a STEP AP203 file [12.15]. Therefore, it is considered appropriate and practicable to use AP203 as the neutral data format.

12.4.1 System Architecture

In order to support web-based applications, STEP AP203 data are converted into XML files. Figure 12.3 shows the architecture of the system that supports STEP product data sharing and exchange using the XML format. Since it can support design activities with different CAD/CAM systems, collaborations on the Internet are enabled. In such a networked environment, any registered user can log onto the Internet to acquire product data that may be generated and posted by his/her fellow designers.

Different design partners may cooperate with each other to design different parts using their own preferred CAD software. As long as the design data can be saved as STEP AP203 files, the converter in the system can convert these STEP files into XML files for them to be web-ready. These XML files can reside on a web server for other designers to access. This system also helps engage customers' involvement and incorporate their requirements at the early stages of the product development.

12.4.2 Overview of the System

Based on the above suggested architecture, a digital manufacturing support system has been developed (Figure 12.4). It integrates the STEP and XML standards. In converting product data from STEP to XML, the late binding approach is used. This ensures the system's interchangeability and expendability. The DTD format gives the data readability and portability. SCHEMAs are used to check data conformity. Relevant database technologies are used to enhance data saving and data sharing.

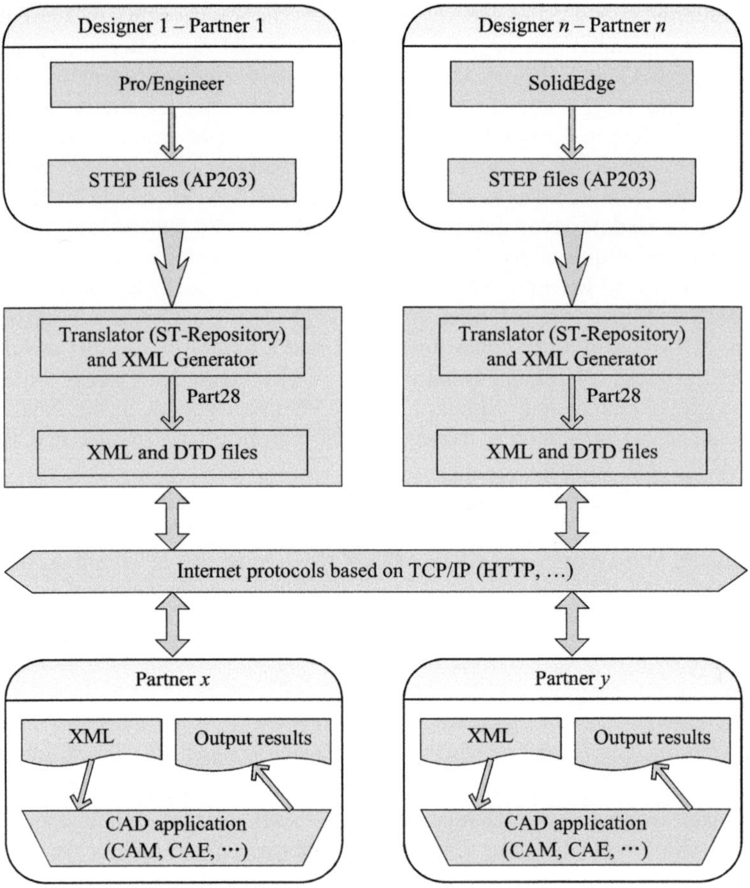

Figure 12.3. Architecture of the system

As shown in Figure 12.4, the system brings together upper-stream and down-stream activities through their corresponding interfaces by a converter in a networked database. At the upper-stream end, the STEP data are generated for any design model by any CAD/CAM system. The converter translates the data into the XML format. The XML files are then saved in the networked database and meanwhile made available on the Internet. The down-stream applications such as real-time dynamic control and ERP may also have access to the necessary data. Some of the main functions of the system are discussed in the next section.

12.4.3 System Functionality

Design data are normally saved as the STEP AP203 format by the proprietary CAD/CAM system. Some pre-processing may be needed before the data are parsed by the converter. Lexical analysis is the next step. The files are read into a buffer area, where every character in the file is scanned. Based on the punctuations (such as

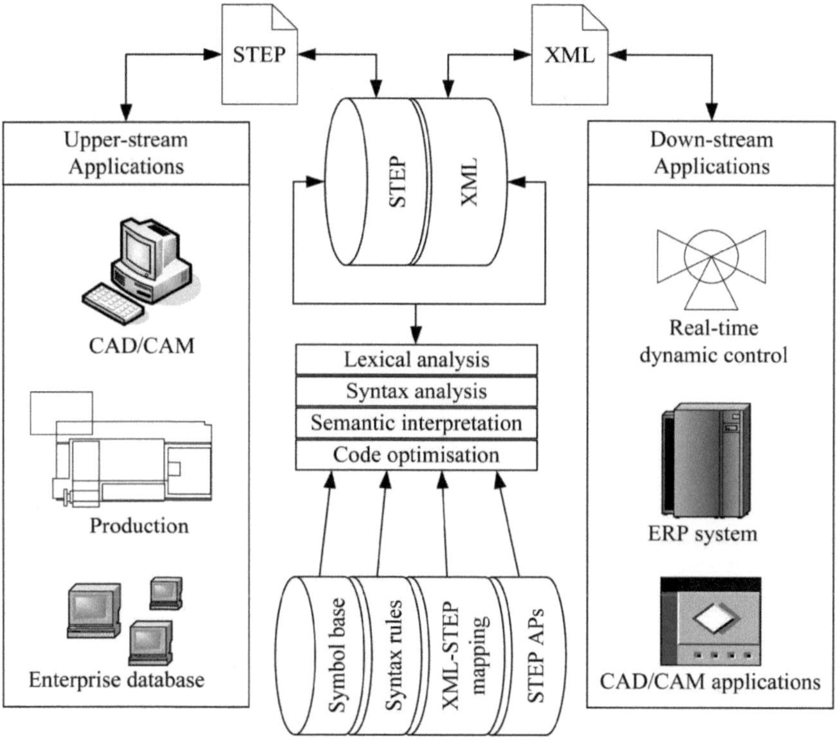

Figure 12.4. Overview of the system

Table 12.1. Keywords and attributes

No.	Keywords	Attributes
1	#1	ID
2	=	ENTITY_START
3	POINT	ENTITY_NAME
4	(	LIST_START
5	1.000000	CONTENT
6	,	SEPERATOR
7	1.000000	CONTENT
8	)	LIST_END
9	;	

space, comma, bracket, semicolon and single quotation marks) stipulated by the corresponding grammar, the file is divided into cellular elements for identification. Strings are identified to be either keywords or attributes. For example, Table 12.1 shows how the keywords and attributes are identified when the following line of a Part 21 file is read,

```
#1= POINT (1.000000, 1.000000);
```

Following the lexical analysis is syntax analysis. The syntax category corresponding to the keywords will be identified. At the same time, the syntax (grammar) tree is generated as shown in Figure 12.5.

By invoking the STEP-XML relationships mapping rules, the corresponding semantic symbols are replaced and the XML format is generated, *e.g.*

```
<entity id="#1" name="point">
  <real_value> 1.000000</real_value>
  <real_value> 1.000000</real_value>
</entity>
```

To enhance the file's readability, the system annotates the XML documents with the STEP terminology. In doing this, the XML document with special formats can be developed. The above example may be therefore changed to

```
<entity id="#1" name="point">
  <attribute name="x">
    <real_value> 1.000000</real_value>
  </attribute>
  <attribute name="y">
    <real_value> 1.000000</real_value>
  </attribute>
</entity>
```

The system keeps the XML data files in the networked database, and makes it available on the Web to support the down-stream applications. Figure 12.6 shows the structure of an XML file generated from a STEP file.

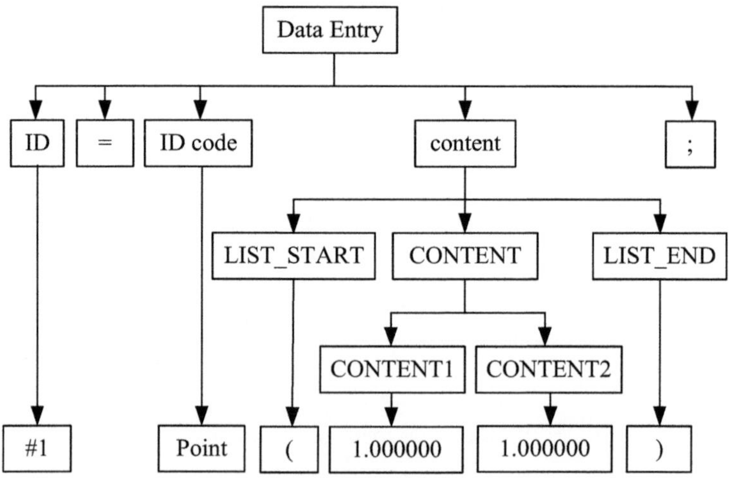

Figure 12.5. Structure of the grammar tree

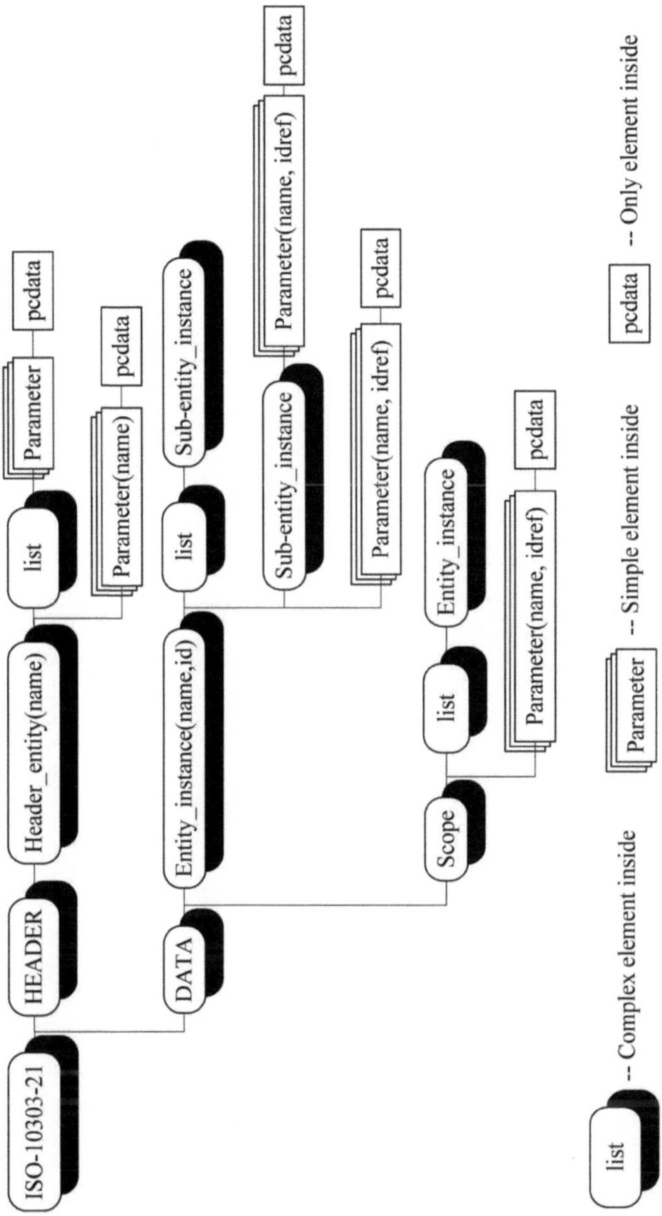

Figure 12.6. Structure of the XML output file

12.4.4 Converter

Figure 12.7 shows the flowchart of the program that converts STEP to XML, and Figure 12.8 shows the flowchart of the program that converts XML back to STEP. Converting a STEP file to an XML file is done by so-called "onion-peeling": *Scope-Entity-Sub-entity*. When converting an XML file back to a STEP file, the key is the detection of keywords, based on which conversions can be carried out in one of the four possible ways, *Special keyword output*, *Scope*, *Header entry* and *Data entry*. Noticeably, the converter is bidirectional.

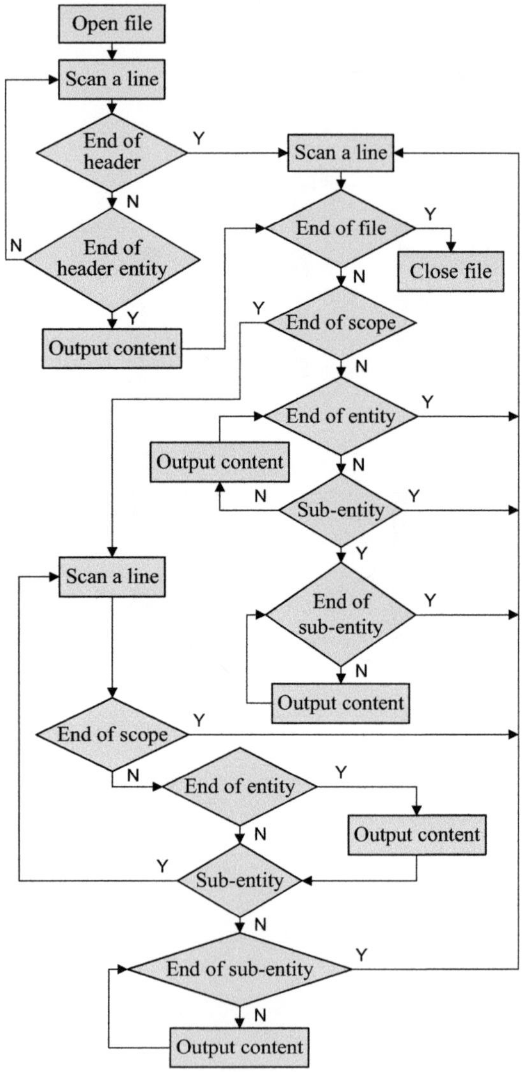

Figure 12.7. Converting STEP to XML

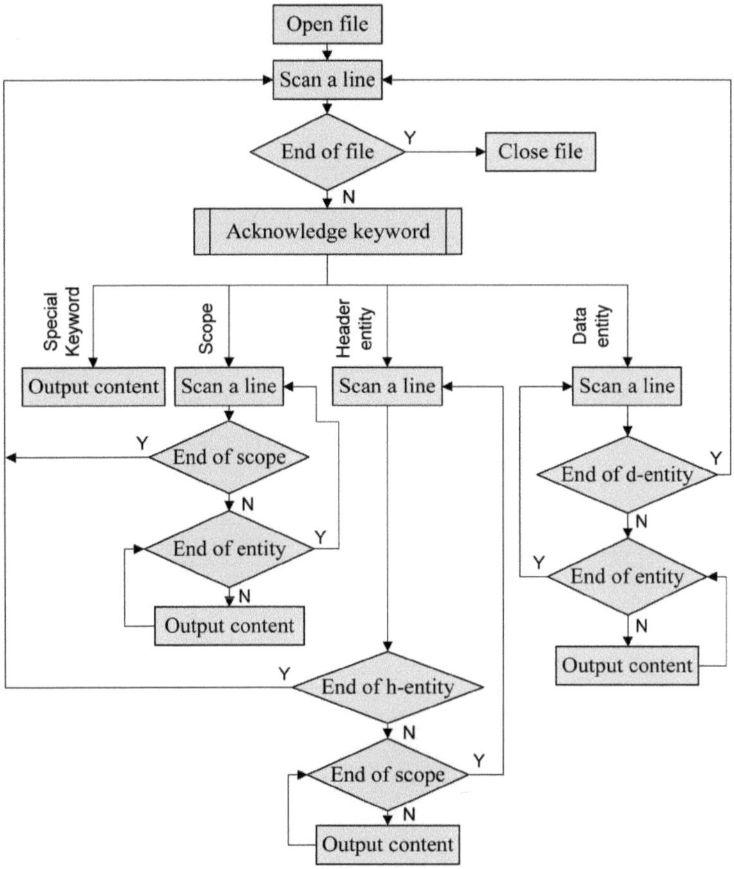

Figure 12.8. Converting XML to STEP

12.4.5 Late Binding Rules

Given an EXPRESS schema, its corresponding data set is defined in the system. The DTD elements defined by the late binding rules can then be carried out. The appendix at the end of this chapter shows the bindings for some of the common data entities, for example, array_literal element, attribute_instance element, author element, authorisation element, bag_literal element, binary_literal element, constant_instances element, and *etc.*

12.4.6 System Interface

The interface of the system (Figure 12.9) has a number of functions. It can load and browse both STEP and XML files, which are displayed in the windows to the right. The interface can also view an XML file in an Internet Explorer window (not shown in the figure) as well as display parts drawings. The main function of the interface is however performing conversions between STEP and XML files. Once a STEP file is converted to an XML file, it can be uploaded to a server or other online systems, so that the data can be shared across the Internet or Intranet.

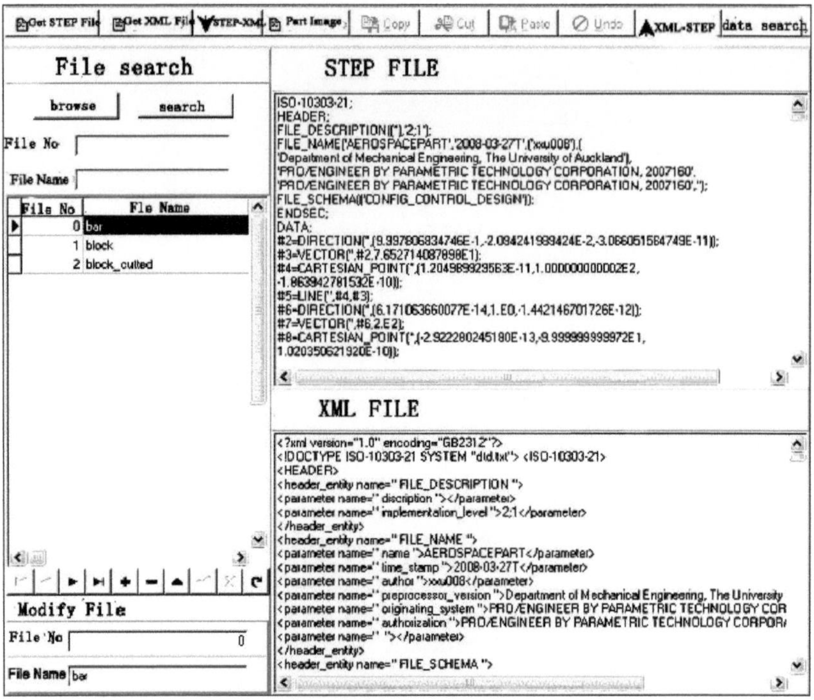

Figure 12.9. The interface of the system

Figure 12.10. An aerospace part created by Pro/Engineer

Take a simple example as shown in Figure 12.10, a test-piece for implementing STEP AP238 in the aerospace industry. It contains some typical features of aircraft components. A Pro/Engineer model has been created and given a name of “AeroSpacePart.prt”. Pro/Engineer has its built-in STEP translator, which is used to convert the model into STEP AP 203 format, named “AeroSpacePart.stp”. This file is then converted to an XML file using the system developed in this research. Figures 12.11 and 12.12 show the beginning parts of the two files.

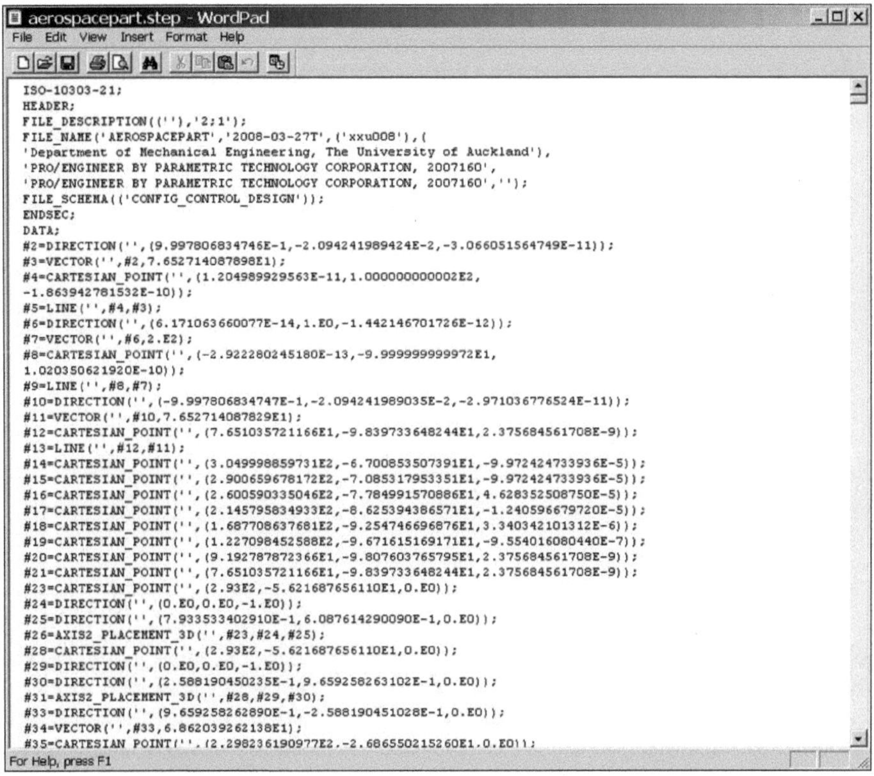


```
ISO-10303-21;
HEADER;
FILE_DESCRIPTION((''),'2;1');
FILE_NAME('AEROSPACEPART','2008-03-27T',('xxu008'),(
'Department of Mechanical Engineering, The University of Auckland'),
'PRO/ENGINEER BY PARAMETRIC TECHNOLOGY CORPORATION, 2007160',
'PRO/ENGINEER BY PARAMETRIC TECHNOLOGY CORPORATION, 2007160','');
FILE_SCHEMA(('CONFIG_CONTROL_DESIGN'));
ENDSEC;
DATA;
#2=DIRECTION('',(9.997806834746E-1,-2.094241989424E-2,-3.066051564749E-11));
#3=VECTOR('',#2,7.652714087898E1);
#4=CARTESIAN_POINT('',(1.204989929563E-11,1.000000000002E2,
-1.863942781532E-10));
#5=LINE('',#4,#3);
#6=DIRECTION('',(6.171063660077E-14,1.E0,-1.442146701726E-12));
#7=VECTOR('',#6,2.E2);
#8=CARTESIAN_POINT('',(-2.922280245180E-13,-9.999999999972E1,
1.020350621920E-10));
#9=LINE('',#8,#7);
#10=DIRECTION('',(-9.997806834747E-1,-2.094241989035E-2,-2.971036776524E-11));
#11=VECTOR('',#10,7.652714087829E1);
#12=CARTESIAN_POINT('',(7.651035721166E1,-9.839733648244E1,2.375684561708E-9));
#13=LINE('',#12,#11);
#14=CARTESIAN_POINT('',(3.049998859731E2,-6.700853507391E1,-9.972424733936E-5));
#15=CARTESIAN_POINT('',(2.900659678172E2,-7.085317953351E1,-9.972424733936E-5));
#16=CARTESIAN_POINT('',(2.600590335046E2,-7.784991570886E1,4.628352508750E-5));
#17=CARTESIAN_POINT('',(2.145795834933E2,-8.625394386571E1,-1.240596679720E-5));
#18=CARTESIAN_POINT('',(1.687708637681E2,-9.254746696876E1,3.340342101312E-6));
#19=CARTESIAN_POINT('',(1.227098452588E2,-9.671615169171E1,-9.554016080440E-7));
#20=CARTESIAN_POINT('',(9.192787872366E1,-9.807603765795E1,2.375684561708E-9));
#21=CARTESIAN_POINT('',(7.651035721166E1,-9.839733648244E1,2.375684561708E-9));
#23=CARTESIAN_POINT('',(2.93E2,-5.621687656110E1,0.E0));
#24=DIRECTION('',(0.E0,0.E0,-1.E0));
#25=DIRECTION('',(7.933533402910E-1,6.087614290090E-1,0.E0));
#26=AXIS2_PLACEMENT_3D('',#23,#24,#25);
#28=CARTESIAN_POINT('',(2.93E2,-5.621687656110E1,0.E0));
#29=DIRECTION('',(0.E0,0.E0,-1.E0));
#30=DIRECTION('',(2.588190450235E-1,9.659258263102E-1,0.E0));
#31=AXIS2_PLACEMENT_3D('',#28,#29,#30);
#33=DIRECTION('',(9.659258262890E-1,-2.588190451028E-1,0.E0));
#34=VECTOR('',#33,6.862039262138E1);
#35=CARTESIAN_POINT('',(2.298236190977E2,-2.686550215260E1,0.E0));
```

Figure 12.11. An aerospace part in STEP AP203 format

12.5 Conclusions

As product developers collaborate over the Internet, different CAD/CAM systems can be used. To enable true collaboration, a neutral data format is a must. STEP emerged as such a standard. In spite of its capability, the conventional implementation methods of STEP (*e.g.* EXPRESS and Part 21) come short of supporting Internet-based communications. To address this problem, XML has been regarded as a popular and effective data format. In fact, EXPRESS or SCHEMA for STEP is familiar to DTD for XML. Therefore, STEP and XML are being regarded as complementary technologies. One of the latest developments in STEP is that of

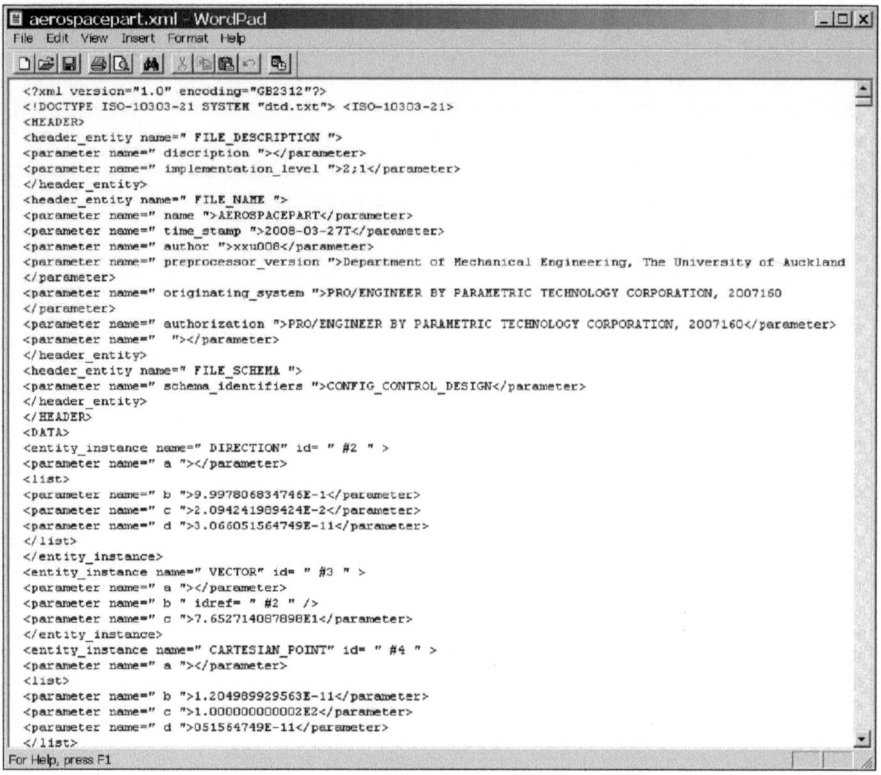


```
<?xml version="1.0" encoding="GB2312"?>
<!DOCTYPE ISO-10303-21 SYSTEM "dtd.txt"> <ISO-10303-21>
<HEADER>
<header_entity name=" FILE_DESCRIPTION ">
<parameter name=" discription "></parameter>
<parameter name=" implementation_level ">2;1</parameter>
</header_entity>
<header_entity name=" FILE_NAME ">
<parameter name=" name ">AEROSPACEPART</parameter>
<parameter name=" time_stamp ">2008-03-27T</parameter>
<parameter name=" author ">xxu008</parameter>
<parameter name=" preprocessor_version ">Department of Mechanical Engineering, The University of Auckland
</parameter>
<parameter name=" originating_system ">PRO/ENGINEER BY PARAMETRIC TECHNOLOGY CORPORATION, 2007160
</parameter>
<parameter name=" authorization ">PRO/ENGINEER BY PARAMETRIC TECHNOLOGY CORPORATION, 2007160</parameter>
<parameter name="  "></parameter>
</header_entity>
<header_entity name=" FILE_SCHEMA ">
<parameter name=" schema_identifiers ">CONFIG_CONTROL_DESIGN</parameter>
</header_entity>
</HEADER>
<DATA>
<entity_instance name=" DIRECTION" id= " #2 " >
<parameter name=" a "></parameter>
<list>
<parameter name=" b ">9.997806834746E-1</parameter>
<parameter name=" c ">2.094241989424E-2</parameter>
<parameter name=" d ">3.066051564749E-11</parameter>
</list>
</entity_instance>
<entity_instance name=" VECTOR" id= " #3 " >
<parameter name=" a "></parameter>
<parameter name=" b " idref= " #2 " />
<parameter name=" c ">7.652714087898E1</parameter>
</entity_instance>
<entity_instance name=" CARTESIAN_POINT" id= " #4 " >
<parameter name=" a "></parameter>
<list>
<parameter name=" b ">1.204989929563E-11</parameter>
<parameter name=" c ">1.000000000002E2</parameter>
<parameter name=" d ">051564749E-11</parameter>
</list>
```

Figure 12.12. An aerospace part in XML format

using XML as the "data carrier" over the Internet. This new technology is rapidly becoming the preferred method for complex design data exchange over the Web. The flexibility and availability of commercial Web/XML tools greatly increases the acceptance, use and spread of STEP. The emerging XML and STEP implementation technologies already show great promise to enable collaborative product development, manufacturing interoperability, and the ultimate product lifecycle management paradigm.

This research has shown that when XML is used to represent STEP information, there can be a host of benefits such as,

- Existing software tools as well as previous research results can be utilised among the collaborators, reducing the time and cost of product development.
- The XML capability supports digital manufacturing.
- As an XML file can also be used as a data structure, there is no need to design or provide an additional internal data structure for data interpretation.

The prototype system developed and presented in this chapter demonstrated the integration of STEP and XML for digital manufacturing. It successfully demonstrated how an EXPRESS early binding XML can be used to capture the EXPRESS data model. The XML-based data is of a high-level as it contains a complete data model. This high-level product data model can support remote applications as well as transfer product data within and among collaborators.

As a key element in the prototype system, the developed STEP/XML converter can convert data in both ways, STEP-XML and XML-STEP. The example shows that the product information over the Internet can be integrated and shared for networked, digital manufacturing. To enhance its practicality, upper-stream and down-stream interfaces have been added. This system can therefore support remote monitoring, real-time control and virtual manufacturing.

References

[12.1] ISO 10303-1: 1994, *Industrial Automation Systems and Integration – Product Data Representation and Exchange – Part 1: Overview and Fundamental Principles*, ISO, Geneva, Switzerland.

[12.2] Marchal, B., 1999, *XML by Example*, Quebec, Canada.

[12.3] Xu, X. and Newman, S., 2006, "Making CNC machine tools more open, interoperable and intelligent," *Computers in Industry*, **57**(2), pp. 141–152.

[12.4] Xu, X., 2006, "Realisation of STEP-NC enabled machining," *Robotics and Computer-Integrated Manufacturing*, **22**(2), pp. 144–153.

[12.5] Xu, X., Wang, H., Mao, J., Newman, S.T., Kramer, T.R., Proctor, F.M. and Michaloski, J.L., 2005, "STEP-compliant NC research: the search for intelligent CAD/CAPP/CAM/CNC integration," *International Journal of Production Research*, **43**(17), pp. 3703–3743.

[12.6] Kemmerer, S. (ed.), 1999, *STEP – The Grand Experience*, NIST Special Publication 939, Gaithersburg, MD, USA.

[12.7] ISO 10303-11: 1994, *Industrial Automation Systems and Integration – Product Data Representation and Exchange – Part 11: Description Methods: The EXPRESS Language Reference Manual*, ISO, Geneva, Switzerland.

[12.8] ISO 10303-22: 1998, *Industrial Automation Systems and Integration – Product Data Representation and Exchange – Part 22: Implementation Methods: Standard Data Access Interface,* ISO, Geneva, Switzerland.

[12.9] ISO 10303-21: 1994, *Industrial Automation Systems and Integration – Product Data Representation and Exchange – Part 21: Implementation Methods: Clear Text Encoding of the Exchange Structure*, ISO, Geneva, Switzerland.

[12.10] ISO 10303-23: 2000, *Industrial Automation Systems and Integration – Product Data Representation and Exchange – Part 23: C++ Language Binding to the Standard Data Access Interface*, ISO, Geneva, Switzerland.

[12.11] ISO 10303-24: 2001, *Industrial Automation Systems and Integration – Product Data Representation and Exchange – Part 24: C Language Binding of Standard Data Access Interface*, ISO, Geneva, Switzerland.

[12.12] ISO 10303-27: 2000, *Industrial Automation Systems and Integration – Product Data Representation and Exchange – Part 27: Java™ Programming Language Binding to the Standard Data Access Interface with Internet/Intranet Extensions*, ISO, Geneva, Switzerland.

[12.13] ISO/CD TS 10303-28 (Edition 1): 2002, *Product Data Representation and Exchange: Implementation Methods: EXPRESS to XML Binding, Draft Technical Specification*, ISO TC184/SC4/WG11 N169, ISO, Geneva, Switzerland.

[12.14] ISO/TS 10303-28 (Edition 2): 2004, *Product Data Representation and Exchange: Implementation Methods: XML Schema Governed Representation of EXPRESS Schema Governed Data*, TC184/SC4/WG11 N223, ISO, Geneva, Switzerland.

[12.15] ISO 10303-203: 1994, *Industrial Automation Systems and Integration – Product Data Representation and Exchange – Part 203: Application Protocol: Configuration Controlled 3D Designs of Mechanical Parts and Assemblies*, ISO, Geneva, Switzerland.

[12.16] Zhu, S., Cheng, D., Xue, K. and Zhang, X., 2007, "A unified bill of material based on STEP/XML," *Lecture Notes in Computer Science (including subseries Lecture Notes in Artificial Intelligence and Lecture Notes in Bioinformatics)*, 4402 LNCS, pp. 267–276.

[12.17] Balakrishna, A., Babu, R.S., Rao, D.N., Raju, R.D. and Kolli, S., 2006, "Integration of CAD/CAM/CAE in product development system using STEP/XML," *Concurrent Engineering: Research and Applications*, **14**(2), pp. 121–128.

[12.18] Kim, H., Kim, H.-S., Lee, J.-H., Jung, J.-M., Lee, J.Y. and Do, N.-C., 2006, "A framework for sharing product information across enterprises," *International Journal of Advanced Manufacturing Technology*, **27**(5–6), pp. 610–618.

[12.19] Chan, S.C.F., Dillon, T. and Ng, V.T.Y., 2003, "Exchanging STEP data through XML-based mediators," *Concurrent Engineering: Research and Applications*, **11**(1), pp. 55–64.

[12.20] Amaitik, S.M. and Kilic, S.E., 2007, "An intelligent process planning system for prismatic parts using STEP features," *International Journal of Advanced Manufacturing Technology*, **31**(9–10), pp. 978–993.

[12.21] Lee, W., Bang, Y.-B., Ryou, M.S., Kwon, W.H. and Jee, H.S., 2006, "Development of a PC-based milling machine operated by STEP-NC in XML format," *International Journal of Computer Integrated Manufacturing*, **19**(6), pp. 593–602.

Appendix: Bindings of Some Common Data Entities

- The array_literal element

```
<!ELEMENT array_literal (binary_literal |
  integer_literal | logical_literal | real_literal |
  string_literal | bag_literal | list_literal |
  set_literal | array_literal | type_literal |
  nested_complex_entity_instance |
  flat_complex_entity_instance |
  simple_entity_instance | entity_instance_ref |
  unknown)*>
```

- The attribute_instance element

```
<!ELEMENT attribute_instance (binary_literal |
  integer_literal | logical_literal | real_literal |
  string_literal | bag_literal | list_literal |
  set_literal | array_literal | type_literal |
  nested_complex_entity_instance |
  flat_complex_entity_instance |
  simple_entity_instance | entity_instance_ref)>
<!ATTLIST attribute_instance express_attribute_name
  NMTOKEN #REQUIRED >
```

- The author element

```
<!ELEMENT author (#PCDATA)>
```

- The authorisation element

```
<!ELEMENT authorisation (#PCDATA)>
```

- The bag_literal element

```
<!ELEMENT bag_literal (binary_literal* |
  integer_literal* | logical_literal* | real_literal*
  | string_literal* | bag_literal* | list_literal* |
  set_literal* | array_literal* | type_literal* |
  (nested_complex_entity_instance |
  flat_complex_entity_instance |
  simple_entity_instance | entity_instance_ref)*)>
```

- The binary_literal element

```
<!ELEMENT binary_literal (#PCDATA)>
<!ATTLIST binary_literal
  notation (hex | base64) #REQUIRED >
```

- The constant_instances element

```
<!ELEMENT constant_instances
  (nested_complex_entity_instance |
  flat_complex_entity_instance |
  simple_entity_instance)*>
```

- The data element

```
<!ELEMENT data (data_section_header?,
  schema_instance)>
<!ATTLIST data data_id ID #REQUIRED >
```

- The data_section_description element

```
<!ELEMENT data_section_description (description*)>
```

- The data_section_header element

```
<!ELEMENT data_section_header
  (data_section_description, data_section_name)>
```

- The data_section_identification_name element

```
<!ELEMENT data_section_identification_name (#PCDATA)>
```

- The data_section_name element

```
<!ELEMENT data_section_name
  (data_section_identification_name, time_stamp,
  author, organisation, preprocessor_version,
  originating_system, authorisation)>
```

- The description element

```
<!ELEMENT description (#PCDATA)>
```

- The entity_instance element

```
<!ELEMENT entity_instance_ref EMPTY>
<!ATTLIST entity_instance_ref entity_instance_idref
  IDREF #REQUIRED >
```

- The enumeration_ref element

```
<!ELEMENT enumeration_ref (#PCDATA)>
```

- The flat_complex_entity_instance element

```
<!ELEMENT flat_complex_entity_instance
  (partial_entity_instance+)>
<!ATTLIST flat_complex_entity_instance
  entity_instance_id ID #REQUIRED >
```

- The integer_literal element

```
<!ELEMENT integer_literal (#PCDATA)>
```

- The ISO-10303-data element

```
<!ELEMENT ISO-10303-data (data+)>
```

- The list_literal element

```
<!ELEMENT list_literal (binary_literal* |
  integer_literal* | logical_literal* | real_literal*
  | string_literal* | bag_literal* | list_literal* |
  set_literal* | array_literal* | type_literal* |
  (nested_complex_entity_instance |
  flat_complex_entity_instance |
  simple_entity_instance | entity_instance_ref)*)>
```

- The logical_literal element

```
<!ELEMENT logical_literal (false | true | unknown)>
```

- The nested_complex_entity_instance element

```
<!ELEMENT nested_complex_entity_instance
  (attribute_instance*,
  nested_complex_entity_instance_subitem*)>
<!ATTLIST nested_complex_entity_instance
  express_entity_name NMTOKEN #REQUIRED
  express_schema_name NMTOKEN #IMPLIED >
<!ATTLIST nested_complex_entity_instance
  entity_instance_id ID #REQUIRED >
```

- The nested_complex_entity_instance_subitem element

```
<!ELEMENT nested_complex_entity_instance_subitem
  (attribute_instance*,
  nested_complex_entity_instance_subitem*)>
```

```
<!ATTLIST nested_complex_entity_instance_subitem
  express_entity_name NMTOKEN #REQUIRED
  express_schema_name NMTOKEN #IMPLIED
  entity_instance_id ID #IMPLIED >
```

- The non_constant_instances element

```
<!ELEMENT non_constant_instances
  (nested_complex_entity_instance |
  flat_complex_entity_instance |
  simple_entity_instance)*>
```

- The organisation element

```
<!ELEMENT organisation (#PCDATA)>
```

- The originating_system element

```
<!ELEMENT originating_system (#PCDATA)>
```

- The partial_entity_instance element

```
<!ELEMENT partial_entity_instance
  (attribute_instance*)>
<!ATTLIST partial_entity_instance express_entity_name
  NMTOKEN #REQUIRED express_schema_name NMTOKEN
  #IMPLIED entity_instance_id ID #IMPLIED >
```

- The preprocessor_version element

```
<!ELEMENT preprocessor_version (#PCDATA)>
```

- The real_literal element

```
<!ELEMENT real_literal (#PCDATA)>
```

- The schema_instance element

```
<!ELEMENT schema_instance (constant_instances,
  non_constant_instances)>
<!ATTLIST schema_instance express_schema_name NMTOKEN
  #REQUIRED >
```

- The set_literal element

```
<!ELEMENT set_literal (binary_literal* |
  integer_literal* | logical_literal* | real_literal*
  | string_literal* | bag_literal* | list_literal* |
  set_literal* | array_literal* | type_literal* |
  (nested_complex_entity_instance |
  flat_complex_entity_instance |
  simple_entity_instance | entity_instance_ref)*)>
```

- The simple_entity_instance element

```
<!ELEMENT simple_entity_instance
  (attribute_instance*)>
```

```
<!ATTLIST simple_entity_instance express_entity_name
  NMTOKEN #REQUIRED express_schema_name NMTOKEN
  #IMPLIED entity_instance_id ID #REQUIRED >
```

- The string_literal element

```
<!ELEMENT string_literal (#PCDATA)>
```

- The time_stamp element

```
<!ELEMENT time_stamp (#PCDATA)>
```

- The type element

```
<!ELEMENT type_literal (binary_literal |
  integer_literal | logical_literal | real_literal |
  string_literal | bag_literal | list_literal |
  set_literal | array_literal | type_literal |
  nested_complex_entity_instance |
  flat_complex_entity_instance |
  simple_entity_instance | entity_instance_ref |
  enumeration_ref)>
<!ATTLIST type_literal express_type_name NMTOKEN
  #REQUIRED express_schema_name NMTOKEN #IMPLIED >
```

13

Pulling the Value Streams of a Virtual Enterprise with a Web-based Kanban System

Hung-da Wan, Sanjay Kumar Shukla and F. Frank Chen

Centre for Advanced Manufacturing and Lean Systems
Department of Mechanical Engineering, University of Texas at San Antonio
One UTSA Circle, San Antonio, TX 78249, USA
Emails: hungda.wan@utsa.edu, sanjay.shukla@utsa.edu, ff.chen@utsa.edu

Abstract
A Kanban system is one of the major enablers of lean manufacturing implementation. With the aid of information technologies, web-based Kanban systems inherit the benefits of using physical Kanban cards and meanwhile eliminate several limitations such as the scope and distance of applied areas, amount and types of information contents, real-time tracking and monitoring, and flexibility for adjustments. With the enhanced functionality, the Kanban system is no longer merely for shop floor control. The web-based Kanban system is ready to bring the "pull" concept into a distributed manufacturing environment to make a virtual enterprise. This chapter presents the functionality requirements and available solutions of web-based Kanban systems. The applications of a web-based Kanban system in various environments, from manufacturing cells to virtual enterprises, are explored. A web-based Kanban system provides great visibility of the production flows in an enterprise system. It can deliver a clearer picture of the up-to-date status of the system as well as the dynamics over time. Using the enhanced information, decision makers will be able to plan and manage production flows of a virtual enterprise more effectively.

13.1 Introduction

Under intense global competition, continuously seeking ways to become leaner is now a crucial task for every company. Lean manufacturing concepts and tools have made significant impacts on various industries. By implementing waste-elimination tools and methodologies, lean manufacturers enjoy various benefits, such as enhanced productivity, product variety, and quality. Even for non-manufacturers, significant results have been reported in various types of operation, such as logistics, healthcare, financial services, construction, and maintenance [13.1].

Among lean principles and tools, the pull concept is the key to reducing work-in-process (WIP) level and revealing hidden waste. A Kanban system is a critical enabler of the pull concept. Using visual signs, a Kanban system simplifies shop floor management and conducts material flows efficiently. Consequently manufacturers can significantly improve leanness by implementing a Kanban system. However, lean improvements within an individual company are limited if the whole supply chain is not integrated. As a result, companies must extend the scope of lean implementation onto an integrated value stream of a supply chain to achieve true leanness [13.2]. Nevertheless, extending the pull system to inter-organisational activities becomes a challenge. In today's competitive market, uncertain customer demand, increased product variety, distributed manufacturing, and increased physical distances between facilities are making the applications of simple Kanban system more complex. The complexity increases incorrect Kanban deliveries, lost cards, and delays. In order to cope with the complexity and accomplish an integrated lean supply chain, an information system that supports the pull concept must be in place for lean logistics. The web-based Kanban system emerges as a natural solution that combines lean principles with advanced information sharing capability [13.3].

On the other hand, consumers' demand patterns are showing a trend of changes. Due to rapid technology development and the information boom, most consumers now long for more product options or services to meet their individual desires. Therefore, the lifecycle of product designs has become shorter and shorter; while the variety of products has increased. In order to survive and thrive in the global competition, manufacturers and service providers must be not only "lean" (*i.e.* efficient and effective) in their operations but also "agile" (*i.e.* flexible and responsive) in their value chains. Through adaptive formation of virtual enterprises, an agile supply chain aims at meeting changing demands in a timely way and eventually achieves mass customisation [13.4].

Many discussions about integrating lean and agile strategies have been made. The question now is how to practically carry out this integration and benefit from it. The key to accomplishing it is information technology: using enhanced information flow to facilitate digital manufacturing and collaborative virtual enterprise. For both lean and agile strategies, information sharing is critical. An advanced information system must be in place to support the timely formation of a virtual organisation and streamline the operations. Since Kanban systems are designed to facilitate workflows and logistics of lean operations, a web-based Kanban system emerges as a perfect candidate to bring about the lean value streams of a virtual enterprise.

The conventional card-based Kanban system lacks tracking and monitoring abilities, which are vital for a supply chain. With the advent of web-based technology, information can be accessed and shared by users anywhere in the world, which can enhance Kanban systems. Through the Internet, web-based technology facilitates multi-media, computer network, distributed computing, *etc.* With these functionalities, a web-based Kanban system can track, monitor, and control the activates in a supply chain. Therefore, a web-based Kanban is a lean tool to streamline customers' demand throughout the organisation and provide the customers what they really want.

This chapter investigates the potential impacts of supporting an agile virtual enterprise using a web-based Kanban system. The functionality requirements of a web-based Kanban system and available solutions are discussed. The infrastructure of the web-based Kanban system for an agile virtual enterprise is introduced. An experimental software program for a web-based Kanban system using PHP+MySQL platform is illustrated. The development process of the experimental program reveals the benefits and limitations of using web-based Kanban system in a virtual enterprise environment. Results of this research show that a web-based Kanban system is applicable to various environments, from manufacturing cells to virtual enterprises. It provides great visibility of the production flows in an enterprise system. With the "built-in" pull concept in the web-based Kanban, operations are streamlined with a customer's demand, and this information is transparent for every designated user. Moreover, with properly developed performance metrics, it can deliver a clearer picture of the up-to-date status of the system as well as the dynamics over time. This enhanced information will enable decision makers to plan and manage production flows in a virtual enterprise more effectively.

13.2 Lean Systems and Virtual Enterprises

13.2.1 Lean Manufacturing Systems

Lean manufacturing has demonstrated significant impacts on various industries. Evolved from the Toyota Production System (TPS) originally, lots of lean tools and methodologies were developed to eliminate non-value-added activities, *i.e.* waste, from all operations [13.5]. In the past century, mass production approaches evolved from Henry Ford's concepts, which successfully satisfied the growing market using abundant resources. In contrast, TPS was the result of striving to survive in a poor condition in Japan right after World War II, where any type of waste was unaffordable [13.6]. The Just-in-Time (JIT) system, focusing on waste reduction and continuous improvement, was introduced to western researchers in the 1970s [13.7]. Later, an MIT research team's publications, *the Machine that Changed the World* [13.8], earned widespread publicity. This team coined the term "lean manufacturing" that precisely reflects the spirit of TPS [13.9]. Meanwhile, Ohno [13.6] introduced the background and concepts of TPS, and Monden [13.5] detailed the practical techniques of TPS in their books. Womack and Jones [13.10] summarised *lean thinking* into five essential steps, *i.e. value, value stream, flow, pull*, and *perfection*. Among them, the "pull" concept aligns production targets throughout the value stream with the end customer's demand and hence minimises inventory and WIP. Therefore, a pull system is the key to smoothing out the flow of a value stream or, in other words, a lean system.

A large amount of tools and techniques have been developed to realise various lean concepts, including all the TPS techniques [13.5] and value stream mapping [13.11]. For the pull concept specifically, the Kanban system is a critical enabler. Kanban is a Japanese term for cards (*i.e.* "ban" in Japanese) that deliver visual information (*i.e.* "kan" in Japanese). A pull system uses Kanbans (cards) to pull material flows based on the demands of downstream workstations. Therefore, the output rate is dictated by the demand for the product [13.7]. As a result, products are produced only when needed, which avoids non-value-added activities (*e.g.* waiting, overproduction, unnecessary material handling, *etc.*) from happening.

Due to the significance of JIT production, extensive investigations of the pull concept have been carried out, such as studies of simulation analysis [13.12], analytical modelling [13.13], and system re-design [13.14]. In general, Kanban-enabled pull systems result in significant productivity improvement compared with a "push" system. The benefits of implementing a pull system include shorter lead time, greater flexibility in response to demand changes, reduced levels of inventory and other wastes, capacity considerations that are restricted by the system design, and inexpensive to implement [13.15]. In terms of production planning, a pull system is reportedly more efficient, easier to control, more robust, and more supportive to improving quality [13.16]. Lean manufacturers using pull systems enjoy a more manageable production environment with much lower WIP and inventory level [13.3].

13.2.2 Lean Supply Chain

Many success stories of lean manufacturers have been reported. However, the pursuit of perfection often meets a bottleneck when the individual company discovers that improvements are constrained by their business partners [13.17]. In a factory, reducing production lead time from three days to three hours results in a more than 90% of improvement. However, in a supply chain where products take three weeks to flow through, the improvement at the factory only accounts for about 6% of the reduction in total lead time [13.2]. In the end, customers still wait for three weeks mostly due to non-value-added activities in the supply chain. Therefore, integrating the lean improvement efforts of each separate silo along the whole supply chain is the key to achieve true leanness.

The main focus of research on the supply chain is to cut the overall cost and maximise overall profit. Creating lean flows throughout the supply chain matches this objective. Jones and Womack [13.18] proposed an approach to extend the value stream mapping technique to cover the whole value chain, from the supplier of raw materials to the end customers. The extended value stream map can visualise the flows and wastes of the overall supply chain, which facilitates the management of lean supply chains. The next question is how to urge the business partners to become lean. Lean supply chain practices are outward and based on full collaboration [13.19]. As more companies in a supply chain become leaner, the peer pressure will result in a self-motivated lean implementation throughout the supply chain.

A lean supply chain must be planned, executed, and designed across the business partners to deliver products of the right design, in the right quantity, to the right place, at the right time [13.20]. Vitasek *et al.* [13.21] defined a lean supply chain as "a set of organisations directly linked by upstream and downstream flows of products, services, finances, and information that collaboratively work to reduce cost and waste by efficiently pulling what is needed to meet the needs of the individual customer." Therefore, in a lean supply chain, all components must be tightly integrated and aligned with end customers' demands. To develop this tight integration, Phelps *et al.* [13.17] suggested a six-step procedure to initiate the establishment of a lean supply chain, which involves selection of supply chain members, current state assessment, value stream mapping for overall and detailed views, timeline chart development, and future state analysis. Rivera *et al.* [13.2] summarised the characteristics of lean supply chains and pointed out four building blocks to facilitate lean flows, which are transparent information, lean logistics, monitored performance, and full collaboration. Among them, the accessibility of information is crucial to facilitate the other building blocks. This conclusion stresses the need for an effective information system, such as a web-based Kanban system, to link the extended value stream.

13.2.3 Agile Virtual Enterprise

Lean manufacturing methods generally work well under stable and repetitive demand. In today's competitive market, the uncertain customer demand, increased product variety, and distributed manufacturing are causing increased complexity of a supply chain and raise the difficulty of being lean. Alternatively, the concept of agile manufacturing provides a different approach from lean manufacturing in the pursuit of success in the marketplace. In the early 1990s, industry leaders figured that the top challenge for business in the 21st century is to provide high-quality, low-cost products, and meanwhile be responsive to customers' specific, unique, and rapidly changing needs [13.22]. The solutions to this challenge were termed "agile manufacturing" in 1991 [13.23]. The main objective is to cope with demand volatility by making changes in an economically viable and timely manner [13.24].

In a typical lean environment, variability should be minimised if it cannot be fully eliminated; however, agile manufacturing aims at a marketplace full of unexpected situations. To accomplish this goal, agile manufacturers have to respond swiftly to demand changes and gain market share before most competitors can react [13.4]. Therefore, being agile is more advantageous than being lean in a volatile, customer-driven environment [13.25]. The agility of an enterprise refers to the capability to proactively establish a virtual organisation in order to "(i) meet the changing market requirements, (ii) maximise customer service level, and (iii) minimise the cost of goods, with an objective of being competitive in a global market and for an increased chance of long-term survival and profit potential" [13.22]. Such an agile system is enabled by strategic planning, product design, virtual enterprise, and automation and information technology [13.22]. Among the four enablers, the virtual enterprise approach is the unique concept that distinguishes agile manufacturing from the other approaches.

A virtual enterprise, based on temporary partnerships, is formed with a group of capable companies when a new and unique demand appears. The partnership is dismissed upon meeting the demand. This approach maximises the flexibility of a supply chain to enhance agility. Goldman *et al.* [13.26] proposed a concept of virtual organisation to promote the virtual partnerships in agile supply chains. A framework for an agile virtual enterprise with a performance measurement model was proposed by Goranson [13.4]. At the shop floor level, the virtual cell uses a similar concept to increase the agility of manufacturing systems [13.27–13.29]. Prince and Kay [13.30] introduced a virtual group approach which is an application of virtual cells with functional layout that combines the lean and agile characteristics. In a different route, Naylor *et al.* in [13.31] proposed a "leagile" system that uses a decoupling point to divide a supply chain into lean upstream section and agile downstream section in order to combine the lean and agile paradigms.

From the literature review, two sets of enablers for lean systems and agile systems have been identified. A comparison is shown in Figure 13.1. Among the four enablers of agile manufacturing, an effective information system is again a crucial component to accommodate all the collaborative activities, such as design, planning, and control. Can a web-based Kanban system designed for a lean supply

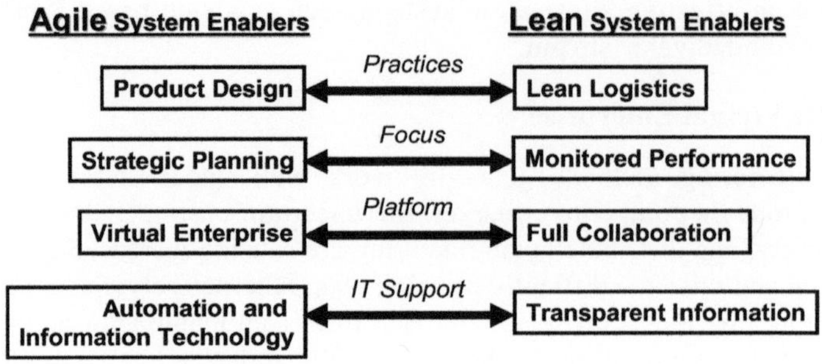

Figure 13.1. Enablers of lean and agile systems

chain be a viable solution for an agile virtual enterprise? Discussions are carried on in the following sections.

13.3 From Kanban Cards to Web-based Kanban

13.3.1 Kanban Systems: The Enabler of Just-in-Time

The Kanban system is one of the most important components of TPS. It facilitates the pull concept of lean manufacturing and typically performs well when demand is repetitive and stable. A Kanban card specifies a job requirement, including product name, item code, card number, *etc*. It is a signal to start production or to move material. In TPS, a dual-card Kanban system provides tight control on production, but it is more difficult to implement. Therefore, a single-card Kanban system is often used as a transition from a push system to the dual-card pull system [13.32].

Although the implementation of Kanban requires only little investment, the simple method has made a significant impact on efficiency and effectiveness of shop floor control. Abundant research on Kanban systems has been performed by previous researchers. In order to properly design a Kanban system, Berkley [13.33] identified 24 elements in the Kanban production system as operational design factors. Philipoom *et al.* [13.34] was one of the pioneers who revealed that a successful Kanban system comprehensively improves throughput and reduces lead time. Karmarkar and Kekre [13.35] found that reduction in container size and increase in number of Kanbans led to better performance. This finding motivated various researchers to develop tools and techniques that can determine the optimal number of Kanbans. Sarker and Balan [13.36] developed an approach to find the optimal number of Kanbans between two adjacent stations. In a flow shop environment, Rshini and Ran [13.37] proposed a recursive paradigm for scheduling the single card Kanban system with dual blocking. Altiok and Shiue [13.38] modelled a multiple product pull system considering setups between different product types. Bitran and Chang [13.39] developed an optimisation model for the multi-stage capacitated assembly line to find the optimal number of Kanbans. Davis and Stubitz [13.40] developed a simulation model for the design of a Kanban system. Finally, Kumar and Panneerselvam [13.32] carried out a comprehensive literature review of Kanban systems that covers previous research in all aspects.

Kanbans are not only used to pull the products. They can also visualise and control the WIP level. A properly designed system can effectively limit the amount of in-process inventory, while co-ordinating the logistics of the system. Therefore, a Kanban system is a manual method to harmoniously manage and control production and inventory within the plant.

A Kanban system can be applied internally within the shop floor or externally between distant facilities to realise JIT delivery in a global supply chain [13.41, 13.42]. Fax and email are often used to dispatch Kanbans among distant sites when delivering physical cards is not efficient enough. For production control, the number of Kanban can be adjusted within a range to meet the capacity requirements. Using demand levelling, the pull system remains stable when demand fluctuates within a certain range. However, the complexity and dynamics of a supply chain reveal the weaknesses of conventional Kanban systems.

13.3.2 Weakness of Conventional Kanban Systems

The conventional card-based Kanban system is known to be simple and effective for manufacturers with stable and repetitive demand. However, with long-distance transportation, time to transport the cards and the hand-offs between two sites create more problems. When the product variety increases or demand fluctuates, the Kanban system needs to adjust the number of cards correspondingly. These issues increase the complexity of the Kanban system and make it difficult to manage. Therefore, more complexity of the actual manufacturing system leads to more problems and difficulties in the management of a physical Kanban system.

In today's competitive market, uncertain customer demand, increased product variety, distributed manufacturing, and increased physical distances between business partners all increase the chance of mistakes in a card-based Kanban system [13.43]. The most common mistake is a lost Kanban card, which leads to material outage, waiting, extra cost, and lower service level [13.44, 13.45]. Mistakes of handling contribute to most of the lost cards, and the problem is amplified when cards are delivered over a long distance or travel time. In addition to the mistakes of operations, visibility is another critical issue when physical Kanban cards travel out of a facility. The conventional card-based Kanban system was designed to enhance the visibility of workflows to a certain extent within a limited shop floor environment. However, the visibility is lost when the cards travel to distant locations [13.44]. A Kanban system often operates autonomously on a shop floor that requires little interference from the management level. It normally controls WIP level strictly, but other performance metrics at the workstation level are not automatically monitored. Therefore, extra efforts on data collection are needed when the performance metrics are requested for further analysis.

Due to the increasing complexity of modern manufacturing systems and supply chains, the problems associated with card-based Kanban applications have become more and more critical. To understand these problems, Wan and Chen [13.3] summarised them into two categories: common mistakes and system weaknesses:

(1) *Common mistakes in the practice of conventional Kanban systems*

- Lost Kanbans
- Incorrect delivery of Kanbans

- Delivering Kanbans with inaccurate information

(2) *Weakness in the system infrastructure of conventional Kanban system*

- Less efficiency in distant delivery
- No visibility in distant delivery
- Limited tracking and monitoring capability
- Limited support for performance measurement
- Limited scalability to handle spikes in demand
- Consuming workforce on managing the cards

To eliminate the weaknesses of conventional Kanban systems, improvements are urgently needed for practitioners to effectively compete in the volatile global market. With the advancement of information technologies, a computerised Kanban system provides an opportunity to address these issues. The integration of information technology and a Kanban system is an important step to improve this lean tool.

13.3.3 Web-based Technology and e-Kanban

In order to address the problems and weaknesses of conventional Kanban systems, information technology is the natural solution. Electronic versions of a Kanban system (*i.e.* e-Kanban) have been developed to minimise the chances of human errors and to enhance the monitoring and tracking capabilities. Currently, e-Kanban systems are usually developed based on three different platforms: existing enterprise resource planning (ERP) systems, electronic data interchange (EDI) connections, or web-based technology [13.44]. Those developed based upon existing ERP and EDI systems require higher levels of investments on software platform and hardware infrastructure. On the contrary, web-based versions of a Kanban system can provide a more affordable and accessible solution for lean practitioners. With the ubiquitous network connections, the visibility of Kanbans and efficiency of deliveries are inherently embedded for individual facility and over long distances. The automated processes can minimise human errors. More importantly, performance of the Kanban-enabled manufacturing system can be analysed and reported through the Internet in real time [13.3]. Based on the application areas, Osborne [13.45] identified four major elements of web-based Kanban, which are the inter-plant, supplier, customer, and internal systems. These four elements cover the logistics of a supply chain from raw material to finished products.

ERP systems were created to be a computerised complete solution for manufacturers. However, the proprietary software infrastructures to encompass this complex system result in a much higher price tag than is affordable by average manufacturers. Due to the distinct routes that ERP and JIT systems took, conventional ERP systems offered limited support for pull-type material flow [13.3]. As lean manufacturing became more popular, major ERP suppliers started to provide add-on modules to integrate Kanban operations into their systems. Nevertheless, these solutions are still costly. On another path, researchers and software providers started to develop fully web-based ERP systems that are more accessible, economic, and user-friendly. Yen *et al.* [13.46] analysed the functionality of ERP systems and concluded that web-based ERP is the solution to support e-business applications. Ng and Ip [13.47] designed a framework of web-ERP to meet the trend of globalisation. Ash and Burn [13.48] investigated the ERP-enabled business-to-business solutions including web-based systems. Verwijmeren [13.49] proposed a web-based software component architecture that encompasses local ERP systems, warehouse management systems, and transportation management system. Helo *et al.* [13.50] proposed a web-based logistics system for agile supply chain management. Tarantilis *et al.* [13.51] constructed a fully web-based ERP system as a total solution for supply chains.

The web-based ERP designers strive to provide an information system for manufactures that covers a wide range of operations, especially the scheduling and planning of all resources. As a core module of web-based ERP or a stand-alone system, the web-based Kanban system focus on the pull-type lean logistics with an emphasis on facilitating both intra-organisational and inter-organisational material flows. A rapidly growing number of commercial web-based Kanban systems have been developed recently by software providers, such as Datacraft Solutions, eBots, Manufactus, SupplyWorks, and an open-source Web Heijunka-Kanban by Bo Ahlberg. These systems have been developed based on web-based infrastructure that provides cross-platform access via web browsers. The systems have different focuses, such as pull production, inventory control, or real-time performance analysis. When integrating with barcode or smart identification systems (*e.g.* RFID, radio frequency identification), the web-based Kanban systems can conduct and monitor production flows and control inventory. As a result, web-based Kanban system becomes a much more economic alternative than a full ERP package and can bring the pull concept into a dynamic supply chain environment.

Despite the growing software market of web-based Kanban system, little research has been carried out. Kotani [13.52] proposed an effective heuristic for changing the number of Kanbans in an e-Kanban system. Cutler [13.43] pointed out three potential cultural shifts when a manufacturer moves from conventional Kanban to e-Kanban. These impacts are: (1) give some control back to the management; (2) enhanced communications and visibility of material status; (3) eliminate humiliation due to lost cards. Web-based Kanban systems inherit the benefits of using physical Kanban cards and meanwhile eliminate several limitations such as the scope and distance of applied areas, amount and types of information content, real-time tracking and monitoring, and flexibility for adjustments. In summary, using web-based technology, the e-Kanban systems are expected to improve the conventional card-based Kanban system by the features listed below [13.3]:

- Automate the data transactions and eliminate human errors.
- Deliver and monitor Kanbans in real time regardless of distance.
- Adjust Kanban quantities to quickly adapt to demand changes.
- Analyse performance and suggest performance targets in real time.

13.4 Building a Web-based Kanban System

Using web-based technology, web-based Kanban systems can meet the needs of managing the production flows of a lean and agile supply chain. This section explores how a web-based Kanban system can be developed and what the functionality requirements are.

13.4.1 Infrastructure and Functionality of a Web-based Kanban System

Several important features of web-based technology, as listed below, facilitate the e-Kanban system to be simple, effective, inexpensive, and easy to use [13.3].

- Cross-platform standard user interface (*i.e.* web browsers)
- Ubiquitous networks
- Improved web-based programming capability

Using web databases and interactive web-based programs, a web server can provide operational functions for e-Kanban systems, including real-time tracking, performance measurement, interactive input/output, dynamic display, *etc*. The cross-platform accessibility allows users from different organisations to be linked together. With the ubiquitous network connections, the visibility of Kanbans and material flows is inherently embedded whether within one facility or over a long distance. The automated Kanban transactions can minimise human errors. More importantly, performance of the Kanban-enabled manufacturing system can be analysed and reported through the Internet in real time.

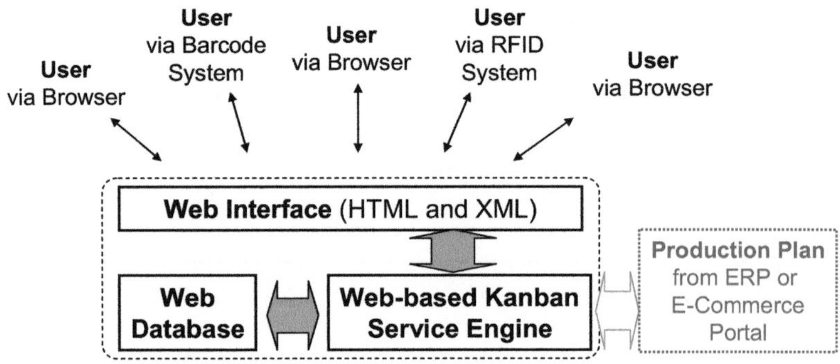

Figure 13.2. Generic infrastructure of web-based Kanban systems

A generic infrastructure of web-based Kanban system is presented in Figure 13.2, which consists of three major components: Kanban service engine, database, and interface. In general, users communicate with the Kanban service engine through the web user interface from various hardware devices and software programs. Kanban service engine contains the core modules to perform the transactions of Kanban, job monitoring, performance measurement, *etc.* A web database stores and provides various data, such as quantity, timeframe, instruction, and rules and procedures. Security of the network connections needs to be ensured through the web interfaces. The production planning module may or may not be included in a web-based Kanban system. Production plans can be imported from existing ERP or e-commerce systems and then executed and controlled by the web-based Kanban system.

The web-based Kanban service engine must be programmed as a dynamic web page. Two fundamental components of dynamic pages are the web-based program and web database, which provide the computational power and data storage and retrieval, respectively. Commonly used web-based programming languages are ASP (Microsoft Active Server Pages), JSP (JavaServer Pages), PERL (Practical Extraction and Report Language), ColdFusion (Adobe), and PHP (PHP: Hypertext Preprocessor). Among them, some send scripts to viewer's browser to be executed by the clients (*e.g.* JSP); the others run server-executed programs (*e.g.* PHP).

As to the web databases, IBM DB2, MySQL, Oracle 9i, Microsoft SQL Server, and Sybase Adaptive Server are among the commonly used commercial packages. The web databases provide online data storage and interact with aforementioned web-based programs to enable the Kanban system functions. An alternative to the relational databases, the Extensible Markup Language (XML), is a structured data representation that facilitates cross-platform applications. XML enjoys greater flexibility; while relational databases are more efficient when variation of data structure is minimal. Hybrids of the two types of database have been developed to embed the advantages of both approaches.

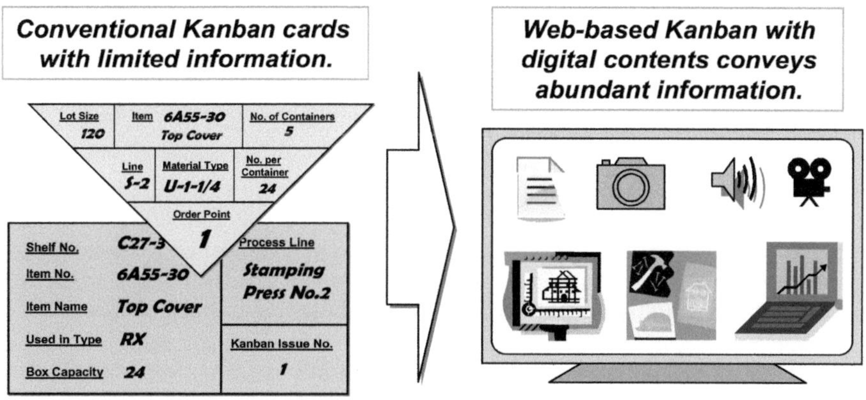

Figure 13.3. Web-based Kanban provides more information than conventional Kanban

The World Wide Web servers provide the web services via the web interface module. With the standard Hypertext Markup Language (HTML), multi-media digital contents can be delivered to users through web browsers. As shown in Figure 13.3, conventional Kanban cards can only deliver very limited information, such as part type, batch size, *etc.* Using the web interface, various types of information can be delivered, including text documents, drawings, pictures, audio, video, *etc.* The ability to convey abundant digital contents opens up a new window of Kanban applications. For example, a production Kanban that triggers a specific job can display work instructions using a combination of engineering drawing, photograph, video clip, and vocal assistance. Thus, a multi-functional operator can pick up the job easily while changing over among product types. This new feature realises several lean principles, including visual systems, standard work, error-proofing, and flexible workforce. This feature is particularly important for agile systems where customer demand changes frequently. The digital contents can be used to guide the assembly and machining processes, display tool and accessory requirements, assist raw material or finished goods visual inspection, and show the transportation routes.

With properly designed digital contents, web-based Kanban is not only a computerised version of the conventional Kanban system but also a new tool for enhancing the effectiveness and efficiency of value streams.

Beside the digital contents made available by the web interface, the interactive dynamic web pages can also embed more guidance and decision-support functions to assist operators perform their jobs. In a dynamic environment like agile virtual enterprise, a "smart and interactive Kanban" can assist operators to identify urgent jobs or reschedule a job based on current situations and operational policies. Work standards can be displayed as performance targets for the operations, and the actual performance can be reported automatically in real time. Finally, the web interface must ensure security of the transactions. Security is always a critical issue that can fail a great system. Through accounts/passwords, secured connections, firewalls, buffer zones, and many other technologies, the web-based Kanban system has to be well protected for its intra-organisational and inter-organisational activities.

In addition to the software infrastructure, peripherals for tracking and control, such as barcode and RFID systems, are also handled by the interface modules. Barcode systems have been used widely in industry to track and control material flows. However, the efficiency is limited due to the piece-by-piece scanning that requires optical contact. On the other hand, RFID is developed to supersede barcode systems. Multiple RFID tags within the range of an antenna can be scanned simultaneously, which greatly enhances the efficiency of material tracking. An RFID system can monitor multiple parts of a final assembly or hundreds of products in a trailer and thus, provides significant visibility of material flows in a supply chain. However, several issues, such as cost and accuracy, must be resolved before it can really supersede barcode. The read-rate is often less than 100%, especially when tagging metal or liquid products [13.53]. The reliability issue is critical for users.

A web-based Kanban system is expected to be a lean management tool that can be used to drive down costs, integrate disparate existing software systems, communicate throughout the supply chain, and eventually achieve customer satisfaction [13.45]. Furthermore, decision makers should make use of the real-time information provided by the Kanban system to guide the lean improvement activities. In order to carry out the expected features, appropriate web-based technologies need to be identified first. In this research, the PHP+MySQL platform is selected to form the infrastructure of an experimental program as introduced in the next section.

13.4.2 An Experimental System Using PHP+MySQL

A web-based Kanban system has been developed by the authors as an experimental program to demonstrate the proposed concept and explore various issues [13.3]. Using PHP+MySQL, the example program is an entirely server-executed web-based program, which requires no installation at the user's end. PHP+MySQL platform is among the most frequently used web-based programming platforms due to its economic advantages. As shown in Figure 13.4, the experimental system consists of three functional modules, *i.e.* system configuration, Kanban operations, and performance monitoring modules. Before operating the system, administrator must configure the program to virtually map the structure of the physical manufacturing system or supply chain onto the cyberspace, including the workstations, product information, routing, *etc.* After the system has been properly configured, orders can be placed to trigger the pull production. The experimental program can level customer demand into smaller lots over time. However, it does not contain the function of production planning.

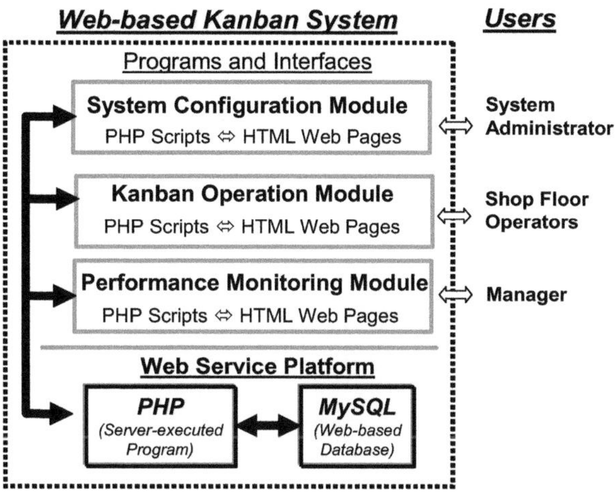

Figure 13.4. Framework of a PHP+MySQL enabled Kanban system

On the operational side of the web-based Kanban system, an operator logged on to the system can only receive jobs and associated information (*e.g.* work instructions) designated to that user account. Jobs are prioritised by preset rules, and the program directs the operators to finish the tasks step by step. The digital Kanbans are assigned based on the end customers' orders, and the demand pulls the production flow throughout the supply chain. At the management level, authorised personnel can use the performance monitoring module to track the Kanbans, monitor time-based performance, and download the data for further analysis.

Some snapshots of the example program are show below to demonstrate the operations and functionalities. Figure 13.5 shows an operation Kanban designated to a workstation, which provides information on quantity, timeframe, work instructions, *etc.* The experimental program allows companies to maintain a limited WIP level at buffer areas that are monitored in real time, allowing business partners to track the status of relevant components along the supply chain. Performance information is also accessible to relevant personnel to facilitate further analysis and decision making. An example of time-based performance of a particular operation is shown in Figure 13.6.

The current experimental program conveys information and material flows. Financial flow has not been included in this system. More software interfaces need to be developed to further enhance the compatibility with other software programs, such as ERP systems.

In order to apply the web-based Kanban system to a supply chain, a management team must be involved in the system configuration and performance monitoring. The

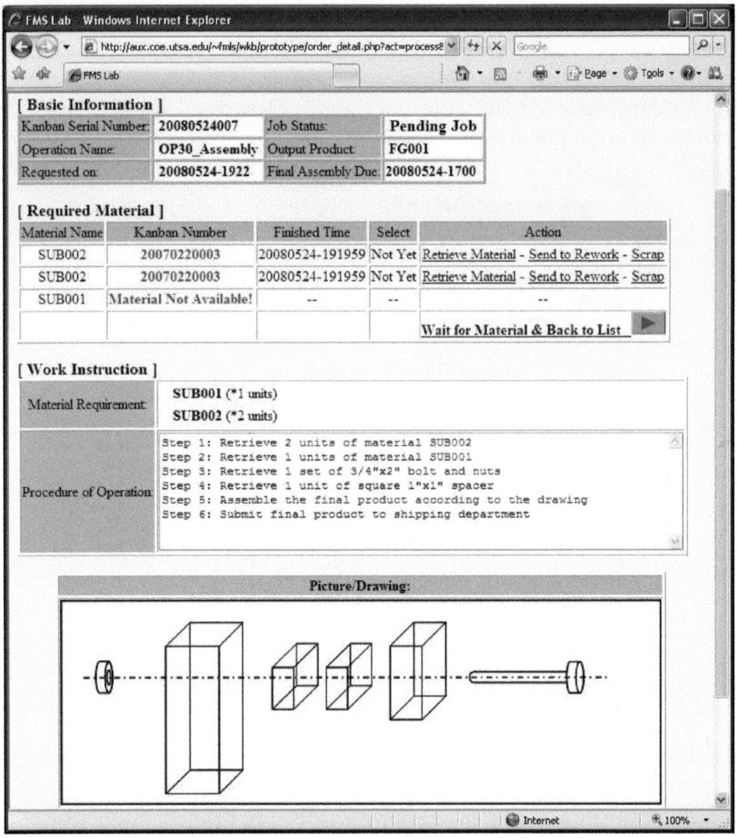

Figure 13.5. Operations module of the example program

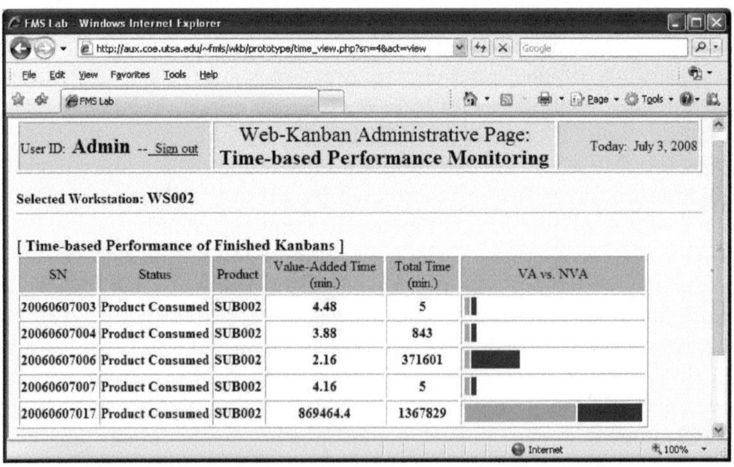

Figure 13.6. Performance monitoring module of the example program

team needs to configure the system with all components (*i.e.* suppliers and distributors) of the supply chain as well as product/process information, bill of material, routing information, *etc*. Since a supply chain is more complex than a single company or shop floor, the configuration of the web-based system for a supply chain is expected to consume much more time and effort. After the system is fully configured, the operation of the Kanban system is similar to that of smaller scopes. On the other hand, for a virtual enterprise environment, participating companies are not constant as typical lean supply chains. The accountability of managing and maintaining the web-based Kanban system becomes an issue. Three different strategies are discussed in the following section.

13.5 Pulling the Value Streams of a Virtual Enterprise

Extending Kanban applications to pull from major suppliers has been done by some renowned lean manufacturers. However, when it comes to an agile virtual enterprise, its dynamic nature challenges the efficacy of conventional Kanban systems. Web-based Kanban systems, on the other hand, have several features that provide the flexibility to support Kanban operations in a virtual enterprise setting. This section discusses the integration of the Kanban system with virtual organisation starting with virtual cells at the shop floor level.

13.5.1 Web-based Kanban for Virtual Cells

Before getting into the agile virtual enterprise, there is a smaller scale virtual organisation, *i.e.* virtual cells, which can be a test bed for the web-based Kanban system at shop floor level. For lean practitioners, cellular manufacturing is a very effective solution for pull production, which tightly integrates a small group of workstations dedicated to certain product families. However, under dynamic demand, it may be difficult to dedicate a manufacturing cell to certain product families. A virtual cell is a temporary "logical grouping of processors that are not necessarily transposed into physical proximity" [13.54]. The formation of virtual cells allows manufacturers to carry out some lean principles to minimise waste and cost. However, the dynamic configuration and physically dispersed layout increase the difficulty of planning and control of a pull system. Therefore, web-based Kanban system becomes an important tool to realise such a system.

Virtual cell is a mass customisation approach at the shop floor level. It increases the flexibility, one of the major characteristics of agility, of a manufacturing system. Using a web-based Kanban system, temporary grouping of workstations can be done easily when customer orders arrive. Workstations can be selected from existing manufacturing lines, cells, or a job shop to form the virtual cells for special projects/ products (Figure 13.7). Upon formation of a virtual cell, the routing of material is determined, and the web-based Kanbans will be delivered to the designated workstations to pull the flow.

Forming virtual cells from existing production lines or cells increases the flexibility of the current system to accommodate more product variety. On the other hand, forming virtual cells in a job shop environment brings in the lean concepts to pull the flows, limit the WIP levels, and enhance the leanness of the value stream. However, there are a few issues associated with virtual cells as listed below:

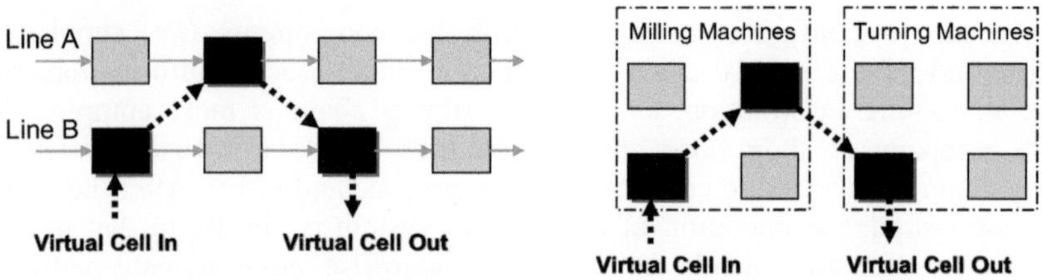

(a) Virtual cell formation in existing production lines

(b) Virtual cell formation in job shops

Figure 13.7. Formation of virtual cells in different environments

- Temporary configurations increase the complexity of the material and information flows on the shop floor.
- Without physically configuring a cellular layout, the distances among workstations may bring down productivity and cause confusion and other problems. Additional material handling is expected.
- The temporary grouping may interrupt the existing production lines or cells, which may lead to chaos if appropriate rules and policies are not in place.
- Progress, performance, and WIP levels are harder to monitor and control if an appropriate report system is not in place.

The web-based Kanban system is expected to minimise the impact of adding virtual cells onto existing manufacturing systems. At a basic level, it is a Kanban system conveying pull signals for jobs. The Kanbans route automatically in the software program and clearly indicate the destination of material flows from each station. This structure realises the concept of cellular manufacturing without physically reconfiguring the layout. When a material handling unit is part of the user groups of the Kanban system, the problems associated with the distant machine layouts and additional material handling requirements can be solved by sending Kanbans to the material handling unit to conduct the material flows. Furthermore, when Kanbans are sent directly from the production planning and control unit to each workstation, the system can operate as a push system, similar to the single-card Kanban system. As a result, a virtual cell with a web-based Kanban system facilitates both fixed and flexible routings and both push and pull manners. Figure 13.8 shows how the web-based Kanban system enables virtual cell operations.

On top of the pull signals, the web-based Kanban system facilitates Kanban tracking, performance monitoring and analysis, and conveyance of enhanced digital contents. It can be an important component of a manufacturing system, but it needs to be properly interfaced with existing programs, such as production planning and ordering systems. Interfacing remains challenging due to lack of standards for data exchange among various proprietary systems. Another challenge is to maintain an accurate database and constantly reconfigure the system to meet the dynamic demand. This issue is inevitable for the information systems to support agile virtual organisations, but it can be alleviated to a certain extent through modular design and better graphical user interfaces. Finally, integrating the web-based Kanban with barcode and RFID systems can automate some manual operations in order to avoid human errors and detect discrepancies to ensure data accuracy and integrity.

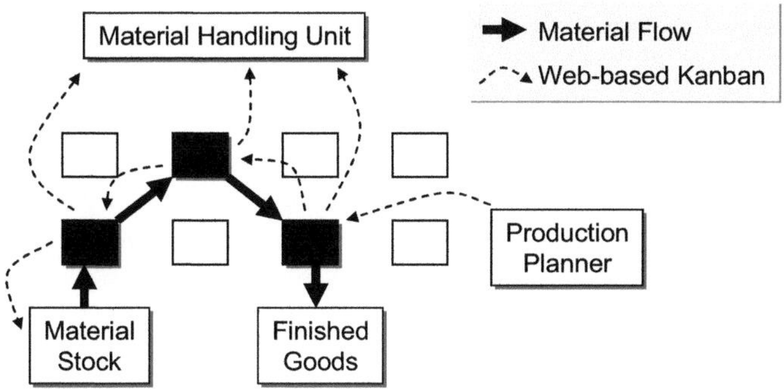

Figure 13.8. A web-based Kanban enabled virtual cell

13.5.2 Cyber-enabled Agile Virtual Enterprise

As discussed earlier in this chapter, information technology is a major enabler for an agile virtual enterprise. Extending the web-based Kanban system to support agile virtual enterprise is the goal of this research. Due to the dynamic nature of the configuration, the issues of virtual cells appear again at the enterprise level. Furthermore, the larger scale leads to more challenges than the shop floor problems. First of all, the requirements and objectives of a virtual enterprise are different from managing a lean or agile shop floor. The objective of forming a virtual enterprise is to team up excellent business partners who are most capable and/or suitable for specific tasks to meet the customer demand. There exists collaboration and also competition between partners. Furthermore, interactions between organisations are established based on mutual trust. The responsibilities and expectations are different from shop floor management. In addition, the business units involved in a virtual enterprise can be much more distant physically, which may cause more problems and higher costs. Differences between the companies, such as company cultures, strategies, and policies, may lead to difficulties in communications and interactions. Therefore, different performance metrics may be required to reflect the real objectives and to guide the operations.

The mission of the web-based Kanban system in a virtual enterprise is to facilitate collaborative production and control. As an information system for the large and complex organisation, it has to be capable of accommodating the amount of data flows and number of users. How to maintain data integrity, create visibility of information, and manage the tasks in such a large volume becomes a challenge. Wan *et al.* [13.55] summarised three types of management strategies for agile supply chain settings as shown in Figure 13.9, and discussed below.

Agile Virtual Enterprise with Few Major Controllers

When there are one or few powerful units in a supply chain, the major controller(s) remain unchanged, while the secondary and other upstream suppliers are engaged dynamically. Many examples of this type of configuration can be found, such as Wal-Mart in the wholesale-retail business, Dell in the computer industry, and GM in the automobile industry. The few major controllers usually face customers directly and enjoy higher power while negotiating with suppliers. In this setting, the web-based Kanban system can be provided and maintained by a major controller in the supply chain. As a result, the system can be better structured and maintained for the core business of the controller, and the accountability of each party can be clearly designated. However, the secondary and third-tier suppliers have relatively weaker power. The accountability and profit sharing may not be perfectly fair in this configuration. On the other hand, this type of configuration inherits more properties of conventional supply chains and lean enterprises. It is less flexible in terms of the formation of virtual enterprises, which may lead to lost business opportunities.

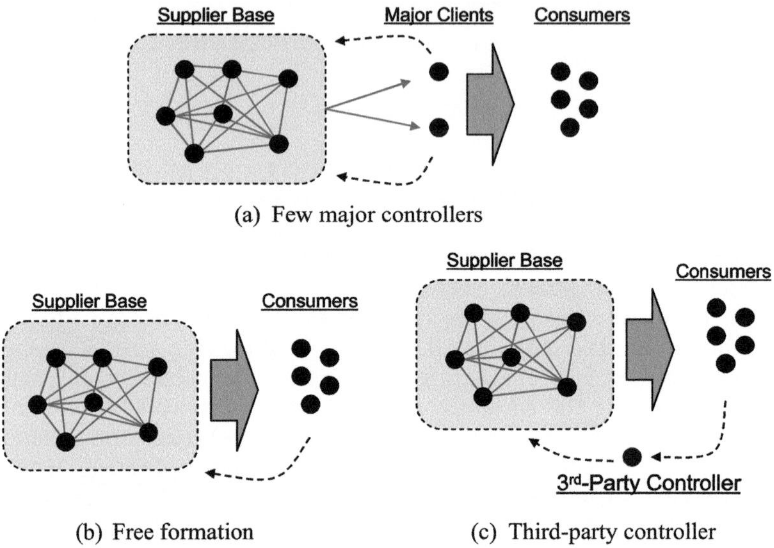

Figure 13.9. Web-based Kanban enabled agile virtual enterprise with three different control strategies

Agile Virtual Enterprise with Free Formation

In the free formation environment, suppliers have equal power in negotiations. A web-based Kanban system for this configuration can be provided by an Internet service provider to host the free-market platform. Without a major controller, an auction module needs to be embedded or linked to the web-based Kanban system. Participants select their suppliers from the auction platform, which determine the upstream routing. Currently, some business-to-business auction web services (*e.g.* alibaba.com in China) resemble the free formation settings, but web-based Kanban has not been involved. The free formation setting enjoys the highest level of flexibility based on commitment of each participant. However, the accountability and profit sharing may not be as clear as the supply chain with major controllers. The auction platform must keep detailed records of the performance and credibility of each participant. Poor performance of any participant would hurt the overall efficiency of the whole virtual enterprise.

Agile Virtual Enterprise with a Third-party Controller

Forming virtual enterprises by a professional third-party controller is another solution. The controller maintains the web-based Kanban system to bridge the end customers with a flexible supplier base. Upon receiving a customer order, suppliers can be recruited by the controller or participate through an auction process. A good example is the trading companies in East Asia that typically take the lead on recruiting suppliers for its customers based on professional judgement. However, using web-based Kanbans throughout a system has not been reported. In this setting, the accountability, profit sharing, and the credibility of suppliers can be better defined and controlled. The controller should possess a certain level of professional knowledge about the industry to ensure the quality of the service. The result is a compromised solution between flexibility and stability compared with the other two strategies. Yet, the controller can actively improve the transparency of information to accommodate lean and agile operations.

13.5.3 Ensuring Leanness of the Agile Virtual Enterprise

Similar to the virtual cells, the formation of a virtual enterprise delivers great flexibility at the supply chain level. The information system plays a critical role to enable and manage the distributed manufacturing setting. In order to enhance the leanness level while ensuring its agility, the distributed system must incorporate lean logistics approaches. Evolved from a lean manufacturing technique, the web-based Kanban system conducts the material flows in a pull manner to minimise non-value-added activities. Upon formation of the virtual enterprise, lean logistics can be implemented with the aid of web-based Kanban system. Some lean logistics approaches mentioned in Rivera *et al.* [13.2] are listed below with discussions on virtual enterprise applications.

- *Just-in-Time Delivery*: With properly configured web-based Kanbans, pull system can be implemented in the virtual enterprise setting with demand levelling to ensure JIT deliveries. Involving material handlers in the Kanban system facilitates a "milk run" system with smaller batches when applicable.
- *Cross Docking*: Using the web-based Kanban system, a distribution centre can be part of the pull system, which transports materials based on Kanban signals to implement cross docking and reduce material handling time.
- *Vendor Managed Inventory (VMI)*: Using web-based Kanban system, the "supermarket" pull systems can be established among the virtual enterprise members regardless the physical locations.
- *Third-party Logistics (3PL)*: As a web-based Kanban system user, a third-party logistics provider can be pulled to transport materials similarly to a material handling unit on a shop floor. As a result, small-batch production or one-piece flow can be realised when transportation cost is affordable.
- *Decoupling Point*: Establishing decoupling points is an important approach of agile supply chains to optimally divide the operations into push and pull segments. In a web-based Kanban setting, high-level Kanbans can be used to manage the flows between push and pull segments, while a separate group of

production Kanbans take care of detailed processes within a segment. The use of decoupling point allows different companies or departments in a virtual enterprise to operate with the most appropriate method (*e.g.* pull or push) to ensure the core competency of each participating organisation.

The lean logistics approaches listed above show the possibility of implementing lean principles in a virtual enterprise. Many other lean tools are also applicable. Some of the approaches may be more difficult to implement depending on various factors of the actual environment. Overall, the goal is to create a user-driven collaborative manufacturing system monitored and controlled by the web-based Kanban system.

13.6 Challenges and Future Research

13.6.1 Challenges of Web-based Kanban in an Agile Virtual Enterprise

The proposed integration of web-based Kanban technology with the agile virtual enterprise provides a practical solution to achieve a lean and agile manufacturing system. However, a few issues associated with the framework and software/hardware implementations remain challenging. In summary, the following issues need to be resolved gradually if not immediately.

- To implement the web-based Kanban system, computer terminals and related devices (*e.g.* barcode reader, *etc.*) must be present at every workstation if the system is being carried out at a detailed level. On the other hand, if the Kanban system is only applied to the macro-level (*e.g.* pulling from a department or a plant), JIT deliveries and other lean principles may not be guaranteed at shop floor level. Visibility also becomes limited.
- Web security is always a critical issue for web-based programs. Appropriate technologies and techniques must be in place to ensure the security of connections, especially for inter-organisational applications. Emergency options should be prepared in case severe problems occur.
- User friendly interfaces have to be established to enhance the leanness of operations involved in the web-based program. Typically, web-based programming is less complicated, but the flexibility is limited compared with other programming environments, such as C++ or Java. For example, creating a web-based reporting module with charts and diagrams similar to MS Excel is already challenging. How to continuously improve the user interface is an important issue.
- Interfacing with other software programs and hardware devices is another challenge. Due to proprietary data format of various systems, many software interfaces may need to be developed to connect with existing systems.
- System setup and maintenance may be the most challenging task. In a stable lean environment, setting up the system for the first time may require some time and effort, but only minor adjustment and maintenance are needed later on. For agile virtual enterprises, every product can be different, which leads to frequent reconfigurations and adjustments. Furthermore, due to the

distance and number of transactions, maintaining the data accuracy and integrity is not easy. How to detect/prevent errors and correct the discrepancies are issues to be addressed. Again, the user interface plays a critical role in the maintenance functions and needs to be well established.

In addition to the abovementioned issues, the users of this system have to clearly define the performance metrics, responsibilities, benefit sharing, and penalties. A secondary platform can be established within the system to facilitate information sharing, interactive communications, and knowledge accumulation. This will result in a learning organisation that grows continuously to meet the market demands.

13.6.2 Conclusions and Future Research

This chapter illustrates an approach to facilitate an agile virtual enterprise with a lean tool, the web-based Kanban system. Several directions and options for carrying out the lean and agile supply chain configuration have been reviewed. Although there are some issues to be addressed, benefits of using a web-based Kanban can be significant. Many software providers have started to develop economic and effective web-based Kanban systems for JIT operations, inventory control, and lean logistics. A rapidly increasing amount of packages are being offered to the market as an affordable alternative to conventional ERP and MES systems. With the increasing needs of customisation and distributed manufacturing, global supply chains have to aim at integrating lean and agile strategies to sustain competitiveness in the market. Web-based Kanban system is an enabler of the integration.

To create a viable web-based Kanban system for agile virtual enterprises, more research effort is necessary. For the software program, building or integrating with an auction platform will enhance the applicability of the system in the formation of partnerships. A module to evaluate performance and credibility of potential partners will provide vital information to the users. Determination of appropriate number of Kanbans in the web-based virtual enterprise settings needs to be studied. Furthermore, software interfaces need to be developed to bridge the web-based Kanban system with existing software and hardware so that the participating companies can be quickly engaged or dismissed. Enhancing security is another major issue that requires continuous effort. To support the flexible configuration, more studies need to be carried out to identify the best method to handle the constantly changing routings, product configuration, and partners. Finally, policies and strategies to define a fair responsibility, performance objectives, costing and profit sharing, and penalties must be investigated and established. The ultimate goal of the research is to create a system to meet customer demand by quickly engaging best-in-class partners to ensure maximum flexibility and responsiveness while embedding lean principles to minimise non-value-added activities.

Acknowledgement

This study has been partially funded by the National Science Foundation (NSF) under the Major Research Instrumentation (MRI) research grant CMMI-0722923.

References

[13.1] Womack, J.P. and Jones, D.T., 2005, *Lean Solutions*, Free Press, New York.

[13.2] Rivera, L., Wan, H., Chen, F.F. and Lee, W., 2007, "Beyond partnerships: the power of lean supply chains," In *Trends in Supply Chain Design and Management: Technologies and Methodologies*, Jung, H., Chen, F.F. and Jeong, B. (eds), Springer, Surrey, UK, pp. 241–268.

[13.3] Wan, H. and Chen, F.F., 2008, "A web-based Kanban system for job dispatching, tracking, and performance monitoring," *International Journal of Advanced Manufacturing Technology*, (in press).

[13.4] Goranson, H.T., 1999, *The Agile Virtual Enterprise: Cases, Metrics, Tools*, Quorum Books, London.

[13.5] Monden, Y., 1998, *Toyota Production System: An Integrated Approach to Just-in-Time*, 3rd Edition, Institute of Industrial Engineers, Norcross, GA.

[13.6] Ohno, T., 1988, *Toyota Production System*, Productivity Press, Portland, OR.

[13.7] Sugimori, Y., Kusunoki, K., Cho, F. and Uchikawa, S., 1977, "Toyota production system and Kanban system: materialisation of just-in-time and respect-for-human system," *International Journal of Production Research*, **15**(6), pp. 553–564.

[13.8] Womack, J.P., Jones, D.T. and Roos, D., 1990, *The Machine that Changed the World*, Macmillan, New York.

[13.9] Womack, J.P., 2004, "Deconstructing the Tower of Babel," Womack's e-letters of October 2004 in *WMEP Articles*, http://www.wmep.org/Articles/towerbabel.aspx.

[13.10] Womack, J.P. and Jones, D.T., 1996, *Lean Thinking: Banish Waste and Create Wealth in Your Corporation*, Simon and Schuster, New York.

[13.11] Rother, M. and Shook, J., 1998, *Learning to See – Value Stream Mapping to Add Value and Eliminate Muda*, The Lean Enterprise Institute, Brookline, MA.

[13.12] Sarker, B.R., 1989, "Simulating a Just-in-Time production system," *Computers and Industrial Engineering*, **16**(1), pp. 127–137.

[13.13] Siha, S., 1994, "The pull production system: modelling and characteristics," *International Journal of Production Research*, **32**(4), pp. 933–949.

[13.14] Spearman, M.L., Woodruff, D.L. and Hopp, W.J., 1990, "CONWIP: a pull alternative to Kanban," *International Journal of Production Research*, **28**(5), pp. 879–894.

[13.15] Sipper, D. and Bulfin, R.L. Jr, 1997, *Production Planning, Control, and Integration*, McGraw-Hill, New York.

[13.16] Hopp. W.J. and Spearman, M.L., 2000, *Factory Physics: Foundations of Manufacturing Management*, McGraw-Hill, New York.

[13.17] Phelps T., Smith M. and Hoenes T, 2004, "Building a lean supply chain," *Manufacturing Engineering*, **132**, pp. 107–116.

[13.18] Jones, D.T. and Womack, J.P., 2002, *Seeing the Whole: Mapping the Extended Value Stream*, Lean Enterprise Institute, Brookline MA.

[13.19] Schonberger, R.J., 2005, "Lean extended: it's much more (and less) than you think," *Industrial Engineer*, December, pp. 26–31.

[13.20] Reeve, J.M., 2002, "The financial advantages of the lean supply chain," *Supply Chain Management Review*, March/April, pp. 42–49.

[13.21] Vitasek, K., Manrodt, K.B. and Abbott, J., 2005, "What makes a lean supply chain," *Supply Chain Management Review*, October, pp. 39–45.

[13.22] Gunasekaran, A. and Yusuf, Y.Y., 2002, "Agile manufacturing: a taxonomy of strategic and technological imperatives," *International Journal of Production Research*, **40**(6), pp. 1357–1385.

[13.23] Sanchez, L.M. and Nagi, R., 2001, "A review of agile manufacturing systems," *International Journal of Production Research*, **39**(16), pp. 3561–3600.

[13.24] Katayama, H. and Bennett, D., 1999, "Agility, adaptability and leanness: a comparison of concepts and a study of practice," *International Journal of Production Economics*, **60–61**(1), pp. 43–51.

[13.25] Yusuf, Y.Y. and Adeleye, E.O., 2002, "A comparative study of lean and agile manufacturing with a related survey of current practices in the UK," *International Journal of Production Research*, **40**(17), pp. 4545–4562.

[13.26] Goldman, S.L., Nagel R.N. and Preiss, K., 1995, *Agile Competitors and Virtual Organisations*, Van Nostrand Reinhold, New York.

[13.27] Kannan, V.R. and Ghosh, S., 1996, "Cellular manufacturing using virtual cells," *International Journal of Operations & Production Management*, **16**(5), pp. 99–13.

[13.28] Babu, A.S., Nandurkar, K.N. and Thomas, A., 2000, "Development of virtual cellular manufacturing systems for SMEs," *Logistics Information Management*, **13**(4), pp. 228–242.

[13.29] Baykasoglu, A., 2003, "Capability-based distributed layout approach for virtual manufacturing cells," *International Journal of Production Research*, **41**(11), pp. 2597–2618.

[13.30] Prince, J. and Kay, J.M., 2003, "Combining lean and agile characteristics: creation of virtual groups by enhanced production flow analysis," *International Journal of Production Economics*, **85**(3), pp. 305–318.

[13.31] Naylor, J.B., Naim, M.M. and Berry, D., 1999, "Leagility: integrating the lean and agile manufacturing paradigms in the total supply chain," *International Journal of Production Economics*, **62**(1), pp. 107–118.

[13.32] Kumar, C.S. and Panneerselvam, R., 2007, "Literature review of JIT-KANBAN system," *International Journal of Advanced Manufacturing Technology*, **32**(3–4), pp. 393–408.

[13.33] Berkley, B.J., 1992, "A review of the Kanban production control research literature," *Production and Operations Management*, **1**(4), pp. 393–411.

[13.34] Philipoom, P.R., Rees, L.P., Taylor, B.W. III and Huang, P.Y., 1987, "An investigation of the factors influencing the number of Kanbans required in the implementation of the JIT technique with Kanban", *International Journal of Production Research*, **25**(3), pp. 457–72.

[13.35] Karmarkar, U.S. and Kekre, S., 1989, "Batching policy in Kanban system," *Journal of Manufacturing Systems*, **8**(4), pp. 317–328.

[13.36] Sarker, B.R. and Balan, C.V., 1996, "Operations planning for Kanbans between two adjacent workstations," *Computers and Industrial Engineering*, **31**(1–2), pp. 221–224.

[13.37] Rshini, B.S. and Ran, C.R., 1997, "Scheduling in a Kanban-controlled flowshop with dual blocking mechanisms and missing operations for part-types," *International Journal of Production Research*, **35**(11), pp. 3133–3156.

[13.38] Altiok, T. and Shiue, G.A., 2000, "Pull type manufacturing systems with multiple product types," *IIE Transactions*, **32**(2), pp. 15–124.

[13.39] Bitran, G.R. and Chang, L., 1987, "A mathematical programming approach to a deterministic Kanban system," *Management Science*, **33**(4), pp. 427–441.

[13.40] Davis, W.J. and Stubitz, S.J., 1987, "Configuring a Kanban system using a discrete optimisation of multiple stochastic responses," *International Journal of Production Research*, **25**(5), pp. 721–740.

[13.41] Levy, D.L., 1997, "Lean production in an international supply chain," *Sloan Management Review*, **38**(2), pp. 94–102.

[13.42] Liker, J.K. and Wu, Y.C., 2000, "Japanese automakers, U.S. suppliers and supply-chain superiority," *Sloan Management Review*, **42**(1), pp. 81–93.

[13.43] Cutler, T.R., 2005, "Dis"card"ing the paper Kanban: electronic web-based Kanbans support lean manufacturing," InMFG, **62**, pp. 30–31.

[13.44] Drickhamer, D., 2005, "The Kanban e-volution," *Material Handling Management*, **60**, pp. 24–26.
[13.45] Osborne, T., 2002, "Internet Kanban delivers just in time," *Illinois Manufacturer Magazine*, March/April, pp. 9–27.
[13.46] Yen, D.C., Chou, D.C. and Chang, J., 2002, "A synergic analysis for Web-based enterprise resource planning systems," *Computer Standards & Interfaces*, **24**(4), pp. 337–346.
[13.47] Ng, J.K.C. and Ip, W.H., 2003, "Web-ERP: the new generation of enterprise resource planning," *Journal of Materials Processing Technology*, **138**(1), pp. 590–593.
[13.48] Ash, C.G. and Burn, J.M., 2003, "A strategic framework for the management of ERP enabled e-business change," *European Journal of Operational Research*, **146**(2), pp. 373–387.
[13.49] Verwijmeren, M., 2004, "Software component architecture in supply chain management," *Computers in Industry*, **53**(2), pp. 165–178.
[13.50] Helo, P., Xiao, Y. and Jiao, J.R., 2006, "A web-based logistics management system for agile supply demand network design," *Journal of Manufacturing Technology Management*, **17**(6), pp. 1058–1077.
[13.51] Tarantilis, C.D., Kiranoudis, C.T. and Theodorakopoulos, N.D., 2008, "A Web-based ERP system for business services and supply chain management: application to real-world process scheduling," *European Journal of Operational Research*, **187**(3), pp. 1310–1326.
[13.52] Kotani, S., 2007, "Optimal method for changing the number of Kanbans in the e-Kanban system and its applications," *International Journal of Production Research* **45**(24), pp. 5789–5809.
[13.53] Saygin, C., Sarangapani, J. and Grasman, S.E., 2007, "A systems approach to viable RFID implementation in the supply chain," In *Trends in Supply Chain Design and Management: Technologies and Methodologies*, Jung, H., Chen, F.F. and Jeong, B. (eds), Springer, Surrey, UK, pp. 3–27.
[13.54] Rheault, M., Drolet, J.R. and Abdulnour, G., 1995, "Physically reconfigurable virtual cells: a dynamic model for a highly dynamic environment," *Computers and Industrial Engineering*, **29**(1), pp. 221–225.
[13.55] Wan, H. and Chen, F.F., 2007, "Enabling agile supply chain with web-based Kanban system," In *Proceedings of 8th APIEMS & CIIE Conference* (on CD-ROM).

14

Agent-based Workflow Management for RFID-enabled Real-time Reconfigurable Manufacturing

George Q. Huang[1], YingFeng Zhang[2], Q. Y. Dai[3], Oscar Ho[1] and Frank J. Xu[4]

[1] Department of Industry and Manufacturing Systems Engineering
The University of Hong Kong, Hong Kong, China
Email: gqhuang@hku.hk

[2] School of Mechanical Engineering
Xi'an Jiaotong University, Xi'an, Shaanxi, China
Email: xjtuzyf@mail.xjtu.edu.cn

[3] Faculty of Information Engineering
Guangdong University of Technology
Goangzhou, Guangdong, China

[4] E-Business Technology Institute
The University of Hong Kong
Hong Kong, China

Abstract

Recent developments in wireless technologies have created opportunities for developing reconfigurable manufacturing systems with real-time traceability, visibility and interoperability in shop floor planning, execution and control. This chapter proposes to use workflow management as a mechanism to facilitate an RFID-enabled real-time reconfigurable manufacturing system. The workflow of production processes is modelled as a network. Its nodes correspond to the work (process), and its edges to flows of control and data. The concept of agents is introduced to define nodes and the concept of messages to define edges. As a sandwich layer, agents wrap manufacturing services (*e.g.* machines, RFID devices and tools) and their operational logics/intelligence for cost-effectively collecting and processing real-time manufacturing data. Some referenced frameworks and architectures of manufacturing gateway, shop-floor gateway and work-cell gateway are constructed for implementing the RFID-enabled real-time reconfigurable manufacturing system. The shop-floor gateway is mainly discussed where three key components (workflow management tools, MS-UDDI and agent-based manufacturing services management tools) are integrated. By means of web service technologies, each agent can be registered and published at MS-UDDI as a web service that can be easily reused and reconfigured as a workflow node according to the workflow of a specific production process through workflow management to a server for reconfigurable goals. The methodologies and technologies proposed in this chapter will allow manufacturing enterprises to improve shop-floor productivity and quality, reduce the wastes of manufacturing resources, cut the costs in manufacturing logistics, reduce the risk and improve the efficiency in cross-border customs logistics and online supervision, and improve the responsiveness to market and engineering changes.

14.1 Introduction

With the increasing competitiveness and globalisation of today's business environment, enterprises have to face a new economic objective: manufacturing responsiveness, *i.e.* the ability of a production system to respond to disturbances that impact upon production goals, and consequently, its ability to adapt to changing production conditions of shop floor level.

Even if many manufacturing companies have implemented sophisticated ERP (enterprise resource planning) systems, the following problems are suffering from:

- Customer orders, process plans, production orders and production scheduling are conducted in separate systems that are not integrated.
- The availability of raw materials is not known at the time of production scheduling. The scheduler has to visit each inventory area to count the available materials before starting scheduling.
- Shop-floor disturbances such as machine breakdowns and maintenance are not fed back and considered when planning the production order, resulting in unbalanced lines.
- The loading level of work orders at specific machine is unknown to the scheduler and production planner, leading to further line unbalances.
- WIP inventories are highly dynamic – changing frequently between production stations or lines.
- Errors and confusions in handling WIP items are common – leaving a piece of hand-written paper form on the stack and leave the pallet wherever a space is spotted.
- Separate personnel are responsible for auditing the materials at the shop floor and warehouse. But the frequencies of such inventory audits are not high enough or consistent with the frequencies required by the production planning and scheduling systems.
- Separate personnel are responsible for entering the shop-floor data into the computer terminals, usually towards the later stages of their shifts. Any left-over data entries would be completed until the next shift.

Therefore, it is essential to adapt advanced manufacturing technologies and approaches (both software and hardware) to cope with the highly dynamic manufacturing requirements. In recent decades, rapid developments in wireless sensors, communication and information network technologies (*e.g.* radio frequency identification – RFID or Auto-ID, Bluetooth, Wi-Fi, GSM, and infrared) have nurtured the emergence of wireless manufacturing (WM), reconfigurable manufacturing system (RMS) as core advanced manufacturing technology (AMT) in next-generation manufacturing systems (NGMS).

The RMS was introduced in the mid-1990s as a cost-effective response to market demands for responsiveness and customisation. The RMS has its origin in computer science in which reconfigurable computing systems try to cope with the inefficiencies of the conventional systems due to their fixed hardware structures and software logic. Here, reconfiguration allows adding, removing or modifying specific process capabilities, controls, software, or machine structure to adjust production capacity in response to changing production demands or technologies.

In order to utilise an RMS, manufacturers must learn to operate effectively in dynamic production environments, which are characterised by unpredictable market demands and by the proliferation of product variety, as well as rapid changes of product and process technologies. The initial idea of reconfigurable computing systems dates from the 1960s [14.1]. This innovative paradigm dissolved the hard borders between hardware and software and joined the potentials of both. Zhao *et al.* [14.2] considered an RMS as a manufacturing system in which a variety of products required by customers are classified into families, each of which is a set of similar products, and which correspond to one configuration of the RMS. Mehrabi *et al.* [14.3] proposes five key characteristics: modularity, integratability, convertibility, diagnosability, and customisation for RMSs. Yigit and Ulsoy [14.4] describes a modular structure to accommodate new and unpredictable changes in the product design and processing needs through easily upgrading hardware and software rather than the replacements of manufacturing services elements such as machines.

Real-time visibility and interoperability are considered core characteristics of next-generation manufacturing systems [14.5]. Pilot projects have recently been implemented and reported (see various whitepapers and reports available at http://www.autoidlabs.com/research archive/ for more descriptions). The progress of wireless technologies such as RFID and Auto-ID applications in the "manufacturing scenario" has been noticeable although limited. As early as the early 1990s, Udoka [14.6] discussed the roles of Auto-ID as a real-time data capture tool in a computer integrated manufacturing (CIM) environment. Early RFID manufacturing applications were briefly quoted in [14.7] and further promoted in [14.8]. Johnson [14.9] presents an RFID application in a car production line. The website http://www.productivitybyrfid.com/ also provides a few links to real-life pilot cases. Chappell *et al.* [14.10] provides a general overview on how Auto-ID technology can be applied in manufacturing. Several relevant whitepapers have been prepared to provide roadmap for developing and adopting Auto-ID-based manufacturing technologies [14.11, 14.12]. More recently, the Cambridge Auto-ID Lab launched an RFID in Manufacturing Special Interest Group (SIG) (http://www.aero-id.org/). However, further investigations are needed if an integrated solution is to be delivered.

The concept of agent has been widely accepted and developed in manufacturing applications because of its flexibility, reconfigurability and scalability [14.13–14.15]. An agent-based concurrent design environment [14.17, 14.18] has been proposed to integrate design, manufacturing and shop-floor control activities. A compromised and dynamic model in an agent-based environment [14.13] was designed for all agents carrying out their own tasks, sharing information, and solving problems when conflicts occur. Some mobile agent-based systems [14.19] have been applied to the real-time monitoring and information exchange for manufacturing control. Jia *et al.* [14.20] proposed an architecture where many facilitator agents co-ordinate the activities of manufacturing resources in a parallel manner. Jiao *et al.* [14.21] applied MAS (multi-agent system) paradigm for collaborative negotiation in a global manufacturing supply-chain network. Besides, in various applications, *e.g.* distributed resource allocation [14.22], online task co-ordination and monitoring [14.23], or supply chain negotiation [14.24], the agent-based approach has played an important role to achieve outstanding performance with agility.

However, to apply real-time manufacturing with RFID technologies in manufacturing enterprises will suffer such difficulties, *e.g.* (1) both enterprises and customs authorities recognise the potential of RFID technologies, but they are not quite clear how to use RFID technologies cost-effectively to achieve shop-floor visibility and traceability and to facilitate the cross-border customs logistics – declaration, inspection and clearance; (2) shop-floor operators and supervisors do not have adequate tools for reconfiguring a manufacturing process, for visualising, monitoring and controlling its execution, and for examining shop-floor bottleneck problems and in turn responding with timely decisions.

Table 14.1. List of key acronyms

Acronym	Description
AMT	Advanced Manufacturing Technology
EAS	Enterprise Application System
FIPA	Foundation for Intelligent, Physical Agents
MAS	Multi-agent System
MS-UDDI	Manufacturing Services UDDI
NGMS	Next-generation Manufacturing Systems
OEM	Original Equipment Manufacturer
PRD	Pearl River Delta
RFID	Radio Frequency Identification
RMS	Reconfigurable Manufacturing System
RTM	Real-time Manufacturing
RTM-SII	Real-time Manufacturing Information Infrastructure
SASs	Shop-floor Application Systems
SOA	Service-oriented Architecture
SOAP	Simple Object Access Protocol
SOSs	Smart Object Services
UDDI	Universal Description, Discovery and Integration
UPnP	Universal Plug and Play
WAS	Work-cell Application System
WFM	Workflow Management
WIP	Work in Progress
WM	Wireless Manufacturing
WSDL	Web Services Description Language
WSIG	Web Services Integration Gateway

In this research, it is particularly challenging to integrate the advantage of RMS, wireless technologies and agent-based methods. On one hand, wirelessly networked sensors facilitate the automatic collection and processing of real-time field data in the manufacturing processes, and reduce and avoid the error-prone, tedious manual activities. On the other hand, an agent-based system enables relevant activities to be more flexible, intelligent and collaborative, especially in a distributed networked environment for building RMSs. Therefore, a concept of agent-based workflow management, using agent theories to wrap manufacturing services, is proposed and applied to reconfigure the production elements according to changing demands of an agent-based and RFID-enabled real-time RMS. This RFID-enabled real-time RMS is a new paradigm for production systems that addresses the need for introducing greater flexibility into a high production environment where product volumes and types change. The acronyms used throughout the chapter are explained in Table 14.1.

14.2 Overview of Real-time Reconfigurable Manufacturing

RFID technologies are applied to develop an easy-to-deploy and simple-to-use shop-floor information infrastructure for manufacturing companies to achieve real-time and seamless dual-way connectivity and interoperability between application systems at enterprise, shop-floor, work-cell and device levels. The role of the proposed technology is shown in Figure 14.1.

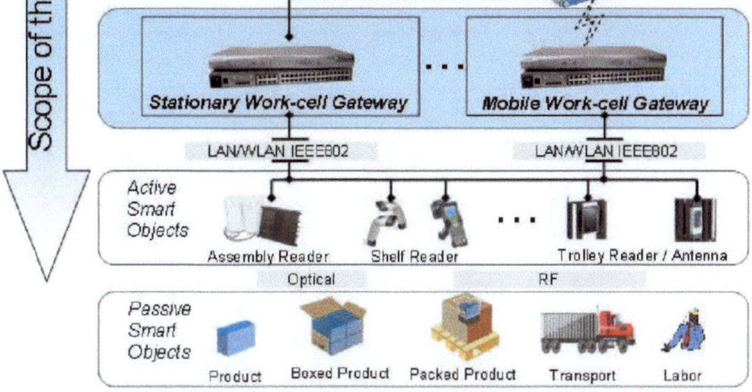

Figure 14.1. Deployment of real-time manufacturing infrastructure

The use of this proposed technology will allow manufacturing enterprises to improve shop-floor productivity and quality, reduce waste of manufacturing resources, cut the costs in manufacturing logistics, reduce the risk and improve the efficiency in cross-border customs logistics and online supervision, and improve the responsiveness to market and engineering changes.

The proposed infrastructure is consistent with the manufacturing hierarchy. That is, a manufacturing factory hosts one or more shop-floor production lines. Each production line consists of many work cells and each work cell is involved with a variety of manufacturing objects such as operators, machines, materials, various containers, *etc.*, and different production lines are often designed to enable different production processes.

Shop-floor-level processes and operations are generally configured from lower-level activities and tasks according to specific manufacturing requirements. Corresponding to these three levels, the real-time manufacturing gateway system (RTM-Gateway in short) proposed in this chapter is composed of the following three core components. RTM-Gateway possesses two-way functionality as shown in Figure 14.2. From low-level shop-floor operations to high-level enterprise decision, RTM-Gateway collects real-time shop-floor data and processes the data into useful contents in mutually understandable formats for EASs. Along the reverse direction, RTM-Gateway receives plans and schedules from EASs and processes these production data into the right contents and formats such as work orders suitable for consumption by shop-floor operators and devices.

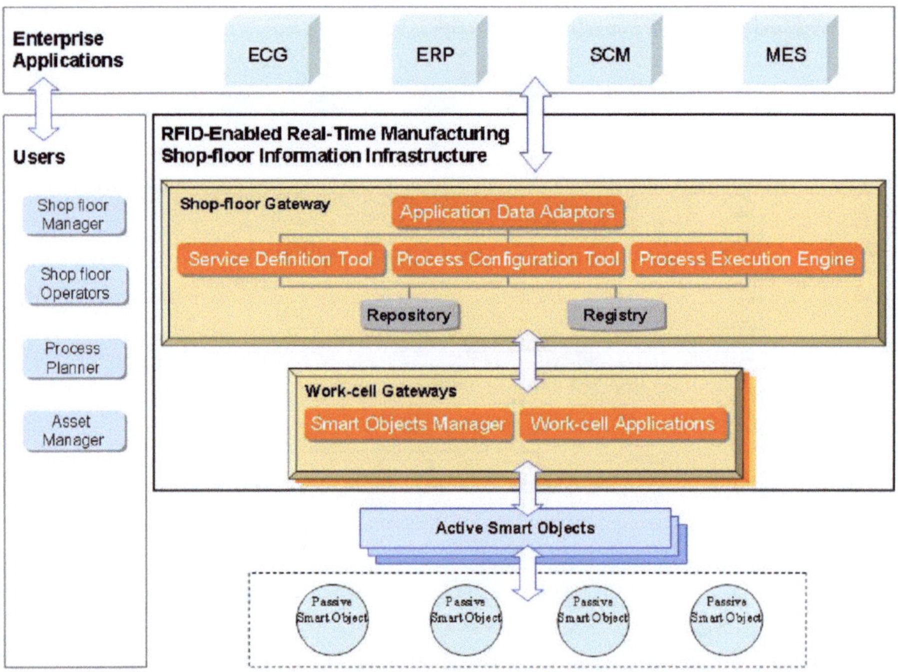

Figure 14.2. Architecture of real-time manufacturing gateway

According to the manufacturing hierarchy, the proposed RTM infrastructure includes the following core components:

- *Shop-floor Gateway*: The shop-floor gateway (SF-Gateway) is at the centre of the overall RTM-SII. Its architecture is based on the service-oriented architecture (SOA). SF-Gateway includes three main components, *i.e.* workflow management tools, MS-UDDI and agent-based manufacturing services management tools, which will be introduced in the next section in detail.

- *Work-cell Gateways*: A Work-cell Gateway (WC-Gateway) acts as a server that hosts and connects all RFID-enabled smart objects of the corresponding work cell. A WC-Gateway has a hardware hub and a suite of software systems. The hub is the server that connects all RFID-enabled smart objects, and the smart objects are represented as software agents in the WC-Gateway operating system within which they are "universal plug and play (UPnP)" and interoperable.
- *RFID-enabled Smart Objects*: The smart objects are those physical manufacturing objects that are made "smart" by equipping them with RFID devices. Those with RFID readers are active smart objects. Those with RFID tags are passive smart objects. Smart objects interact with each other through wired and/or wireless connections, creating what is called an intelligent ambience. In addition, smart objects are also equipped with their specific operational logics, and with data memory and processing functions. Therefore, smart objects are able to sense, reason, act/react/interact in the intelligent ambience community.

14.3 Overview of Shop-floor Gateway

Figure 14.3 shows the overall framework of the SF-Gateway. Following the SOA, enterprise application systems (EASs), shop-floor application systems (SASs), equipments and smart object services (SOSs) in a manufacturing company can all be considered as manufacturing services. SF-Gateway manages the lifecycle of these manufacturing services. At the definition stage, all manufacturing services are deployed and installed on their servers and then registered at the SF-Gateway MS-UDDI, readily available for the next lifecycle stage – *configuration*. Process planners use SF-Gateway's configuration facilities to search and set up suitable WASs and SOSs for configuring a specific manufacturing process. The result from the configuration stage is a manufacturing process represented as a workflow between work cells, ready for the next lifecycle stage – *execution*. While the actual execution of work-cell application systems is carried out at their gateway servers, SF-Gateway provides tools for shop-floor managers to monitor and control the status and progress of the execution of the manufacturing process. Real-time data are handled centrally at the SF-Gateway repository. SF-Gateway is composed of three major components, which are described as follows.

14.3.1 Workflow Management

This component includes four modules for: (a) definition, (b) reconfiguration, (c) execution, and (d) repository.

- *Definition* module is responsible for defining the workflow according to specific production processes.
- Process planner use *reconfiguration* module to search and choose the suitable agents (*e.g.* work cell) through MS-UDDI for producing specific production

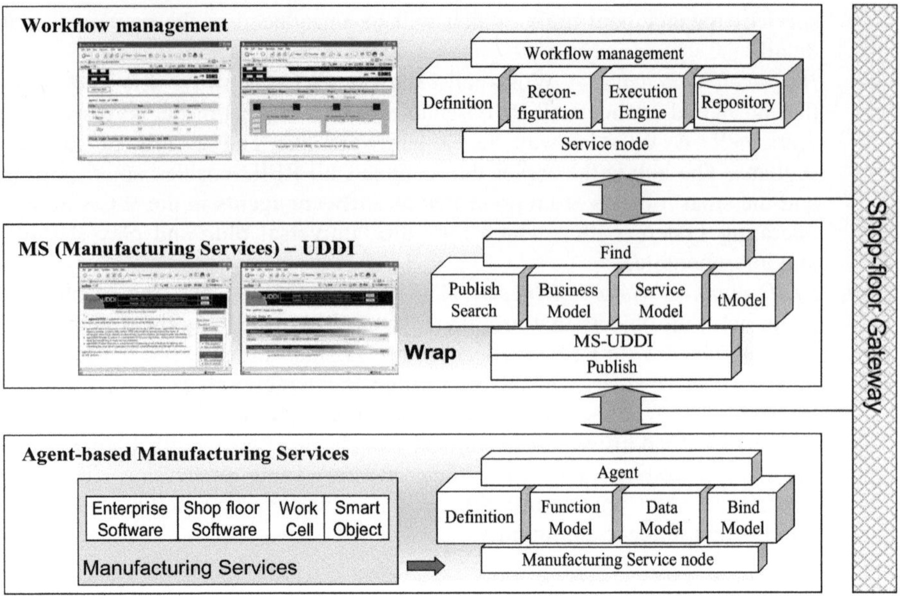

Figure 14.3. Overview of shop-floor gateway

processes. The data configuration operators between output data and input data of chosen agents are also executed in this module.

- *Execution* module provides tools for shop floor managers to monitor and control the status and progress of the execution of the manufacturing process while the actual execution of agents are carried out at their manufacturing services.
- The data generated from the lifecycle stages are maintained in the *repository*. Real-time data are handled centrally at the repository. At each lifecycle stage, the SF-Gateway provides services (facilities) for target users and service consumers.

14.3.2 Manufacturing Services UDDI

Manufacturing services UDDI (MS-UDDI) performs functions similar to those of standard UDDI that is a platform-independent framework for describing services, discovering businesses, and integrating business services through the Internet. MS-UDDI includes four main modules, which are (a) publish and search, (b) business model, (c) service model, and (d) tModel.

- *Publish* module is used to issue agents as web services that can be easily found and communicated by using WSDL (Web Services Description Language) and SOAP (Simple Object Access Protocol), while the *search* module is a web-based GUI (graphic user interface) for users to discover web services published at MS-UDDI and perform the services according to the WSDL documents and binding information.

- *Business model* is contained in a *businessEntity* structure, which contains information about the business that has published the service, such as business name, description, contacts and identifiers.
- *Service model* describes a group of web services that are contained in a *businessService* structure. The *businessService* contains information about families of technical services. It groups a set of web services related to either a business process or group of services. Each *businessEntity* might hold one or several *businessServices*.
- *tModel* describes the specifications for services. To invoke a web service, one must know a service's location and the kind of interface that the service supports. A *bindingTemplate* indicates the specifications or interfaces that a service supports through references to the specification information. Such a reference is called a *tModelKey*, and the data structure encapsulating the specification information is called a *tModel*.

14.3.3 Agents-based Manufacturing Services

This component includes four modules, which are (a) definition, (b) function model, (c) data model, and (d) bind model.

- *Definition* module is responsible for defining the corresponding agents to wrap the manufacturing services deployed and installed on each WC-Gateway server and then their general information is published in a standard format at the MS-UDDI. When an agent is defined and published, the agent represents the behaviour, activity and function of the manufacturing services. Generally, the manufacturing services of WC-Gateway include hardware and software, *e.g.* equipment, workstation, smart objects and corresponding software (driver of smart object, shop floor and enterprise software).
- *Function model* is used to implement the behaviour of the WC-Gateway that is wrapped by its agent. In this research, each function includes one activity. Accordingly, the behaviour of each manufacturing service is a logic and sequence relationship among these functions. For example, for an RFID smart manufacturing service, a *write string 'x' to tag '0001'* behaviour of its agent will be (1) read tags, (2) find tag '0001' and (3) write 'x' to this tag; here, read tags, find tag and write tag are functions of the RFID smart manufacturing service.
- *Data model* describes the basic input and output data standard of the WC-Gateway that is wrapped by its agent. The data model adopts XML-based schema that can be easily edited, transformed and extended. It is stated that this data model only defines the static structure of the input and output of agents. The dynamic output data is stored for sharing at the Shop-floor Gateway when the agent is running.
- *Bind model* provides information required to invoke a web service (agent) published at the MS-UDDI. Once the agent has been configured to the production process, the bind model will automatically record the relationship between the agent and the process. It is useful for reconfiguring various production processes.

14.4 Overview of Work-cell Gateway

A WC-Gateway has a hardware hub and a suite of software, which acts as a server that hosts and connects all RFID-enabled smart objects of the corresponding work cell and provides work-cell applications for operators to conduct, monitor and control their operations. Basically, there is a smart object manager (SOM) responsible for co-ordinating all smart objects in their lifecycle, including definition, installation, and configuration. Apart from smart objects, the gateway also hosts the work-cell applications that are built-in services providing integrated real-time information of work cells (*i.e.* WIP, the status of workstation, *etc.*) such that essential functions could be performed on the shop floor, such as WIP tracking, throughput tracking, capacity feedback, and status monitoring. Moreover, the gateway could be wired or wirelessly connected to the enterprise network and hence there are two types of WC- Gateway, a stationary WC-Gateway and a mobile WC-Gateway, and allowed to be configured in terms of the manufacturing environment. Also, there are a variety of channels provided for Auto-ID devices to be connected to a WC-Gateway computer. In general, the common ways of wireless connections are Bluetooth, ZigBee, and Wi-Fi, *etc.*, whereas the methods of wired connections include USB, and serial ports, *etc.* The overall architecture of WC-Gateway is shown in Figure 14.4. The key components of WC-Gateway and smart objects are described below.

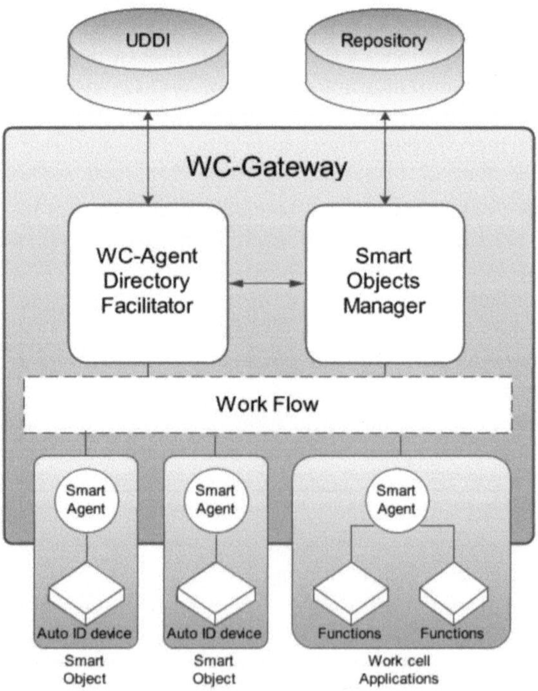

Figure 14.4. Architecture of work-cell gateway

- *Smart Object*: A smart object is a manufacturing object that is made "smart" by equipping it with a certain degree of intelligence: memory, computing power, communicating ability, and task-specific logics. Therefore, smart objects are able to sense, reason, and interact. Auto-ID approaches such as RFID and barcode will be deployed to create an intelligent ambience where smart object interact with each other.
- *Agent-based Smart Object Manager*: An SOM manages smart objects and facilitates their operation. Also, it manages all real-time events and related information from smart objects. The most important features of smart object manager is to allow smart objects to be "universal plug and plays" (UPnP) and communicate with other objects. Since a number of smart agents work together, the WC-Gateway becomes a multi-agent system (MAS) that is compliable with the standard of the Foundation for Intelligent, Physical Agents (FIPA) [14.25], and hence smart agents in the gateway must be FIPA compliant. Therefore, SOM can directly manage the smart agents of the smart objects without being concerned with the problem of communication protocols or devices incompatible with the gateway during the lifecycle of smart objects.
- *Real-time Work-cell Application*: Operators at the work cell use the real-time application to accomplish, monitor and control their work tasks. It is considered as middle lower-level manufacturing applications, and designed and developed according to specific logic requirements from manufacturing processes. In the WC-Gateway, a number of built-in functions are designed and developed locally, including WIP tracking, throughput tracking, status of WC-Gateway monitoring, and devices configuring, *etc.*
- *Work-cell Agent DF*: A work-cell agent directory facilitator (WC-Agent DF) is a component under FIPA specification for agent management. It provides yellow pages for finding services provided by other internal agents in the WC-Gateway. Typically, smart agents can register, deregister, modify and search services in WC-Agent DF and its actions would automatically update into UDDI by SOM. For instance, a smart agent can register a service for retrieving data from an RFID reader in the WC-Agent DF. Such a request is then passed to the SOM and made as tModel before being registered in the UDDI repository. Therefore, other agents or applications may use this service to obtain data of the particular RFID device by means of the web services technology.

14.5 Agent-based Workflow Management for RTM

14.5.1 Workflow Model

As opposed to conventional process models, agents-based workflow management (WFM) model is about the processes and manufacturing resources used by the workflow but are not assigned to actual objects. The agents-based WFM model for SF-Gateway is simply a series of production process flows that can be executed by the correct agents according to the actual situation.

Figure 14.5 shows the agent-based WFM framework, which is used to plan and control the flow of production processes, data and control, and execute any process node of the workflow from the optimal agent in the registered agents. There are two basic elements in the agent-based WFM model: process and agent. A process corresponds to a generic piece of production task, which can be assigned to a certain manufacturing resource. As mentioned above, the agent wraps the corresponding function of the specific manufacturing resource, *e.g.* a work cell, which can execute and finish the specific production process. In other words, each agent can be regarded as a manufacturing resource, *e.g.* work cell.

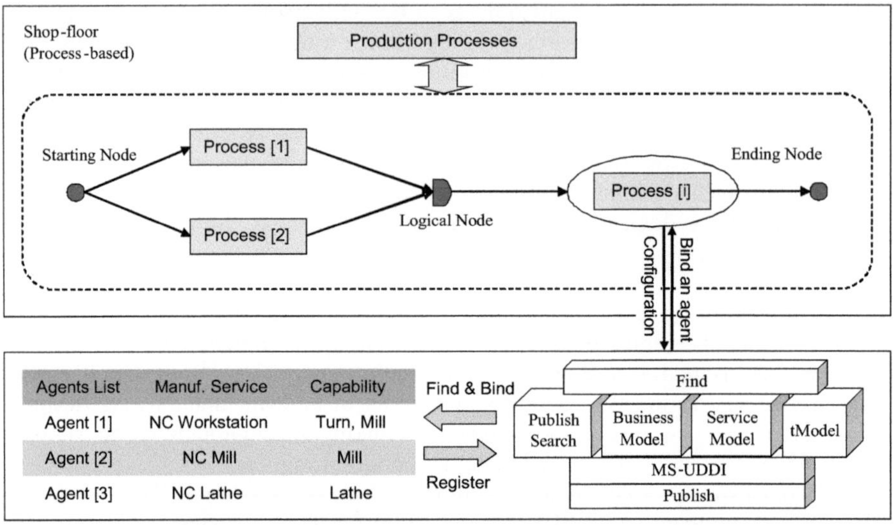

Figure 14.5. Agent-based workflow management model

At the level of SF-Gateway, the WFM is mainly concerned with the co-ordination of distributed agents of work cells. Several features are incorporated. First, the production process network model is adopted as the workflow model. The process nodes represent complex processes or simply process, and the logical nodes represent the trigger condition. Edges represent the logical relationships between production processes, *i.e.* the flows of control and data. Second, the proposed system builds on the concept of agents proposed in the preceding section. An agent represents the work package in the workflow. All the agents involved in a workflow share the same repository and the repository becomes a common working memory. This sharing information ensures the traceability of the decisions at different stages by recording them in a decision tree in terms of the contents of the decisions, the decision makers, and precedence decisions, *etc.* Finally, all interactions are delegated to their agents. Agents are only one of the two main constructs in the workflow network model. That is, agents are only used to define the work as nodes of the network model of production process workflow. Relationships between nodes are separately defined in terms of flows. Without flow definitions, agents still do not know where inputs are obtained from and outputs are sent to. The separation of flow definition from work definition provides opportunities to reuse agents for different production projects once they are defined for RTM. No further changes are necessary when agents are used for other production projects. What the project team needs to do is to choose the agents according to the different production project and define the flows of control and data between agents to suit specific requirements.

14.5.2 Workflow Definition

At the workflow definition stage, two work modes are needed. One is the editing mode where a process planner defines the agent-based workflow for a specific production project. The other is the executing mode where a manager monitors and controls the progress of executing a production workflow. Workflow definition in turn involves the "work" definition and the "flow" definition.

A production project consists of a number of processes or activities. Each process is defined by an agent. This agent can be selected from pre-defined agents in the MS-UDDI. Alternatively, a new agent may be defined and published for this specific work package using agent definition facilities of MS-UDDI. MS-UDDI is also contacted to manage the definition details of agent templates.

Two types of flows are identified in this WFM. They are flow of precedence and flow of data. The flow of precedence and logic node between work-cell agents defines their dependencies. For example, supposing a simple hypothetical product consists of two components, B and C. Component B is outsourced and component C is produced at work-cell 2. Finally, components B and C are assembled to form product A at work-cell 3. Accordingly, the production of A is decomposed into three production processes that can be depicted by a directional network-topology mode. Here, Agent 1 represents a "delivery B" work, Agent 2 represents a "producing C" work, and Agent 3 represents an "assembling A" work. As shown in Figure 14.6, Agent 3 can only start its work after Agents 1 and 2 complete their works under the and logical condition. Agents 1 and 2 may work simultaneously.

The flow of data refers to the situation where agents share their property data. Some outputs from an agent may be the inputs to other agents. Such relationships

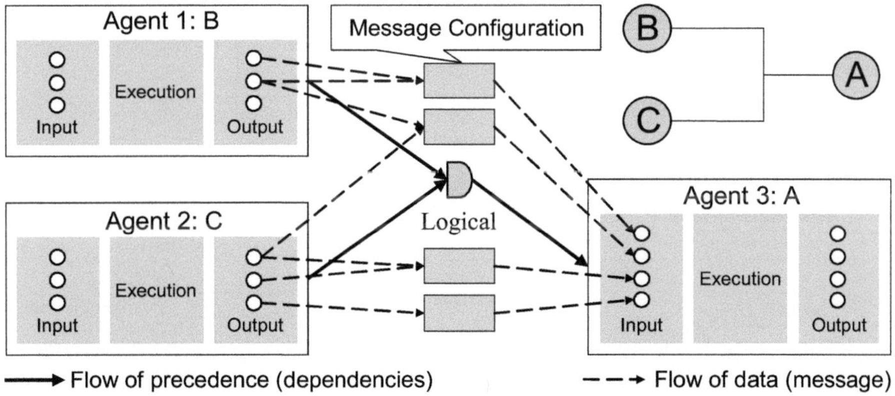

Figure 14.6. Two types of workflow: control flow and data flow

can be easily defined similarly to the way that relationships are defined between data tables in a relational database. Flows of data can be compared to messages widely used in an MAS for communication. And the message configuration tool configures where inputs are obtained from and outputs are sent. For example, Figure 14.6 combines some output items of Agents 1 and 2 as the input items of Agent 3 according to the real requirements.

Flows of data, or message passing, are triggered by the flow of precedence and logical condition. For example, under the "and" condition, if Agents 1 and 2 have not finished their work, flows of data associated with Agent 3 will not be processed.

14.5.3 Workflow Execution

Once the workflow is fully defined, it can be executed as seen in Figure 14.7. During the execution, nodes in a workflow are translated to the corresponding agents. Each agent will invoke its manufacturing services (*e.g.* work cells, smart objects, *etc.*) of the real manufacturing environment to enable their intelligent management of the manufacturing process. Explorers are provided to operators, managers and supervisors for monitoring and controlling the workflow execution lifecycle. The users can simply follow the logic and execute the production project. At the SF-Gateway, the shop floor manager as a user can have a clear overview of the progress of a production project; while at the WC-Gateway, the operators of work cells can use this facility to check if the conditions of their tasks are met before they can be started. The general procedure of executing a workflow is as follows:

- The agents use SOA framework to connect to the web server where SF-Gateway is deployed.

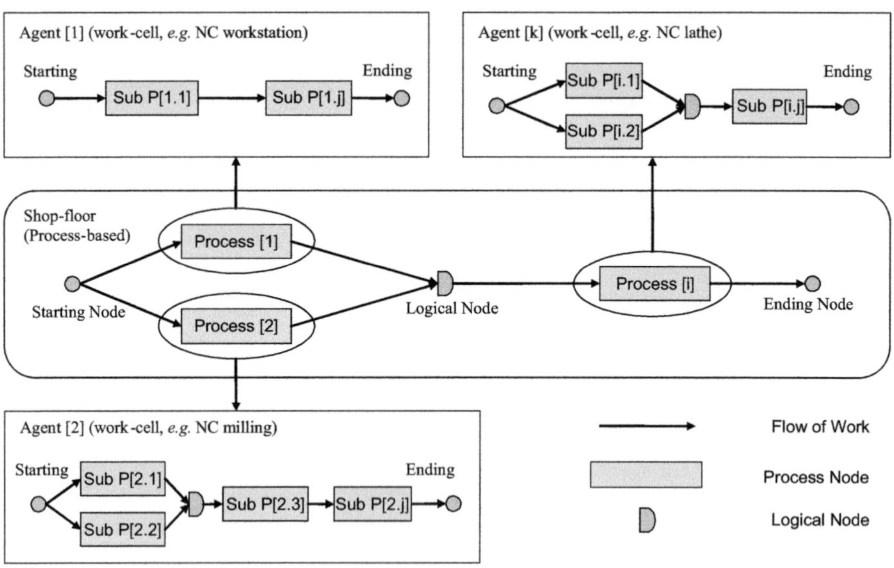

Figure 14.7. Agent-based execution of a workflow model

- XML-based workflow is then automatically downloaded to and manually activated at the corresponding agents.
- Repository is contacted to retrieve the workflow model defined in advance.
- The first agent in the workflow is activated. The agent is executed according to the procedure discussed in the preceding section. Its incoming messages defined as flows of data associated are fired. Therefore, this agent knows from where its input data come.
- After preparing its input data, repository is contacted to save the input/output and other data of the agents.
- Execution engine notifies all agents about the changes.
- The agent is prompted if the output is accepted or a backtracking is necessary.
- Upon completion, the control is passed over to subsequent agents.
- This process repeats until the last agent in the workflow is completed.

14.6 Case Study

14.6.1 Re-engineering Manufacturing Job Shops

Configuration of a Representative Manufacturing Job Shop

The research is motivated through early collaboration with a few OEM (original equipment manufacturer) companies operating in the Pearl River Delta (PRD) region of southern China. Figure 14.8 shows a job shop hypothetically formulated for this study. It is much simplified for ease of understanding and discussion. It is representative among shop floors of several companies that have been studied. This

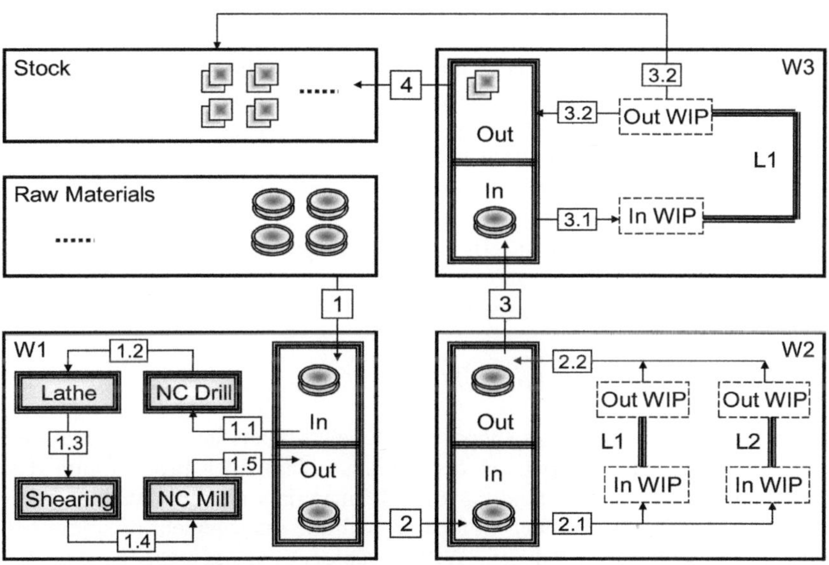

Figure 14.8. A representative manufacturing plant and job shops

generalisation also helps retaining the anonymity of companies. Among these manufacturing companies, parts are typically manufactured from raw materials in different areas of several workshops. Parts and components are then assembled into finished products in corresponding assembly lines. Most individual parts normally undergo a sequence of production operations. These operations may take place at the same or different physical locations. Although assembly operations take place in reconfigurable lines where workstations are movable to suit particular products, workstations for part fabrication are grouped according to their functions. In other words, shop floors for part manufacturing are typically of functional layout.

Re-engineering Job Shops with Wireless Manufacturing Technology

As reviewed previously, job shops suffer from shortcomings widely known in textbooks. Many of them are solved through a change from functional layouts to flexible cells or lines if such change is justified and deemed to be appropriate. Such a re-engineering approach is, however, not preferred by the companies that we investigated although they are aware of the advantages of flexible reconfigurable and cellular manufacturing. Several reasons are evident. This is partly due to its historical way of running manufacturing operations and partly due to the very high product variety that requires extremely high manufacturing flexibility. Another contributing factor to adopt functional layout is the difference in the costs of different manufacturing equipment used in workstations. Some are far more expensive than others and require special cares for maintenance and production scheduling. Finally, shop-floor operators do not have the necessary skills and empowerments to manage themselves in the cell or line. Substantial training is necessary. In summary, all these factors are somewhat quite specific characteristics of the OEM industries in the PRD region of China.

Instead, forward-thinking managements are looking into RFID technology for improving shop-floor productivity and quality. A common belief is shared that better flows and traceability of materials and visibility of information are fundamental to performance improvement. This approach is not only technically but economically viable with the recent developments in RFID technology. The flexibility of job shops is retained without or with little layout changes. The workers' working habits and existing IT systems are also retained if they prove to be already effective and efficient. Ultimately, real-time traceability and visibility will overcome the shortcomings of job shops briefly mentioned previously, meet challenges in managing shop-floor WIP inventories, and solve some typical shop-floor problems summarised in the next two sub-sections.

Challenges in Managing Shop-floor WIP Inventories

The aim of re-engineering job shops with wireless manufacturing technology but without changing their layouts is to maximise shop-floor productivity. This is achieved by improving the handling of WIP materials and minimising the errors involved in handling WIP items. Compared with the management of stocks for raw materials and finished products, WIP inventory management is much more complicated. Reasons are obvious. First, WIP stocks move along the production lines according to the planned processes. Second, the status of WIP stocks change from one workstation to another. Third, the capacities of buffers provided for WIP stocks in production lines are normally small and therefore their uses must be optimised. Fourth, the numbers of WIP material items travelling between different workstations of the production line may be different. Fifth, there are separate departments specifically looking after raw material stocks and finished product stocks with dedicated computer software tools. In contrast, shop floors have to look after production orders and schedules as well as WIP inventories.

14.6.2 Definition of Agents and Workflow

Agents Definition

In order to define agents for the case study, the first important thing should be to wrap the manufacturing services as UPnP component. Two difficulties must be overcome when developing UPnP and interoperable agents. One is that manufacturing services are often developed by third parties that may use different environments and standards. The other is that different manufacturing services at the same agent have different functionalities and therefore have different functional and configuration properties.

To overcome the above two difficulties, Huang and Zhang [14.26] applied the concept of agents to wrap proprietary device drivers of heterogeneous smart objects. A referenced model of wrapping an assembly work cell is shown in Figure 14.9. Because agents share the same data and implementation models, they are therefore

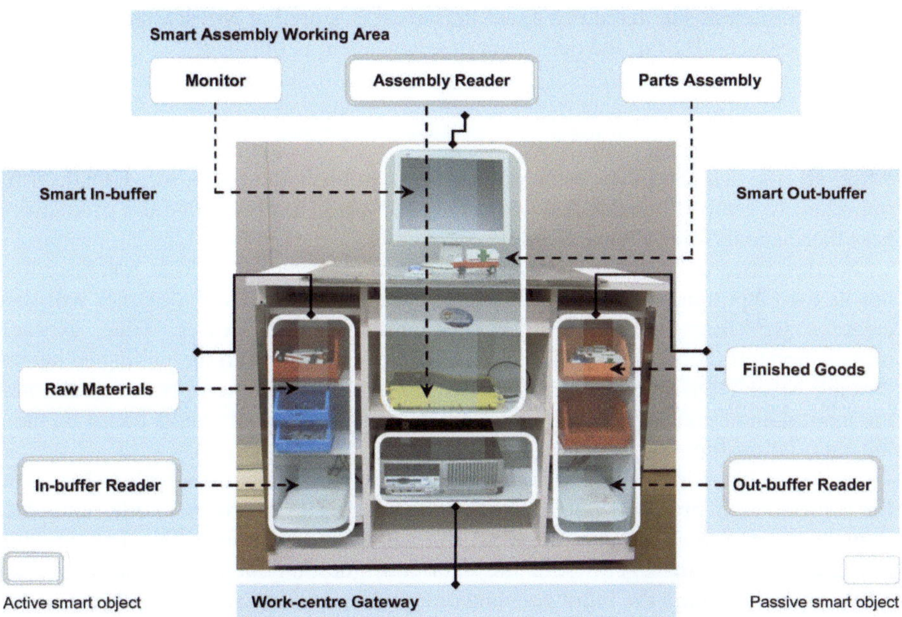

Figure 14.9. Smart assembly station as a stationary smart work cell

interoperable. This research adopts and extends this approach in building the WC-Gateway. Details can be found in [14.26].

Each agent must be registered and published at the MS-UDDI so that they can be found and used by the WFM according to the specific production processes. Main stages for registering and publishing for agents are explained briefly as follows:

- *Register basic information*: The supervisor logins to MS-UDDI and uses definition explorer to specify general description of a new agent. It should be pointed out that each manufacturing service (*i.e.* a smart object) has been registered beforehand. The registered data records the basic information such as to which physical device a specific smart object is linked. This information is available for the agent definition explorer; therefore, the supervisor needs to select the corresponding smart object from the registered list.
- *Define businessEntity*: Each agent works as a web service; therefore, the businessEntity of the agent should be published so that other systems or agents can find it easily and know what services it provides. The businessEntity definition contains information about the business name, description, contacts and identifiers. For each agent, in this MS-UDDI, there is one and only one businessEntity.
- *Define service information*: For the reason that each agent may provide several services for different systems or agents, the supervisor can create the business services for each agent after the businessEntity is defined. The *businessService* definition contains information about the service name, description, access point (WSDL address) and identifiers.
- *Finished*: Through the above steps, the definition of an agent is almost completed. The resulting agent definition is stored at MS-UDDI in an XML document. By far, the definition of a new agent is finished.

Workflow Definition

After all the agents have been defined and published at the MS-UDDI, it is important to establish the workflow on the shop-floor level to solve the problem of how these agents communicate and work together.

As described in previous section, each agent has been wrapped as a service and has its own input and output data during its execution. The CAPP designer will first create a workflow for producing specific production processes. Then, at each process node of the workflow, the optimal agent is chosen from the published agents through MS-UDDI according to their capabilities. Finally, the operator configures the input data of each agent and the output data of the relevant agents based on their logical relationship.

For example, Figure 14.10 illustrates the main flows about defining a workflow for a production process. Once an agent is selected to a process node, an XML segment is created. The XML segment stores the binding information between the process node and the agent. And the data configuration stage will also create an XML segment to bind the input data and output data of these agents.

These processes result in an XML-based workflow definition file. The XML-based configuration file is created at the definition stage with default settings and

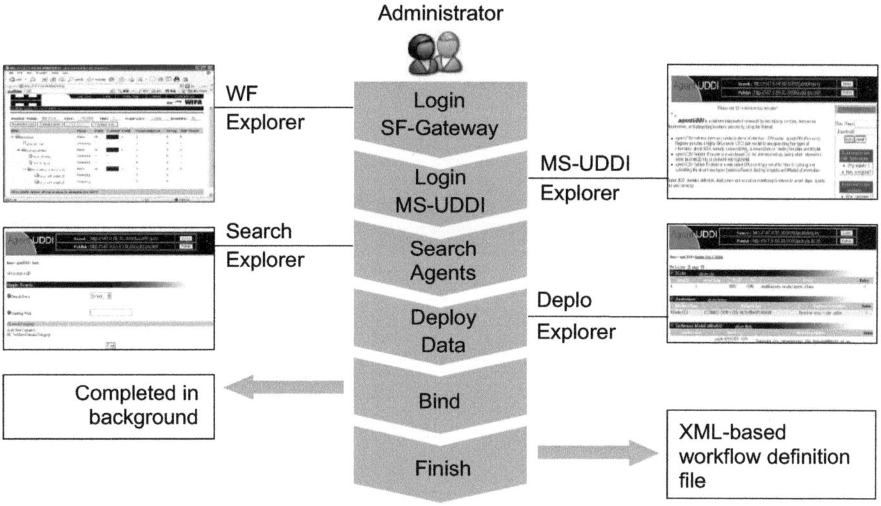

Figure 14.10. Workflow definitions in SF-Gateway

can be dynamically updated after its creation. The XML-based certificate of the agents is also initially created and updated subsequently depending on the situation. By far, the definition of workflow has been finished.

14.6.3 Facilities for Operators and Supervisors

Two groups of shop-floor operators are common in many PRD manufacturing companies. They are production operators at workstations, and internal logistic operators who are responsible for moving WIP items across the shop floor. All these personnel and their supervisors/managers are tagged through their staff cards that are readable by RFID readers. Accordingly, two groups of information and decision explorers are provided.

Let us first consider the group of production operators and their supervisors. Figure 14.11(a) shows the facilities, called explorers. Production operators at workstations can use the Workstation Explorer to check the receipt and despatch of WIP items in the incoming and outgoing buffers, respectively, in addition to their primary tasks specified in the operation sheets. This is only possible if the computers associated with the workstations are networked. Production supervisors use the Line /Shop Explorer to oversee the WIP statuses of all workstations and WIP inventories.

Let us now consider the other group of internal logistic operators and their supervisors. Similar to the facilities provided for the production operators and their supervisors, two explorers are devised for logistic operators and supervisors as shown in Figure 14.11(b). Logistic supervisors work with production supervisors to prepare and/or receive WIP logistics tasks (WIP material requisition orders) according to production orders at workstations and shop floors. Such decisions are automatically issued by the IT system in normal cases. In the meanwhile, logistic supervisors can use the Logistics Explorer to monitor and control the execution of WIP material requisition/logistics orders.

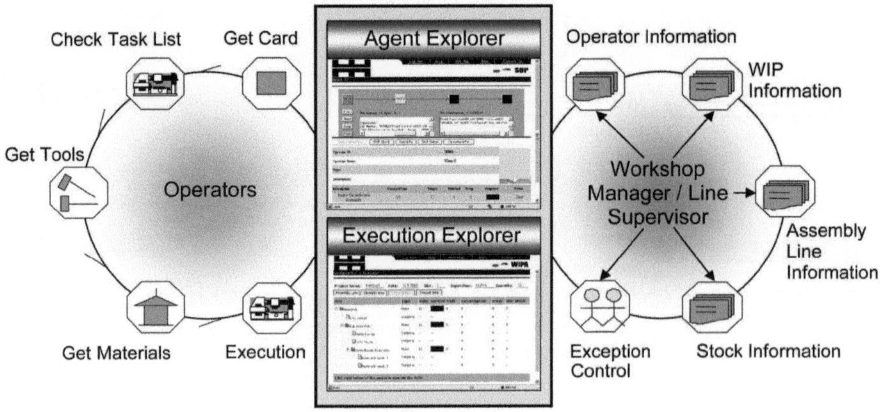

(a) Facilities for production operators and their supervisors

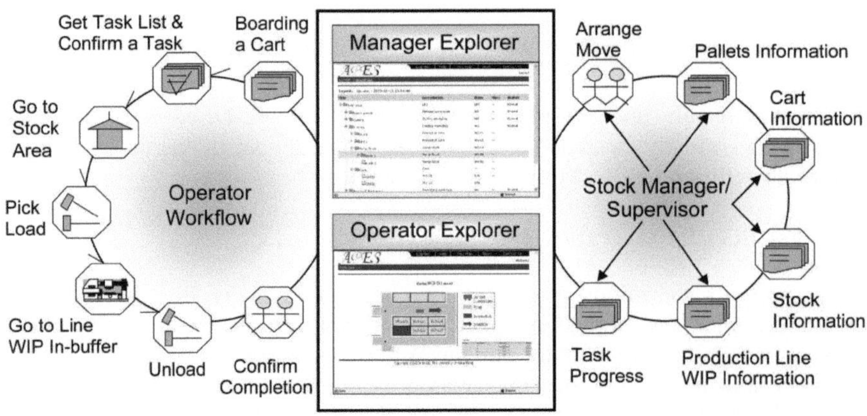

(b) Facilities for internal logistics operators and their supervisors

Figure 14.11. Overview of assembly execution and control through explorers

Internal logistic operators are primarily responsible for choosing and executing the logistics orders to move WIP items between shop-floor inventories and buffers of production workstations. In this regard, a Move Explorer is provided. A more comprehensive explanation on how the Move Explorer assists the logistic operator is given in the next section.

14.6.4 WIP Logistics Process

This section illustrates how the proposed agent, workflow and RFID technology can improve the shop-floor WIP inventory management. The discussion is focused on a situation where an internal logistics operator chooses and follows a logistics order to move WIP items from a shop-floor inventory area to a buffer of a production workstation. The setting is shown in Figure 14.12. The cycle of completing a typical logistics task includes three phases. They are: (a) get on board a cart to start the task; (b) move to the source location and pick up the WIP items; and (c) leave the source location, move to the target workstation, and unload WIP items. The process of moving WIP pallets from the outgoing buffers of an operational workstation to a shop-floor WIP stock area is more or less similar to that described above, thus omitted here.

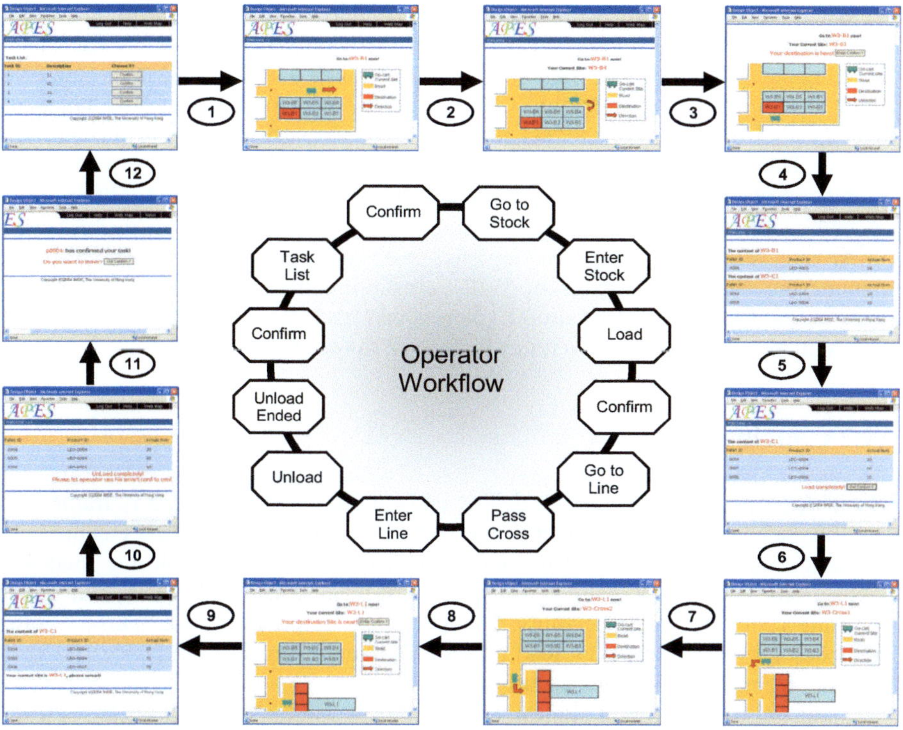

Figure 14.12. Moving WIP items from a WIP inventory to a workstation buffer

At the initial location, the logistics operator gets on board a smart trolley (cart), and logs onto the system using the staff card. At this moment, a binding is established between the operator and the trolley. A list of internal WIP logistics orders is displayed on the screen of the Move Explorer. The order at the top of the list is normally selected for execution. The detail of the location where WIP items are fetched is then displayed in the Move Explorer. The system through the display indicates the way to get to this source location.

On the way, the reader on the trolley reads in tag information along the corridor. The display prompts and directs the operators towards the source location. For example, "turn right" is indicated on the screen when the trolley reads the tag mounted at location *A*. When the trolley arrives in the right aisle, a confirmation is given to the operator. When the trolley is near the source location (in the reading range), the screen indicates to the operator to stop and locate the exact position of the pallets.

Once the cart arrives at the right WIP locations, the operator confirms the arrival and starts picking up the WIP pallets according to the task specification. The Move Explorer will prompt the operator when all pallets are loaded onto the cart. The operator moves from the source location towards the target workstation. The Move Explorer provides necessary navigation on this journey, just as described above for the journey from the initial cart location to the source location.

When the smart cart with the right WIP items moves close to the target workstation, it tracks the tags laid out around this workstation. Once the cart enters the workstation's in-buffer, the logistics operator unloads the WIP items onto the smart locations of the workstation's in-buffer. The new locations of WIP items are then updated through the wireless network from the cart's reader to the backend system. The receipt of this batch of WIP items is confirmed by the workstation production operator. The entire logistics task has now been completed. The logistics operator starts the next cycle of a new task.

At each stage of a shop-floor logistics task, the logistics supervisor is able to monitor the changes of status of WIP items through the Logistics Explorer. In contrast, the Line and Workstation Explorers can track the changes of WIP items at the production lines and workstations but not on the carts.

14.7 Conclusions

This chapter presented an easy-to-deploy and simple-to-use SF-Gateway framework that integrates the concept of agents into workflow management. RFID technologies are used to achieve real-time manufacturing data collection, enable the dual-way connectivity and interoperability between high-level EASs and SASs, and create real-time visibility and traceability throughout the entire enterprise. The proposed methodologies and framework lead to advanced applications of enterprises for real-time and reconfigurable manufacturing.

There are two important contributions in this research. One contribution is using the concept of agent to wrap the manufacturing services and their operational logics/intelligence for cost-effectively collecting and processing real-time manufacturing data; the other contribution is the integration of the agent concept into workflow management. Production processes at shop-floor level are represented as workflows. At the workflow definition stage, the shop-floor manager achieves reconfigurable manufacturing by configuring agents for a production process. At the workflow execution stage, real-time manufacturing is achieved by detecting events and processing data of smart objects at all work cells according to the business and operational rules captured in the workflows.

Acknowledgements

We are most grateful to various companies who provided technical and financial supports to this research. Financial supports from the HKU Teaching Development

Grants (TGD), Seed Fund for Applied Research, and HKSAR ITF grant are also gratefully acknowledged. Discussions with fellow researchers in the group are gratefully acknowledged.

References

[14.1] Radunovic, B., 1999, "An overview of advances in reconfigurable computing systems," In *Proceedings of the 32nd IEEE International Conference on System Sciences*, Hawaii, pp. 1–10.

[14.2] Zhao, X., Wang, J. and Luo, Z., 2000, "A stochastic model of a reconfigurable manufacturing system, Part 1: a framework," *International Journal of Production Research*, **38**(10), pp. 2273–2285.

[14.3] Mehrabi, M.G., Ulsoy, A.G. and Koren, Y., 2000, "Reconfigurable manufacturing systems: key to future manufacturing," *Journal of Intelligent Manufacturing*, **11**, pp. 413–419.

[14.4] Yigit, A.S. and Ulsoy, A.G., 2002, "Dynamic stiffness evaluation for reconfigurable machine tools including weakly non-linear joint characteristics," *Proceedings of the Institute of Mechanical Engineers*, Part B, **216**, pp. 87–100.

[14.5] Huang, G.Q., Zhang, Y.F. and Jiang, P.Y., 2008, "RFID-based wireless manufacturing for real-time management of job shop WIP inventories," *International Journal of Advanced Manufacturing Technology*, **36**(7–8), pp. 752–764.

[14.6] Udoka, S.J., 1992, "The role of automatic identification (Auto ID) in the computer integrated manufacturing (CIM) architecture," *Computers and Industrial Engineering*, **23**(1–4), pp. 1–5.

[14.7] Brewer, A., Sloan, N. and Landers, T., 1999, "Intelligent tracking in manufacturing," *Journal of Intelligent Manufacturing*, **10**(3–4), pp. 245–250.

[14.8] Li, Z.K., Gadh, R. and Prabhu, B.S., 2004, "Applications of RFID technology and smart parts in manufacturing," In *Proceedings of ASME 2004 Design Engineering Technical Conferences and Computers and Information in Engineering Conference*, Salt Lake City, USA, DETC2004–57662.

[14.9] Johnson, D., 2002, "RFID tags improve tracking, quality on Ford line in Mexico," *Control Engineering*, **49**(11), pp. 16–16.

[14.10] Chappell, G., Ginsburg, L., Schmidt, P., Smith, J. and Tobolski, J., 2003, "Auto ID on the line: the value of Auto ID technology in manufacturing," *Auto ID Centre*, CAN-AutoID-BC-005.

[14.11] Harrison, M. and McFarlane, D., 2003 "Whitepaper on development of a prototype PML server for an Auto ID enabled robotic manufacturing environment," available on-line at http://www.ifm.eng.cam.ac.uk/automation/publications/w_papers/cam-autoid-wh010.pdf.

[14.12] Chang, Y., McFarlane, D. and Koh, R., 2003, "Whitepaper on methodologies for integrating Auto ID data with existing manufacturing business information systems," available on-line at http://www.ifm.eng.cam.ac.uk/automation/publications/w_papers/cam-autoid-wh009.pdf.

[14.13] Sikora, R. and Shaw, M.J., 1998, "A multi-agent framework for the coordination and integration of information systems," *Management Science*, **44** (11), pp. 65–78.

[14.14] Macchiaroli, R. and Riemma, S., 2002, "A negotiation scheme for autonomous agents in job shop scheduling," *International Journal of Computer Integrated Manufacturing*, **15**(3), pp. 222–232.

[14.15] Maturana, F.P., Tichy, P., Slechta, P., Discenzo, F., Staron, R.J. and Hall, K., 2004, "Distributed multi-agent architecture for automation systems," *Expert Systems with Applications*, **26** (1), pp. 49–56.

[14.16] Tan, G.W., Hayes, C.C. and Shaw, M., 1996, "An intelligent-agent framework for concurrent product design and planning," *IEEE Transactions on Engineering Manufacturing Management*, **43**(3), pp. 297–306.

[14.17] Krothapalli, N. and Deshmukh, A., 1999, "Design of negotiation protocols for multi-agent manufacturing systems," *International Journal of Production Research*, **37**(7), pp. 1601–1624.

[14.18] FIPA, 2002, "Foundation for intelligent physical agents," *http://www.figa.org.*

[14.19] Shin, M. and Jung, M., 2004, "MANPro: mobile agent-based negotiation process for distributed intelligent manufacturing," *International Journal of Production Research*, **42**(2), pp. 303–320.

[14.20] Jia, H.Z., Ong, S.K., Fuh, J.Y.H., Zhang, Y.F. and Nee, A.Y.C., 2004, "An adaptive upgradable agent-based system for collaborative product design and manufacture," *Robotics and Computer-Integrated Manufacturing*, **20**(2), pp. 79–90.

[14.21] Jiao, J.R., You, X. and Kumar, A., 2006, "An agent-based framework for collaborative negotiation in the global manufacturing supply chain network," *Robotics and Computer Integrated Manufacturing*, **22**(3), pp. 239–255.

[14.22] Lee, W.B. and Lau, H.C.W., 1999, "Multi-agent modelling of dispersed manufacturing networks," *Expert Systems with Applications*, **16**(3), pp. 297–306.

[14.23] Wang, L., Sams, R., Verner, M. and Xi, F., 2003, "Integrating Java 3D model and sensor data for remote monitoring and control," *Robotics and Computer-Integrated Manufacturing*, **19**(1-2), pp. 13–19.

[14.24] Wu, D.J, 2001, "Software agents for knowledge management: coordination in multi-agent supply chains and auctions," *Expert Systems with Applications*, **20**(1), pp. 51–64.

[14.25] Greenwood, D., Lyell, M., Mallya, A. and Suguri, H., 2007, "The IEEE FIPA approach to integrating software agents and web services," In *Proceedings of International Conference on Autonomous Agents*, Honolulu, Hawaii.

[14.26] Huang, G.Q., Zhang, Y.F. and Jiang, P.Y., 2007, "RFID-based wireless manufacturing for walking-worker assembly islands with fixed-position layouts," *Robotics and Computer Integrated Manufacturing*, **23**(4), pp. 469–477.

15

Web-based Production Management and Control in a Distributed Manufacturing Environment

Alberto J. Álvares[1], José L. N. de Souza Jr.[1], Evandro L. S. Teixeira[2] and Joao C. E. Ferreira[3]

[1] Universidade de Brasília, Dep. Engenharia Mecânica e Mecatrônica
Grupo de Automação e Controle (GRACO)
CEP 70910-900, Brasília, DF, Brazil
Emails: alvares@alvarestech.com, leojunior@unb.br

[2] Autotrac Commerce and Telecommunication S/A
Department of Hardware Development (DDH)
CEP 70910-901, Brasília, DF, Brazil
Email: evandro.teixeira@autotrac.com.br

[3] Universidade Federal de Santa Catarina, Dep. Engenharia Mecânica
GRIMA/GRUCON, Caixa Postal 476
CEP 88040-900, Florianopolis, SC, Brazil
Email: jcarlos@emc.ufsc.br

Abstract

This chapter presents a methodology for web-based manufacturing management and control. The methodology is a part of a WebMachining system, which is based on the e-manufacturing concept. The WebMachining virtual company encompasses three distributed manufacturing systems, all of them located in different cities in Brazil, *i.e.* a flexible manufacturing cell (FMC) at GRACO/UnB (Brasília), a flexible manufacturing system (FMS) at SOCIESC (Joinville), and a lathe at UFSC (Florianopolis). The methodology includes planning, scheduling, control, and remote manufacturing of components. A user (customer) uses the manufacturing services provided by the WebMachining virtual company through the Internet, in order to execute operations and processes to design and manufacture the components. The proposed methodology integrates engineering and manufacturing management through an enterprise resource planning (ERP) software that previews which of the three systems will produce the ordered component, and this decision is based on parameters related to each of the three systems. After the decision, the ERP system will generate the production schedule. Also in this work, the implementation aspects of a web-based shop floor controller for the FMC at GRACO/UnB are presented. The FMC consists of a Romi Galaxy 15M turning centre, an ASEA IRB6 robot manipulator, a Mitutoyo LSM-6100 laser micrometer, an automated guided vehicle (AGV), and a pallet to store the blank and finished components. The functional model, which depicts the modules and their relationships in the web-based shop floor controller, serves as a basic model to implement the real system. After that, the proposed implementation architecture based on the object-oriented technology is presented.

15.1 Introduction

Production planning and control (PPC) is concerned with managing the details of what and how many products to produce and when, and obtaining the raw materials, components and resources to produce those products. PPC solves these logistics problems by managing information [15.1]. PPC aims at guaranteeing that the production occurs efficiently and effectively, and that products are manufactured as required by the customer [15.2]. This requires that the manufacturing resources are available in an adequate amount, time, and level of quality.

Production planning and control systems support the efficient management of material flows, the use of manpower and equipment, the co-ordination of the internal activities with the supply and expediting activities, and communication with the customers for its operational necessities. PPC systems help administrators in the function of decision making [15.3].

According to MacCarthy and Fernandes [15.4], there are different systems used for PPC, and some of them are PBC (period batch control), OPT (optimised production technology), and PERT (Program Evaluation and Review Technique)/ CPM (Critical Path Method), and Kanban (a signalling system to trigger some action, which historically uses cards to signal the need for an item). Because of this diversity, the choice of which PPC system is the most adequate for different situations is very important. No PPC system can be considered the solution for all cases, since in order to work with different reasoning to meet diverse necessities and demands, many times it is necessary to use more than one PPC system.

In these circumstances, a methodology is proposed in this chapter for the web-based manufacturing management and control of the *WebMachining* virtual company, whose shop floor is composed of three distributed manufacturing systems located in different cities in Brazil: an FMC GRACO/UnB (Brasília), an FMS SOCIESC (Joinville), and a lathe UFSC (Florianópolis). The proposed methodology includes the development of an ERP system and the integration of this ERP system with other engineering module (CAD/CAPP/CAM). In the engineering module, two component development environments are used: WebMachining [15.5] and CyberCut [15.6]. The ERP system is developed for the web, thus allowing customers to input its orders of components anywhere, without having the equipment and software for carrying out the product development cycle. The methodology also allows the company employees to connect remotely to the system and perform activities from any place (Figure 15.1).

For the implementation of the methodology, tele-manufacturing is used, which is a part of the electronic-manufacturing concept [15.7]. The customer uses the manufacturing services via the web to execute the operations and the necessary processes, designing and manufacturing the desired component efficiently and with flexibility, using computational tools for the development of the product lifecycle.

This work also presents the implementation aspects of a web-based shop floor controller for the FMC at GRACO/UnB. The FMC consists of a Romi Galaxy 15M turning centre, an ASEA IRB6 robot manipulator, a Mitutoyo LSM-6100 laser micrometer, an AGV, and a pallet to store the blank and finished components. The functional model, which depicts the modules and their relationships in the web-based shop floor controller, serves as a basic model to implement the real system.

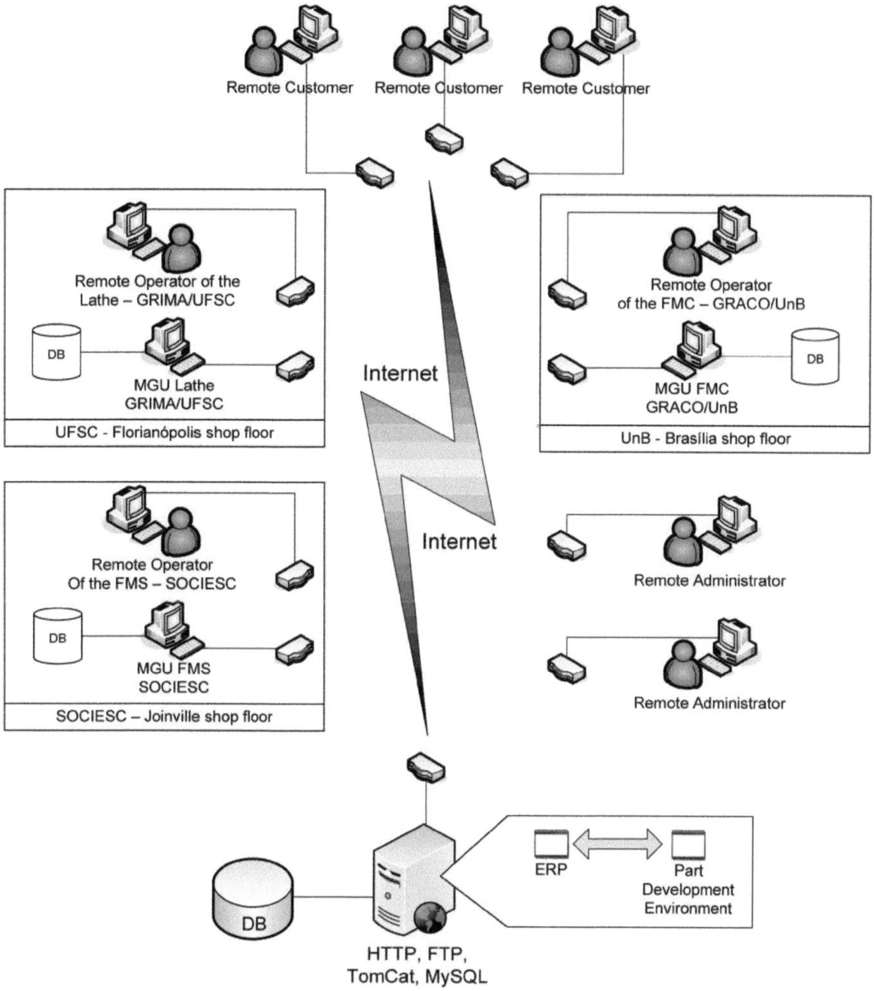

Figure 15.1. Remote access to the system

After that, the proposed implementation architecture based on the object-oriented technology is presented.

15.2 Overview

15.2.1 ERP Systems

With the advances in information technology (IT), companies started to use computer systems to support their activities. Generally, in each company, some systems were developed to meet specific requirements of the diverse business units, factories, departments and offices. Thus, information was fragmented among different systems. The main problems of this fragmentation are the difficulty in getting consolidated information, and the inconsistency of stored redundant data in more than one system. ERP systems solve these problems by including, in just one integrated system, functionalities that support the activities of different companies [15.8].

Because of the evolution from MRP (manufacturing resource planning) systems to ERP systems, it is possible to include and to control all the company processes, without the redundancies found in the previous systems. Information is displayed in a clearer way, immediately and safely, providing a greater control of the business, which includes its vulnerable points, such as costs, financial control and supplies.

15.2.2 Electronic Manufacturing (e-Mfg)

IT, especially the network communication technology and the convergence of wireless and the Internet, is opening a new domain for building the future manufacture environments called e-Mfg (electronic-manufacturing), using labour methods based on collaborative e-Work (electronic work) [15.9], especially the activities developed during product development in integrated and collaborative CAD/CAPP/CAM environments. In essence, e-Work is composed of e-activities (electronic-activities), *i.e.* activities based and executed by the use of information technology. These e-activities include v-Design (v for virtual), e-Business, e-Commerce, e-Manufacturing, v-Factories, v-Enterprises, e-Logistics, and similarly, intelligent robotics, intelligent transport, *etc.*

E-Mfg can be considered as a new paradigm for these computer systems based on global environments, network-centred and spatially distributed, enabling the development of activities using e-Work. This will allow product designers to have easier communication, making possible sharing and collaborative design during product development, as well as tele-operation and monitoring of the manufacturing equipment [15.5].

15.2.3 WebMachining Methodology

The design portion of the *WebMachining* methodology is based on the synthesis of design features, *i.e.* union of features for turning operations and subtraction of features for milling operations [15.5]. The methodology has the purpose of allowing the integration of the collaborative design activities (CAD), process planning (CAPP) and manufacturing (CAM planning and CAM execution). In order to achieve this, it uses as design reference the manufacturing features model defined by part 224 of STEP – Standard for Exchange of Product Model Data (ISO 10303) [15.10], and more specifically the taxonomy of form features for cylindrical components defined by CAM-I [15.11].

The procedure begins in the collaborative modelling of a component using features in the context of remote manufacturing via the web, in a client-server computer model. Some of the data that are generated by the system include the geometric and feature-based model of the component (detailed design), the process planning with alternatives (*WebCAPP* module), and a NC program. Then, tele-operation of the CNC lathe is carried out (*WebTurning* module). The methodology can be applied to the manufacture of both cylindrical and prismatic components.

15.2.4 CyberCut

CyberCut is a web-based design-to-manufacturing system, developed by Brown and Wright [15.6], consisting of the following major components:

- Computer-aided design software written in Java and embedded in a web page. This CAD software is based on the concept of destructive solid geometry (DSG); that is, by constraining the user to remove entities from a regularly shaped blank, the downstream manufacturing process for the component is inherently incorporated into the design.
- A computer-aided process planning system with access to a knowledge base containing the available tools and fixtures.
- An open-architecture machine tool controller that can receive the high-level design and planning information and carry out sensor-based machining on a Haas VF-1 machine tool.

According to Brown and Wright [15.6], by providing access to the CyberCut CAD interface over the Internet, any engineer with a web browser becomes a potential user of this on-line rapid prototyping tool. A remote user would be able to download a CAD file in some specified universal exchange format to the CyberCut server, which would in turn execute the necessary process planning and generate the appropriate NC code for milling. The component could then be manufactured and shipped to the designer. The engineer could have a fully functional prototype within a matter of days at a fraction of the cost of in-house manufacturing.

15.3 PROMME Methodology

PROMME is a methodology to provide the means to manufacture and control management in a distributed manufacturing environment, and is applied to the virtual WebMachining company, whose shop floor is formed of three distributed manufacturing systems. ERP software was developed in order to carry out manufacturing management, enabling the receipt of customer orders, management functions, CAD/CAPP/CAM integration, and component manufacture in one of the three manufacturing systems.

15.3.1 Distributed Shop Floor

The shop floor of the virtual WebMachining company, as described previously, is formed of three distributed manufacturing systems: FMC GRACO/UnB, FMS SOCIESC and Lathe UFSC. The UnB (Brasilia-DF) system is an FMC composed of a CNC turning centre, an industrial robot, an AGV, pallets of components, laser micrometer, management unit (MGU) and an audio/video monitoring system, as shown Figure 15.2. In the FMC-UnB system, the components with non-concentric features can be manufactured in the 3-axis Romi Galaxy 15M CNC turning centre.

The SOCIESC system (Joinville), shown in Figure 15.3, is an FMS composed of a Feeler CNC lathe, a Feeler CNC machining centre, an ABB 2400 robot, and an automated storage and retrieval system (AS/RS).

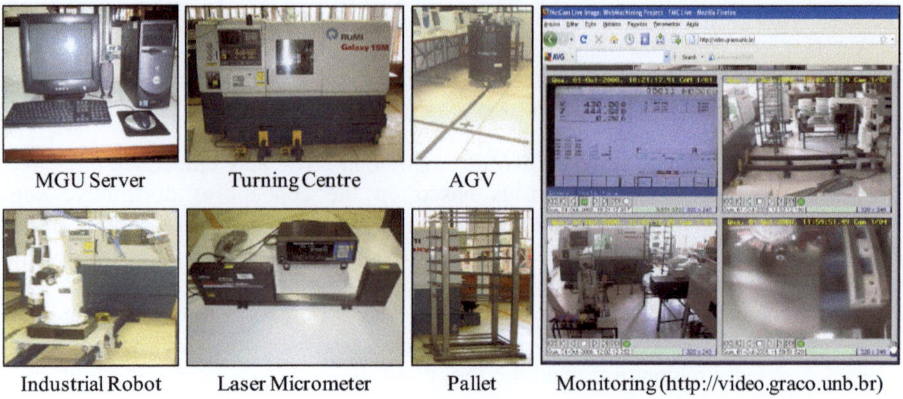

Figure 15.2. FMC GRACO/UnB in Brasilia (http://www.graco.unb.br)

Figure 15.3. FMS SOCIESC (http://www.grima.ufsc.br/sociesc/fms2/FMS2.htm)

The third system, located at UFSC (Florianópolis), is composed of only a Romi CNC lathe. This system makes only components with concentric features, and in this case the presence of an operator to feed the lathe is necessary.

15.3.2 ERP Manufacturing

ERP Manufacturing is a web-based system, written in Java (http://java.sun.com) and JavaServer Pages (JSP – http://java.sun.com/products/jsp), which enables the management of the virtual company. The management is composed of the control of user accesses through the Internet, integration with CAD/CAPP/CAM modules, the computer-aided production module (CAP), integration with the management units of the distributed shop floor, and the management activities of the company. All these modules are described below.

Institutional Module

The institutional module is where the employees of the WebMachining company perform the administrative and operational activities of the company. The managers are responsible for registering new employees, excluding or modifying a register of an employee, modifying values of the production cost calculation of each shop floor, visualising monthly profits and expenses, and visualising the systems' production by using Gantt graphs.

Each shop floor has operators who have functions such as registering suppliers, updating supply of tools, requesting the purchase of materials, registering monthly expenses, getting the daily production of the manufacturing system.

Commercial Module

After having access to the site of the virtual WebMachining company, the customer enters the commercial module, registers, performs system log-in, and then the page with the customer menu is available. On this page, the customer can input a new work order, see the work order status, modify or cancel a work order and modify its registered data. This is the first stage in the production process of the company, and one of the most important. It is at this stage that the customer registers information of the priority, the component type and the batch size.

The customer priority can be determined based on the production time (*e.g.* if the batch must be manufactured in the shortest possible time, thus the work order becomes more expensive), or on the production cost (*e.g.* the time is not the most important factor, but the final batch price).

In the proposed methodology, there are two types of components that the customer must inform the system: prismatic or cylindrical. This definition is the first information that is used by the system for decision making, since only the FMS-SOCIESC is capable of producing prismatic components.

Integration with the Component Development Environment

The component development environment is composed by the WebMachining and CyberCut systems. In this work, WebMachining is used for the design of cylindrical components, whereas CyberCut is used for prismatic components.

After the preliminary pieces of information about the work order are registered, the ERP shows to the customer, via a servlet, a CAD interface with one of the two tools, depending on the component type. In this interface, the customer designs the component, which is then sent to the process planning module (WebCAPP) [15.12] that is responsible for including in a database the information about the machining operations, the time of each machining operation, the list of tools to be used in the manufacturing process, and the NC program to be sent to one of the three management units. These pieces of information are crucial in the decision making about where the component will be manufactured.

In order to help to provide the necessary information for machining, *workingsteps* are used in the WebCAPP module in the WebMachining methodology, which are high-level machining features that can be associated to parameters of processes. A workingstep contains the following information: cutting tool, cutting conditions, machine-tool functions and machining strategy associated with the cutting tool movement. The workingsteps form part of the ISO 14649 (STEP-NC) standard [15.13, 15.14], which is a new model for data transfer between CAD/CAM systems and CNC machines, aiming at replacing the ISO 6983 (G code) standard. STEP-NC eliminates the disadvantages of the ISO 6983 standard, once it specifies machining processes instead of the cutting tool motion, and it uses the object-oriented technology.

Computer Aided Production

Production planning in PROMME is divided into two parts: decision making, which determines which shop floor will be responsible for manufacturing the component, and production scheduling, which is composed of the daily production plan, and by the component families formation.

Decision Making

Decision making is carried out considering basically the production capacity of the shop floor and customers priorities. The algorithm below describes how a decision is made.

Decision Making Algorithm

1. If (Component Type = prismatic)
    1.1. start CyberCut
    1.2. If (Priority = cost)
        1.2.1. calculate the production cost and the maximum production time
    1.3. Else, calculate the production cost and the minimum production time
2. Else If (Component Type = cylindrical)
    2.1. start WebMachining
    2.2. compare the required work order cutting tools with the manufacturing systems tools
    2.3. If (Priority = cost)
        2.3.1. verify the sending type
        2.3.2. calculate the sending cost + production cost for each system, and get the lowest
        2.3.3. calculate the maximum production time
    2.4. Else
        2.4.1. verify which system has the least number of work orders to be produced
        2.4.2. calculate the production cost and the minimum production time

The total production time of the work orders is calculated based on the time of machining operations available in the database, and on the tools setup time. It is assumed that the shop floors are available 16 hours a day, 7 days a week. According to the production times of the components, scheduling is performed each day.

The calculation of the minimum production time is made by analysing the work orders in the database. All work orders have a completion date, *i.e.* the date on which manufacture will be concluded. Every time the system has to preview when the work order will be completed, a search in the database is made to know the next date available to manufacture the batch. The search result is shown to the customer, and he/she decides if it confirms the manufacture of the batch. The batch will only be scheduled if the customer confirms production of the work order.

The maximum manufacturing time, which is the same for all shop floors, is one month after the date the work order entered the system. As in the previous case, the batch will only be scheduled if the customer confirms manufacture of the work order.

Production Scheduling

The master production schedule groups the work orders, which were created by the customers and included in the database, to their destination shop floor. The master schedule provides the knowledge about which shop floor the work order is.

With regard to the formation of the component families, the algorithm of rank order proposed by King [15.15] was applied. In this case, the components that have the same tool characteristics are grouped into the same family. This prevents a new setup of tools being made every time that a new component is processed. The components are grouped using the list of tools included in the database by the component development environment.

Production scheduling is performed in the decision-making process because the foreseen date for manufacturing completion must be shown to the customer for him/her to confirm. The work order will only be included in the master production schedule if the customer accepts the foreseen date. The sequence in which the component families are produced in a day does not matter. The important thing is that the work orders are manufactured before the date that was shown to the customer.

Integration with the Management Units of the Distributed Shop Floors

The integration with the MGUs is made remotely, via a database. All the work orders to be produced are in the master production schedule. The operators on each shop floor connect via the Internet with the web server and get the information about the work orders to be done in the day. The MGUs are responsible for the success of the production. They are responsible for controlling the pieces of equipment, and also for the production scheduling on the shop floor.

15.4 System Modelling

15.4.1 IDEF0

According to the IDEF0 standard (http://www.idef.com), IDEF0 may be used to model a wide variety of automated and non-automated systems. For new systems, IDEF0 may be used first to define the requirements and specify the functions, and

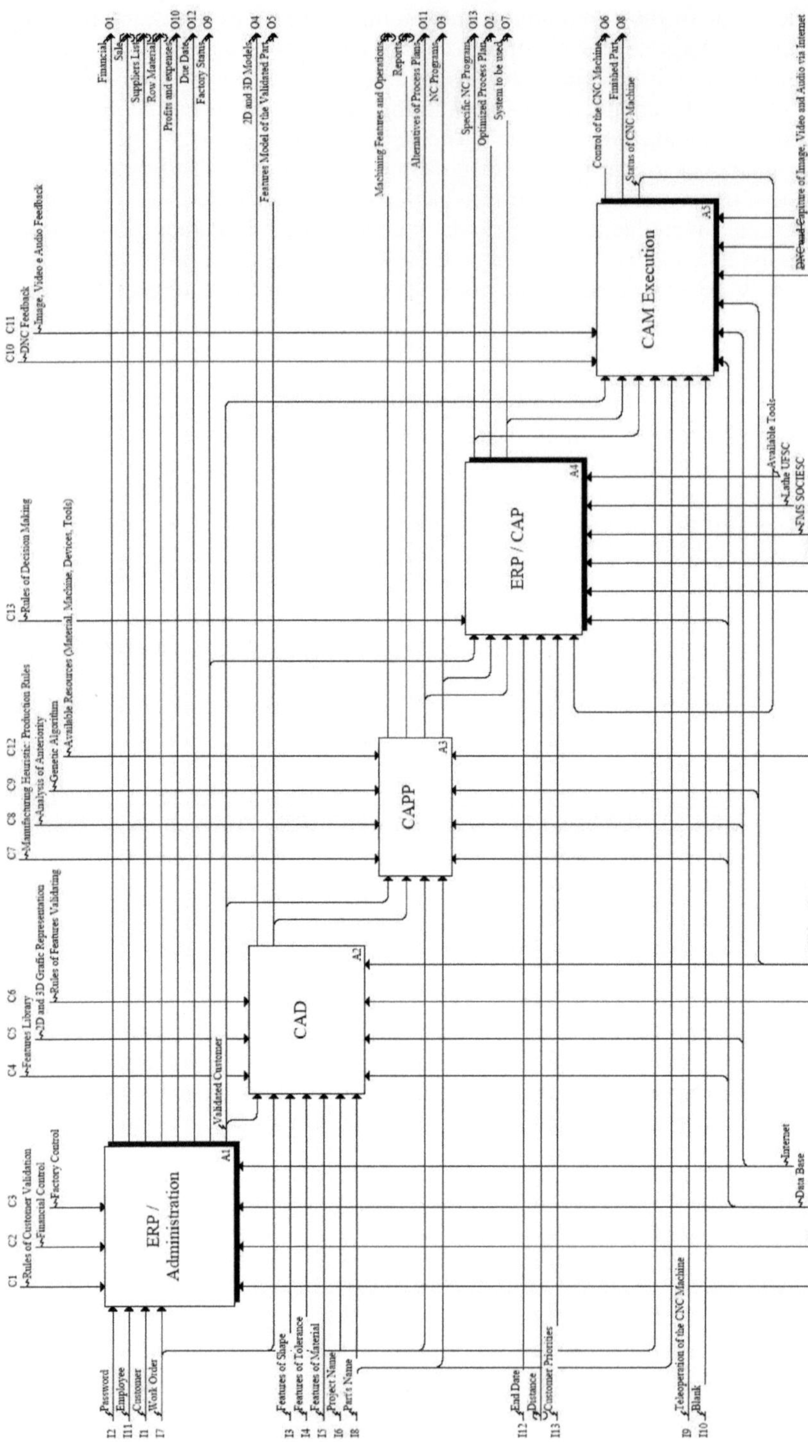

Figure 15.4. IDEF0 modelling of PROMME methodology

then to design an implementation that meets the requirements and performs the functions. For existing systems, IDEF0 can be used to analyse the functions that the system performs and to record the mechanisms (means) by which these are done. The IDEF0 methodology also prescribes procedures and techniques for developing and interpreting models, including ones for data gathering, diagram construction, review cycles and documentation. Figure 15.4 shows the IDEF0 modelling of the PROMME methodology.

15.4.2 UML

In the field of software engineering, the Unified Modelling Language (UML– http://www.uml.org) is a standardised specification language for object modelling. It is a general-purpose modelling language that includes a graphical notation used to create an abstract model of a system, referred to as a UML model. There are many diagrams to model a system.

In UML, a package diagram depicts how a system is split up into logical groupings by showing the dependencies among these groupings. As a package is typically thought of as a directory, package diagrams provide a logical hierarchical decomposition of a system. Figure 15.5 illustrates the package diagram of the PROMME methodology.

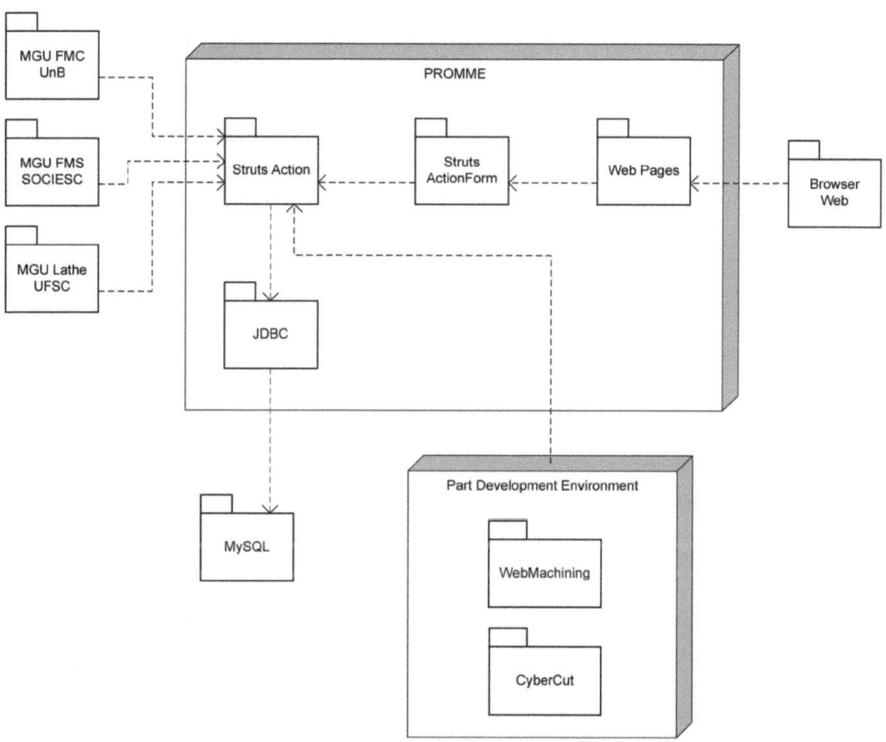

Figure 15.5. UML package diagram – PROMME architecture

15.5 Web-based Shop Floor Controller

A web-based shop floor controller (WSFC) for the FMC at GRACO/UnB (in Brasilia) was also implemented, and it uses WWW resources to perform the remote manufacture of components. The FMC receives instructions from the controller and converts them onto the operations necessary to manufacture the components. The web-based shop floor controller (WSFC), as a computer system, should meet the following requirements: (a) it supports production planning; (b) it should have functions to verify the availability of the production resources allowing the instruction loading on the workstations; and (c) it should control and monitor the production process reacting on abnormal condition that can hinder the fulfilment of the activities established previously on the production planning.

15.5.1 Communication within the Flexible Manufacturing Cell

In order to describe the implementation of the control for the FMC at GRACO/UnB in Brasilia, it is important to visualise the FMC communication structure, describing the method used by the human operator to access the FMC resources (Figure 15.6).

The turning centre (Romi Galaxy 15M) communication is established by an Ethernet interface, using the TCP/IP protocol, linked to the programming library (FOCAS1). The API FOCAS1 drivers and programming library provide the communication and programmable access to a PC-based CNC system [15.16]. This programming library supplies about 300 function calls that can be implemented in customer applications.

The Ethernet/radio system is used to establish the AGV communication. This system possesses a Proxin RangeLan2 interface to connect the robot on the computer network and to communicate with the bridge server. The server connects the robot on the local network using the TCP/IP protocol [15.17]. This configuration mode provides access to the main network mechanisms and patterns (ftp, telnet, TCP/IP, sockets) and the robot can be operated as a network workstation.

The micrometer communication is established by means of an RS232C interface. The communication process is restricted to 23 programmed commands defined by the micrometer manufacturer (Mitutoyo). These commands provide both remote programming of geometric parameters (diameter and tolerances) of a component feature and the conditions in which the inspection will be performed.

The material handling communication is limited to 13 digital I/Os (7 inputs and 6 outputs). This constraint resulted in the design of a dedicated interface to establish indirect communication with the robot based on a CNC/PMC/Robot controller architecture. This was carried out in a partnership with the manufacturer of the CNC turning centre.

15.5.2 Web-based Shop Floor Controller Implementation

The implementation architecture of the controller should encompass the main functionality that the real system should offer (from the human operator to the workstations). The implementation architecture was built based on object-oriented technology.

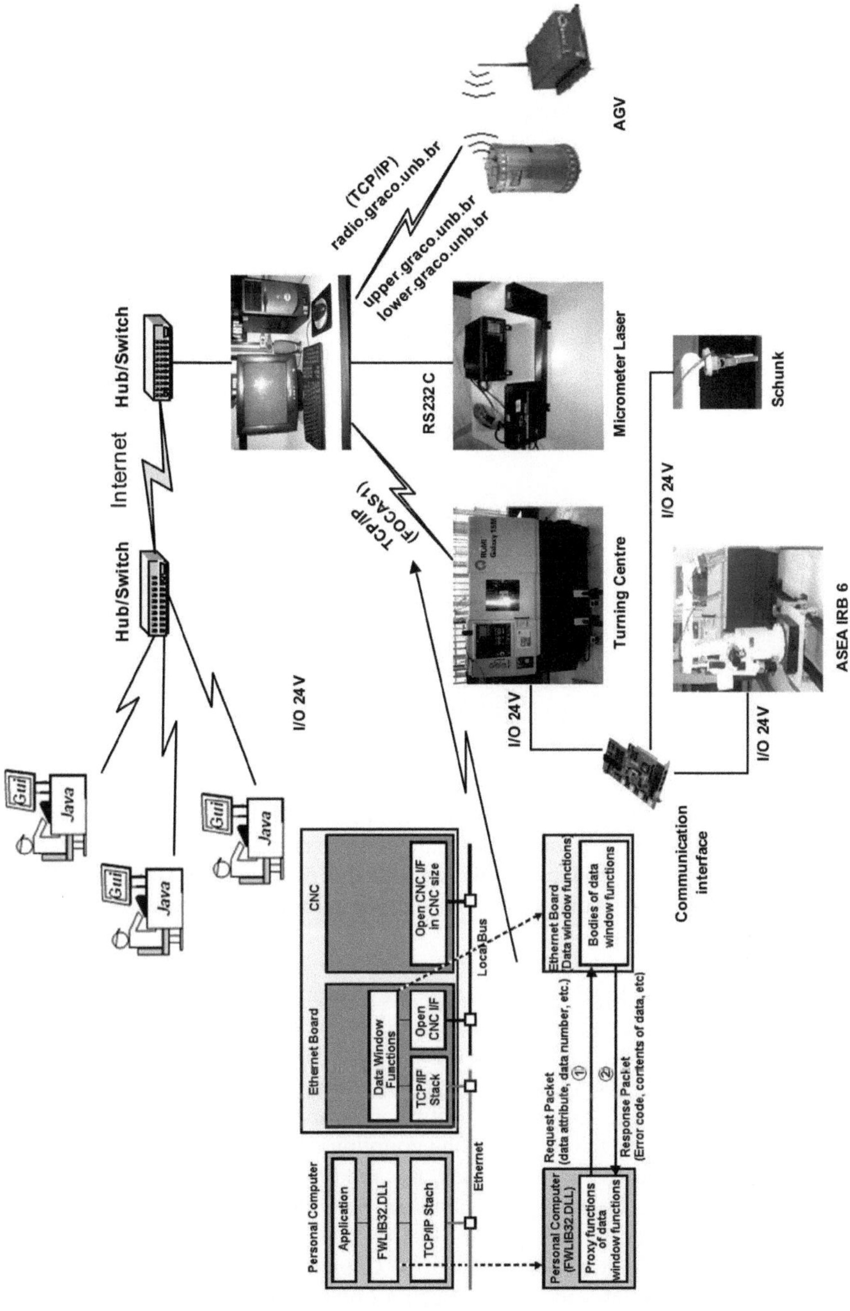

Figure 15.6. Communication structure in the FMC at GRACO/UnB in Brasilia

Implementation Architecture – Package and Component Diagram

The package and the component diagrams of UML were used to design and to document the implementation architecture. The WSFC modules have their responsibilities distributed in different packages. One package is a basic mechanism used to organise and classify the elements of a group [15.18]. Class, interfaces and components that possess similar functionality were grouped in package. Figure 15.7 shows the WSFC package diagram, where the *Initialisation* package has only one class (the *Initialisation* class). This class has only one method (the main method) invoked every time that the WSFC is initialised. The main method is responsible for instantiating the WSFC *NavigatorController*.

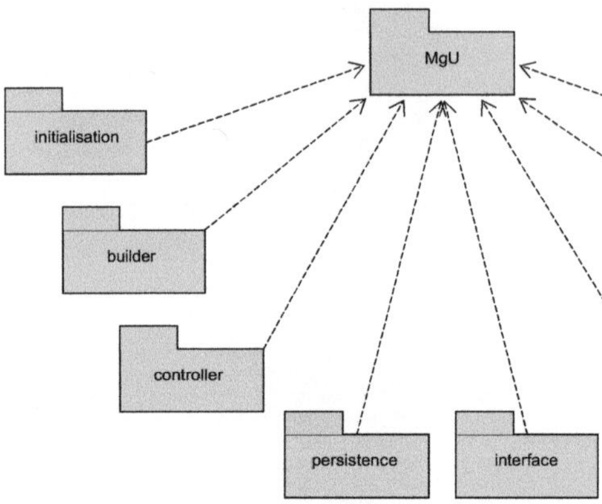

Figure 15.7. Package diagram of web-based shop floor controller

The package *Controller* groups all the *Controller* classes, which encapsulate the logical approach of the system. It can be classified as *FrameController* or *LayerController* class. One *FrameController* class listens to every user interaction with the GUIs, formatting and encapsulating user information to be processed, while the *LayerController* class manages the system navigability and the service changes among the software layers.

BuilderScheduler, *BuilderDispatcher* and *BuilderMonitor* are the main classes stored in the *Builder* package. These classes are responsible for building the WSFC modules and their interconnections. The build process of the WSFC consists of instantiating all the *FrameController* and *LayerController* objects that compose the WSFC modules, and connects them by means of relationships.

The *Interface* package groups the entire interfaces used in the WFSC. One interface is a mechanism used to reduce the coupling degree among objects. When a software layer is connected by interfaces, the modification of one layer does not expand to the others. Thus, this mechanism provides easier expansibility and maintenance of the system.

The *Command* package groups the entire *Command* class. The instance of a *Command* class encapsulates any request as an object, and consequently the object that invokes the operation does not need to know how the request should be executed. For example, when the *SQLInsertWorkOrder* (one Command object) is initialised, it receives one *WorkOrderData* object, which encapsulates the *WorkOrder* attributes such as due date, priority, *etc*. After the *execute* method is invoked, a string containing the SQL instruction is created and used to insert one new *WorkOrder* on the database.

The *View* package aggregates the entire GUIs developed for the WSFC. These GUIs provide users a means to interact with the workstations. Some customer objects, such as *DateChooseButton*, are built considering the principle to modularise the information (encapsulates all the common functionalities), besides preventing code repetition.

The *Persistence* package contains the entire classes that establish the communication with the external devices. The instances of *DBConnection*, *MicrometerConnection*, *TurningCenterConnection*, *RobotConnection* and *AGVConnection* classes establish the connection with the database for the micrometer, the robot, the turning centre and the AGV, respectively.

Figure 15.8 shows the component diagram of the WSFC based on the client-server architecture. The GUIs and the upper-level functionalities are available on the client module, while the lower-level functionalities (*e.g.* direct connection with the workstations) are available on the server module. The communication between these modules is established by sockets, using the TCP/IP communication protocol.

The client (any remote user) connects to the FMC server through the following URL: http://webfmc.graco.unb.br/mgu/mgu.jnlp. This link points to the JNLP archive, one XML (eXtensible Markup Language) document, that specifies which JAR archive file from the client module should be downloaded to the remote computer user. When all the files are download (specified on the XML document), the WSFC is ready to be executed.

Besides incorporating all the advantages offered by Java applets (*i.e.* to execute an application via web without installing it, *etc.*), the use of the Java Network Launch Protocol (JNLP - http://java.sun.com/products/javawebstart/download-spec.html) technology allows the incremental download of the application. It means that every time the application is to be executed on the client computer, only the modified JAR archives from the web server will be downloaded.

Implementation under Distributed Architecture

To provide remote access to the workstations (via web), maintaining the portability that the system should offer, the web-based shop floor controller was implemented in a client-server architecture. The system was designed in two modules: the client and server modules. Figure 15.9 shows the component diagram of WSFC designed under the client-server architecture.

The client module encapsulates the upper-level functionality that does not access directly the operating system resources, providing system portability, and communicates with the logical controller of the server module. In this module, the instances of *SchedulerController*, *DispatcherController* and *MonitorController* encapsulate the upper-level functionalities (*requestMasterPlanningScheduling()*, *sendWorkstationCommands()*, *verifyWokstationStates()*, *etc.*), and establish the communication with the workstation's logical controller by means of their respective interfaces.

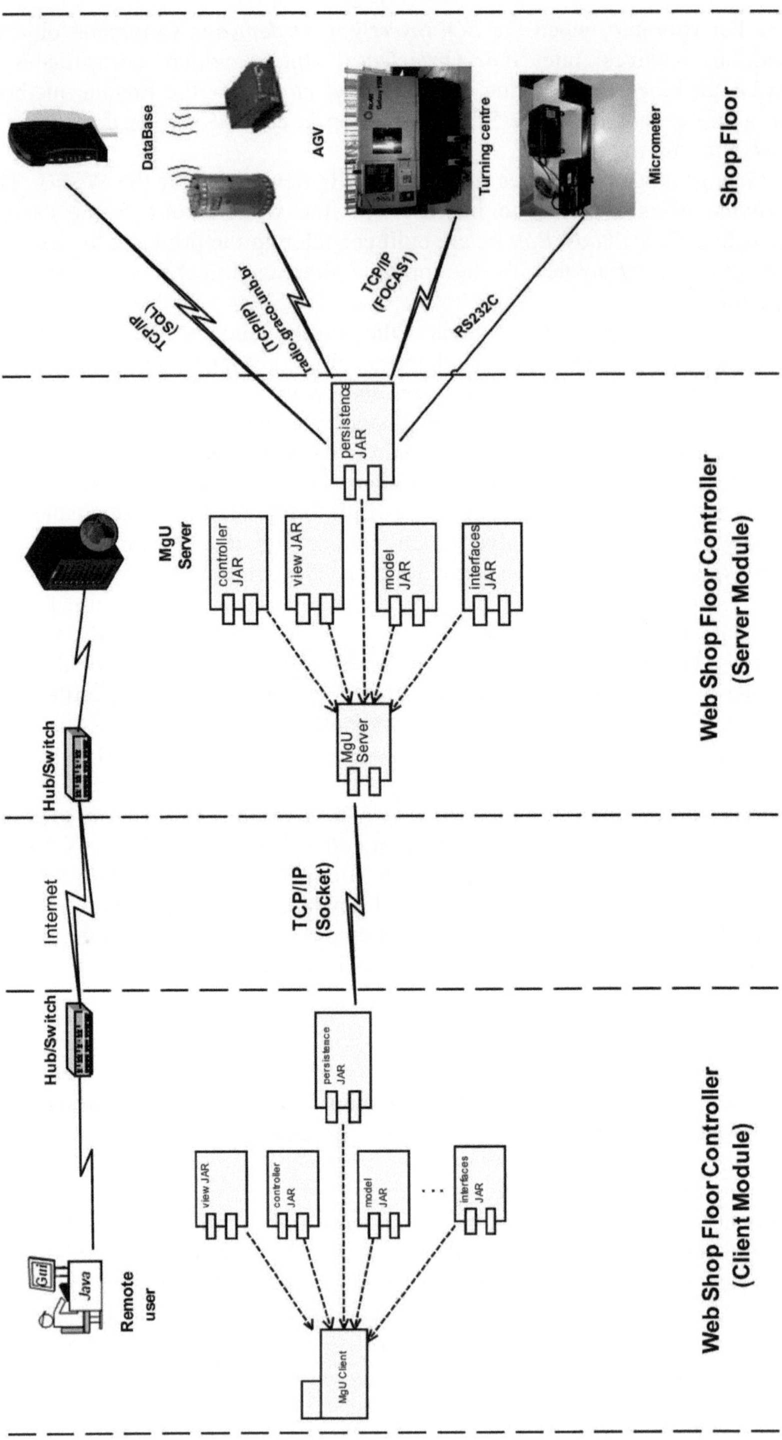

Figure 15.8. Component diagram of web-based shop floor controller

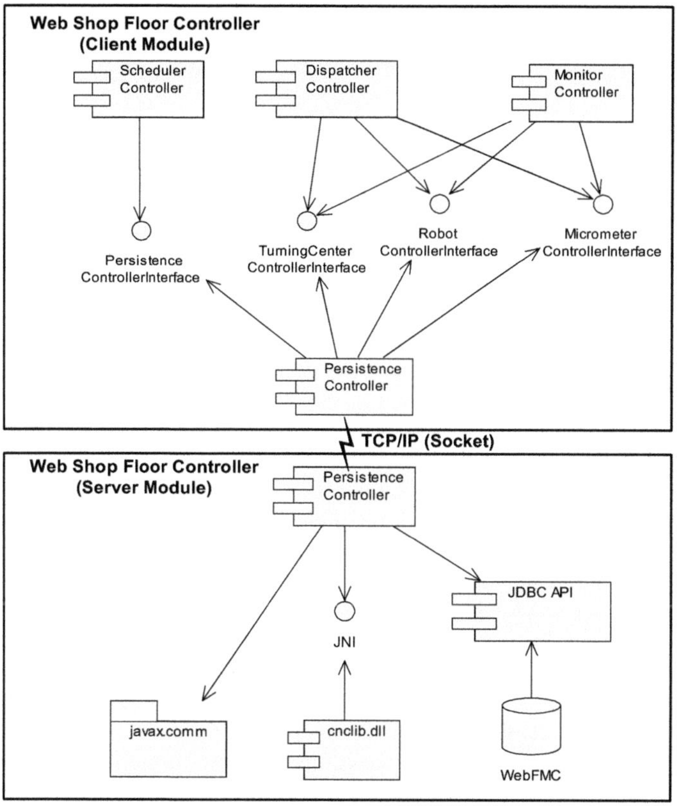

Figure 15.9. Component diagram of client-server architecture

The server module, designed under the multithread environment, implements the upper-level functionalities offered to the client module and establishes the direct communication with the workstation controllers. In this module, the communication with the CNC turning centre is established using the JNI technology (Java Native Interface – http://java.sun.com/j2se/1.5.0/docs/guide/jni/index.html). JNI allows the code being executed in the JVM (Java virtual machine) to interact with other applications and libraries written in different programming languages, such as C, C++, *etc.*

Afterwards, by means of these interfaces, the WSFC (written in Java language) was able to communicate with the library developed in C++ (cnclib.dll) that encapsulates some programming functions offered by API FOCAS1 (used for CNC communication).

The Java Communication is the API (application programming interface) used to establish the communication with the micrometer. This API provides serial port access (via RS232 interface) and parallel port access (IEE-1284). To access the WebFMC database, Java Database Connectivity (JDBC - http://java.sun.com/products/jdbc/overview.html) was used. This API is an industrial standard established to connect the Java technology and some databases, using Structured

Query Language (SQL). It is necessary to maintain the Java portability and to permit to change the database without modify the server module from the WSFC.

15.5.3 Results

Inspection and Production Planning

An inspection plan is included to the information used to plan, control and monitor the inspection of the components. This plan is composed by a set of inspection programs previously recorded on the database. Each program has the component geometry information (diameter and tolerance of each feature that will be inspected), as well as the inspection conditions (unit system, reference, scale, *etc.*).

The GUI *InspectionPlan* (Figure 15.10(a)) was implemented in order to provide the possibility to add, edit or delete the inspection programs of the inspection plan. Each inspection program can be used to group the geometric information of up to ten features, and consequently the same production program can be used to inspect several components without modifying the current micrometer program, since the inspection conditions are the same.

Master production scheduling (MPS) is added to the work orders recorded in the database. One work order has attributes such as priority, due date, process time, *etc.*, which will be used by other WSFC modules. The GUI *ProductionPlan* (Figure 15.10(b)) provides the possibility to add, edit or delete the work orders from the database. Each work order has an attribute called "status", which informs the system about the situation of the work order (*i.e.* "to produce", "in production", or "produced"). Therefore, in order to schedule production, the operator should select the work orders that will be produced, setting the work order status attribute to "in production".

Scheduling Production and Dispatching

After concluding the production plan, the next step consists of establishing the sequence in which the work orders will be manufactured. The scheduling method adopted in this work is based on priority rules [15.19]. Figure 15.11(a) shows the GUI *GanttGraph* used to provide the necessary support to generate the operation sequence for the work orders selected from the production plan.

When an operator selects the manual mode for scheduling, a JDialog will show the planned work orders, and the human operator can schedule the work orders manually. On the other hand, if the automatic mode is selected, the automatic scheduling algorithm will verify the programming method (*forward* or *backward*) and the sequence rule (*priority*, *earliest due date*, *first in first out*, or *shortest processing time*) chosen to schedule the work orders.

After determining the task list (*i.e.* the work order operations sequence), the human operator should dispatch the task list to the workstation. This action is composed of two phases: (a) verification of the workstation status, and (b) loading of the task list on the workstations. The GUI *VerifyWorkstationStatus* (Figure 15.11(b)) was implemented to provide the necessary operation interaction to verify the workstation status.

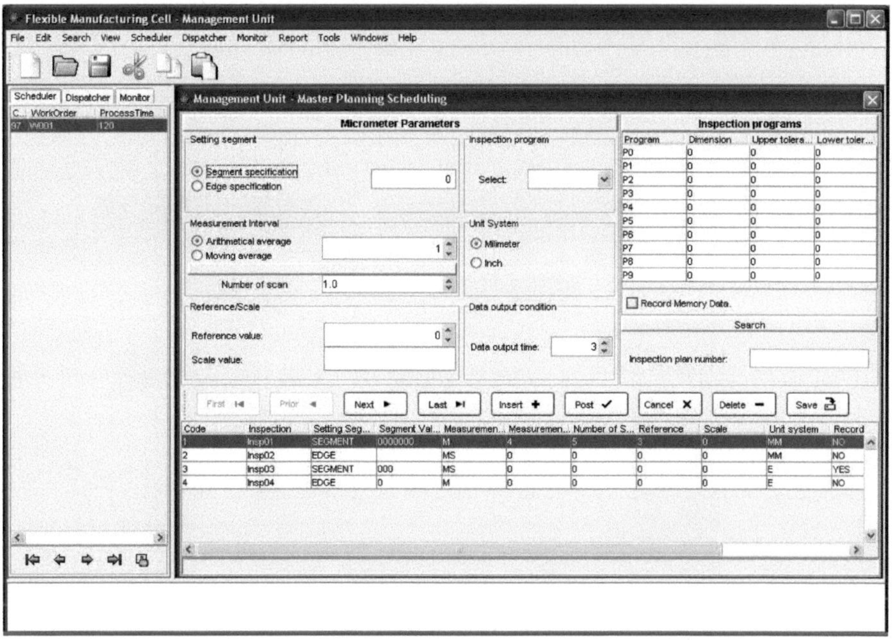

(a) GUI InspectionPlan

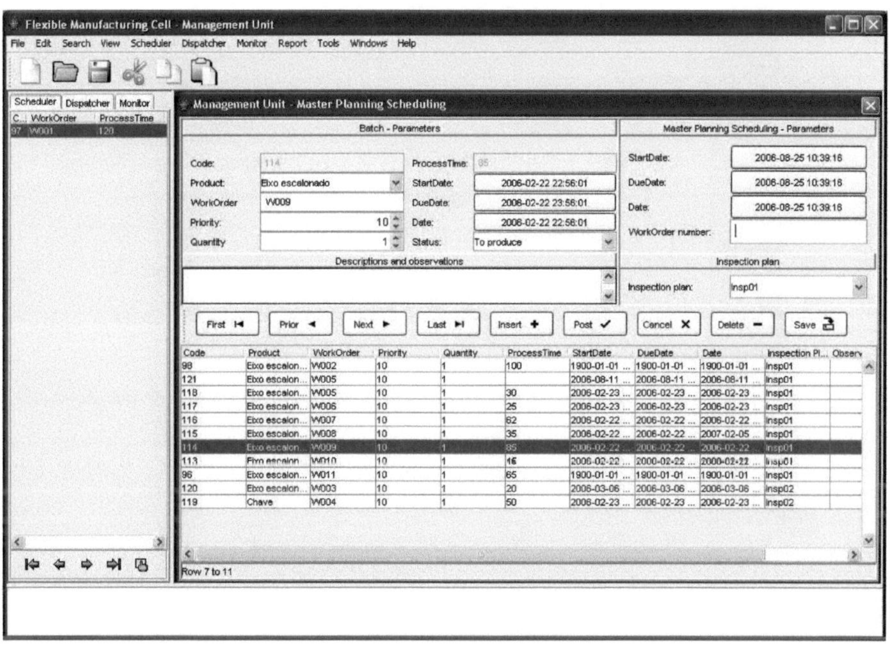

(b) GUI ProductionPlan

Figure 15.10. GUI *InspectionPlan* and GUI *ProductionPlan*

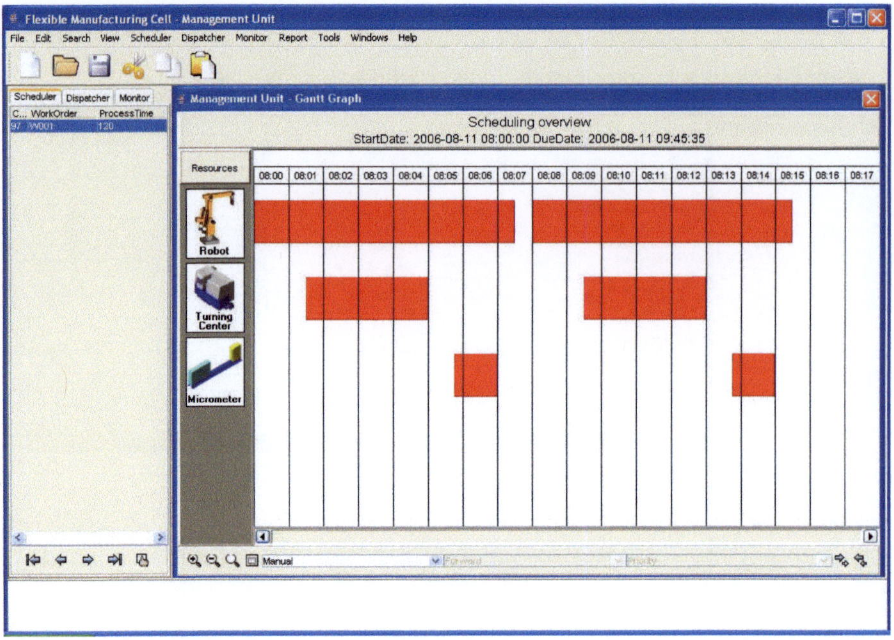

(a) GUI GanttGraph

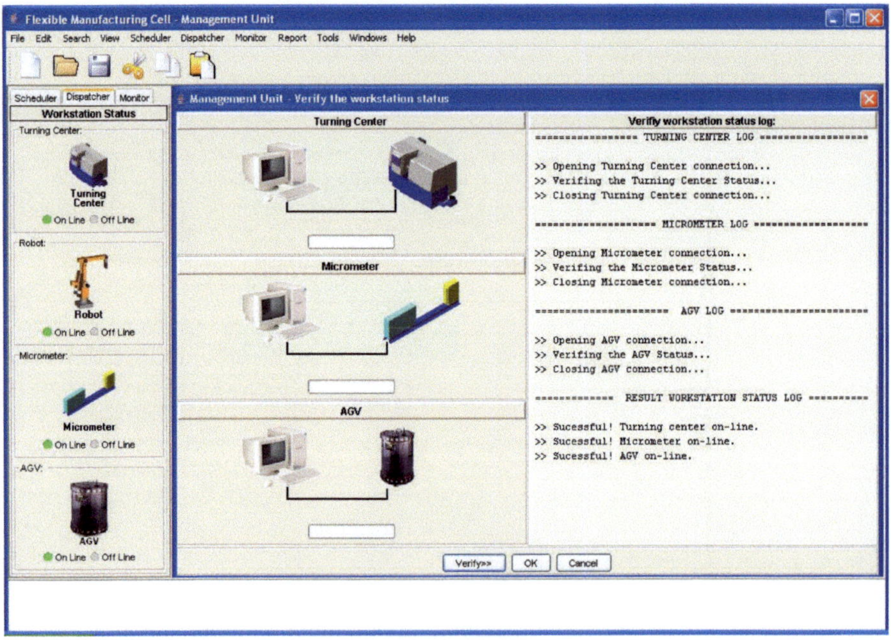

(b) GUI VerifyWorkstationStatus

Figure 15.11. GUI *GanttGhaph* and *VerifyWorkstationStatus*

The WSFC communicates with each workstation in order to verify whether it is available to receive a task upload. While the WSFC is checking the workstation status, a report (log) of the executed command is shown on right side of the GUI. If any fault occurs during the verification process, the human operator cannot dispatch the task list to the workstations.

Production Monitoring and Quality Control

The *VirtualMonitorFrame* (Figure 15.12(a)) provides the virtual monitoring of the workstations. The tab *monitor* (on the left-hand side) shows the PMC tags from the CNC turning centre allocated to the workstation integration. In the centre, the virtual images (from the workstations) of manufacturing a component are presented. On the right-hand side, a JPanel shows the report (log) of each occurred event on the shop floor.

The GUI *QualityControl* provides the human operation interaction with the quality control process. The statistical method selected to control the process is the pre-control [15.20], which is composed of three steps: to qualify the process, operation, and sample frequency. The inspection of a component starts with the positioning of the manufactured component on the micrometer's read unit. After positioning the component on the micrometer and the programmed inspection time (*DataOutputTimer*) expires, the micrometer sends the inspection result to the WFSC via the RS232 interface.

The program number is the geometric information from the inspected feature (diameter and tolerance). If the judgment criteria were activated, the micrometer process unit evaluates the inspection result and checks if the measured value is within the tolerance limits (pre-defined). The result can be –NG (if the measured value is lower than the lower tolerance limit), GO (if the measured value is within the tolerance limits) and +NG (if the measured value is larger than the upper tolerance limit). Figure 15.12 shows the virtual monitoring of a real inspection.

15.6 Conclusions

The proposed PROMME methodology contains a concept for web-based manufacturing management that encompasses a web based system, an ERP software written in Java language, and the presence of distributed manufacturing systems located in different places.

Control is performed using an e-manufacturing concept that integrates the remote access of users (customers and employees), customer orders, engineering activities (CAD/CAPP/CAM), a distributed shop floor, and sales. This integration allows a customer to execute operations and the required processes to design and produce the components with a high amount of efficiency and flexibility, without possessing the necessary pieces of equipment. The integration also allows the employees to carry out company activities remotely.

With regard to the WSFC, it is a computer system that uses Internet resources to promote the remote manufacturing of components. Besides the portability and the remote access via the Internet, the WSFC schedules, controls and monitors the activities on the shop floor. The first WSFC prototype can be executed at

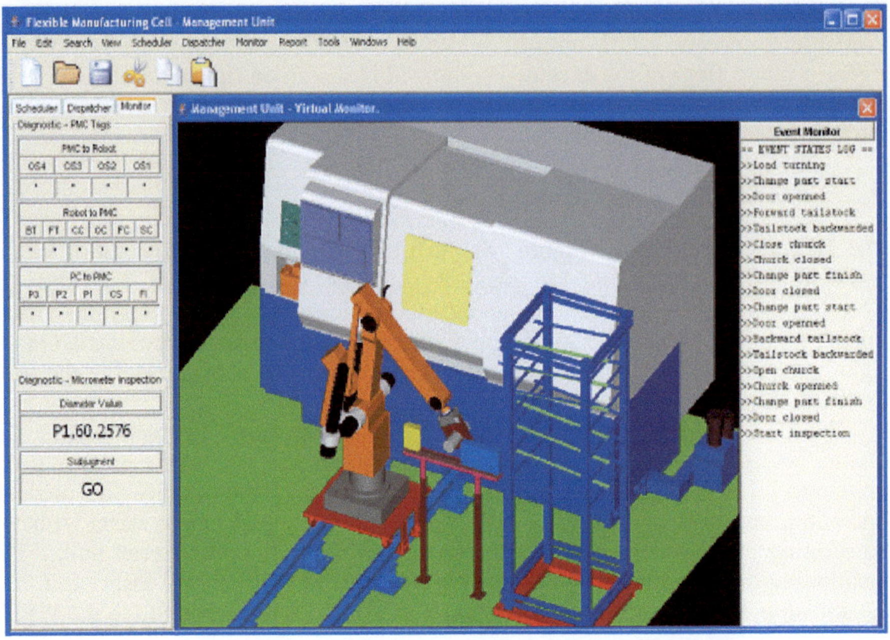

(a) Virtual monitoring

(b) Real inspection

Figure 15.12. Virtual monitoring of real inspection

http://webfmc.graco.unb.br/mgu/mgu.jnlp. Although many functions are not implemented yet, the proposed algorithms and the supplied UML diagrams have laid a solid foundation for accomplishing further implementation in the future.

These UML diagrams were designed based on object-oriented technology, and therefore can be applied to any object-oriented programming languages. The package and component diagrams serve as a model in order to facilitate future changes.

The implementation, based on Java technology, enables the WSFC to be executed over the Internet without the need for users to install any applications, except the Java Runtime Environment (JRE), which must be installed to support the Java 2 platform.

The implementation, based on client-server architecture, encapsulates services that call the operating system functions (*e.g.* CNC communication and micrometer communication) on the server side. Therefore, the portability inherited from the Java technology is maintained, and the WSFC can be executed on different operating systems (at client side).

References

[15.1] Groover, M.K., 2001, *Automation, Production Systems, and Computer-Integrated Manufacturing*, 2nd edn, Prentice-Hall, New Jersey.

[15.2] Slack, N., Chambers, S. and Johnston, R., 2004, *Operations Management*, 4th edn, Prentice-Hall, Harlow, UK.

[15.3] Vollman, T.E., Berry, W.L. and Whybark, D.C., 1997, *Manufacturing Planning and Control Systems*, 4th edn, McGraw Hill, New York.

[15.4] McCarthy, B.L. and Fernandes, F.C., 2000, "A multi-dimensional classification of production systems for the design and selection of production planning and control systems," *Production Planning & Control*, **11**(5), pp. 481–496.

[15.5] Álvares, A.J. and Ferreira, J.C.E., 2008, "A system for the design and manufacture of feature-based parts through the Internet," *International Journal of Advanced Manufacturing Technology*, **35**(7–8), pp. 646–664.

[15.6] Brown, S.M. and Wright, P.K., 1998, "A progress report on the manufacturing analysis service, an Internet-based reference tool," *Journal of Manufacturing Systems*, **17**(5), pp. 389–401.

[15.7] Malek, L.A., Wolf, C. and Guyot, P.D., 1998, "Telemanufacturing: a flexible manufacturing solution," *International Journal of Production Economics*, **56–57**, pp. 1–12.

[15.8] Davenport, T.H., 1998, "Putting the enterprise into the enterprise system," *Harvard Business Review*, **16**(4), pp.121–131.

[15.9] Nof, S.Y., 2004, "Collaborative e-Work and e-Mfg.: the state of the art and challenges for production and logistics managers," *International Federation of Automatic Control, 11th IFAC Symposium on Information Control Problems in Manufacturing* (INCOM), Salvador, Brazil, April 5–7.

[15.10] National Institute of Standards and Technology, 1996, *Part 224 – Mechanical Product Definition for Process Planning*, Document ISO TC184/WG3 N264(T7), USA.

[15.11] CAM-I, Deere & Company, 1986, *Part Features for Process Planning*, Moline, Illinois.

[15.12] Álvares, A.J., Ferreira, J.C.E. and Lorenzo, R.M., 2008, "An integrated web-based CAD/CAPP/CAM system for the remote design and manufacture of feature-based cylindrical parts," *Journal of Intelligent Manufacturing*, **19**(6), pp. 643–659.

[15.13] International Standards Organisation TC 184/SC 1, 2003, *Industrial Automation Systems and Integration – Physical Device Control – Data Model for Computerised Numerical Controllers – Part 121: Tools for Turning Machines*, ISO/DIS 14649-121.

[15.14] International Standards Organisation TC 184/SC 1, 2003, *Industrial Automation Systems and Integration – Physical Device Control – Data Model for Computerised Numerical Controllers – Part 12: Process Data for Turning*, ISO/DIS 14649-12.

[15.15] King, J.R., 1980, "Machine-component grouping in production flow analysis: an approach using a rank order clustering algorithm," *International Journal of Production Research*, **18**(2), pp. 213–232.

[15.16] GE Fanuc, 2006, *Application Development – FOCAS1 (Drivers & Programming Libraries)*, http://www.geindustrial.com/cwc/products?pnlid=2&id=cnc_mec_39.

[15.17] Álvares, A.J., Andriolli, G.F., Dutra, P.R.C., Sousa, M.M. and Ferreira, J.C.E., 2004, "A navigation and path planning system for the Nomad XR4000 mobile robot with remote web monitoring," *ABCM Symposium Series in Mechatronics*, **1**, pp. 18–24.

[15.18] Booch, B., Rumbaugh, J. and Jacobson, I., 1998, *The Unified Modelling Language User Guide*, Addison-Wesley Object Technology Series, USA.

[15.19] Starbek, M., Kušar, J. and Brezovar, A., 2003, "Optimal scheduling of jobs in FMS," *CIRP Journal of Manufacturing Systems*, **32**(5), pp. 419–425.

[15.20] Steiner, S.H., 1998, "Pre-control and some simple alternatives," *Quality Engineering*, **10**(1), pp. 65–74.

16

Flexibility Measures for Distributed Manufacturing Systems

M. I. M. Wahab and Saeed Zolfaghari

Ryerson University, Toronto, ON M5B 2K3, Canada
Emails: wahab@ryerson.ca, zolfaghari@ryerson.ca

Abstract
Currently, enterprises operate in a tremendously competitive environment characterised by a number of changed business conditions including the trend to global and transparent markets, the rise of mass customisation, and reduced product lifecycles. Competence in the optimal use of information and communication technologies supporting a global co-operation of enterprises will be a key feature to remain competitive in the present market. To meet these requirements, manufacturing systems control has moved away from traditional centralised approaches and has focused on the development of a spectrum of distributed manufacturing systems, which have capability to adapt internal as well as external uncertainties. In the literature, several configurations/architectures of distributed manufacturing systems have been discussed. Those systems have numerous characteristics such as easy to remove and introduce new manufacturing equipment, easy to introduce new products, easy to reconfigure the system and its control, and so forth. Even though some of the suggested configurations/ architectures seem promising, to the best of our knowledge, none has fully investigated how a given distributed manufacturing system could be capable of coping with uncertainties that influence its performance. In order to fill this niche, we will study the performance measure of distributed manufacturing systems. This will help enterprises to evaluate alternative configurations/architectures of distributed manufacturing system and choose the one to meet their goal.

16.1 Introduction

Manufacturers around the world are facing a fast growth in competition in a global market. Intense competition at home and abroad forces companies to adopt new strategies that enable them to compete globally. Short lifecycles for products, emerging technologies, access to cheaper labour and proximity to customers are among the reasons for enterprises to go global. To gain competitive advantage, enterprises are constantly looking for ways to be more productive and at the same time more responsive to changes in the market. As a result, manufacturing enterprises are moving towards architectures that allow them to integrate their operations with those of their customers and suppliers through a partnership that is known as a distributed manufacturing system [16.1, 16.2]. Implementation of distributed manufacturing systems poses new challenges to organisations that include decentralised operations and decision making, as manufacturing units will be operating in distributed geographical locations.

In the literature, several configurations/architectures for distributed manufacturing systems have been discussed [16.3–16.6]. Those systems have numerous characteristics such as easy to remove and introduce new manufacturing equipment, easy to introduce new products, easy to reconfigure the system and its control, and so forth. Even though some of the suggested configurations and architectures seem promising, to the best of our knowledge, none has fully investigated how a given distributed manufacturing system could be capable of coping with uncertainties that influence its performance. The uncertainties can come from different sources that may include availability of materials and resources, changes in local regulations, and most importantly the demand for products in different geographical locations. If the demand for a product is constant, it would be much easier for organisations to assign the product to locations with the constantly highest demand. In reality, however, product demands fluctuate substantially over time and from one location to another. This may force enterprises to revisit their product allocation decisions to better meet market fluctuations. Such revisions could result in transferring products from their original manufacturing sites to new locations. These transfers are costly options and many enterprises may not have the necessary infrastructures to furnish such moves. Those who are able to do so have obviously greater flexibility in their system to respond to market changes.

In this chapter, we study the performance measure of distributed manufacturing systems. This will help manufacturing enterprises to evaluate alternative configurations/architectures for distributed manufacturing systems and choose the one that meets their goal best. The proposed model has distinctive features that take into considerations demand uncertainty, routing flexibility and network flexibility. As in the literature, flexibility is measured between 0 and 1, where "0" indicates that the system has the lowest flexibility and "1" the highest flexibility. The purpose of this relative measure is to select the system with the highest flexibility among alternative systems. In the following sections, these features are explained in detail and numerical examples are given.

16.2 Routing Flexibility

Routing flexibility is one of the fundamental flexibilities addressed in the manufacturing systems [16.7]. If a manufacturing system has a high routing flexibility, then during a breakdown, repair, or maintenance of a machine, products can be easily re-routed to other machines that can process the particular product. Considering flexible manufacturing, several definitions of routing flexibility have been presented in the literature. For instance, Bernado and Mohamed [16.8] define it as a system's ability to continue producing a given part mix despite disturbances. Das and Nagendra [16.9] define it as a system's ability to manufacture products via a variety of different routes. Stecke and Raman [16.10] define routing flexibility as the measure of the alternative paths that a part can effectively follow through a system for a given process plan. For more definitions one can refer to [16.11] and [16.12]. In a distributed manufacturing system, routing flexibility refers to the ability of the distributed manufacturing system to continue routing a given product mix to alternative manufacturing systems despite uncertainty in the system.

Several alternative measures for routing flexibility have been proposed in the literature. For example, routing flexibility is measured as the average number of possible routes in which a product can be processed in the given manufacturing system [16.13, 16.14]; it is also measured as the ratio of possible number of paths to the total number of part types in the manufacturing system [16.15]. Routing flexibility has also been measured based on entropy derived from thermodynamics [16.16–16.18]. In this chapter, we also use an entropy approach to measure the routing flexibility of distributed manufacturing systems.

Routing flexibility has been studied in the literature since the 1980s. From the definitions and measures provided in the literature, it can be seen that routing flexibility is measured in terms of three dimensions: range, cost, and time. Most of the measures are based on the range dimension, which is the number of alternative routes for a product in the given manufacturing systems. Measures based on the range can be found in [16.8, 16.13, 16.15–16.18]. Even though different routes have different costs to process a product, the cost dimension has not yet been included in the measure of routing flexibility in the literature. The time dimension is also considered in the measure of routing flexibility (*e.g.* [16.15]). A routing flexibility measure based on only one dimension does not provide a comprehensive measure. Therefore, in this chapter, which is the first study to address flexibility measures in distributed manufacturing systems, we incorporate all three dimensions of routing flexibility.

Consider a distributed manufacturing system that can produce several different products. Figure 16.1 depicts the alternative manufacturing systems that can process a given product. We consider a distributed manufacturing system that has a total of I products and J manufacturing systems. For each product, there exists a number of manufacturing systems that it can be assigned to; however, each manufacturing system may have different technology and may require a different cost and time to process a product. Let c_{ij} represent the cost to process product $i \in I$ in manufacturing system $j \in J$, and t_{ij} represent the required time to process product $i \in I$ in manufacturing system $j \in J$. Depending on the available technology at a

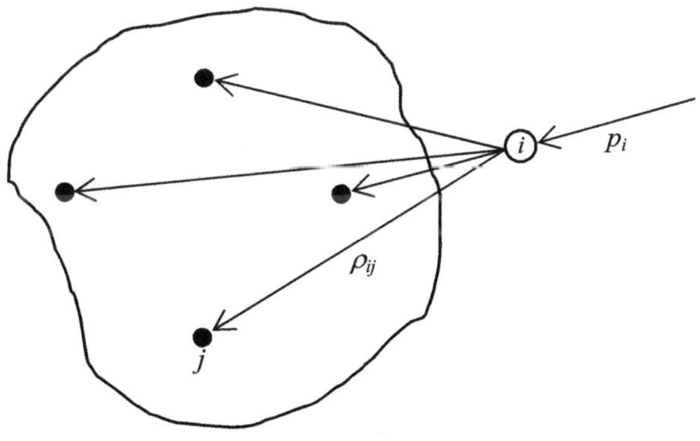

Figure 16.1. Alternative assignments of a product in a distributed manufacturing system

manufacturing system, a product can be assigned to a manufacturing system to be processed. To account for the possible assignments that a given product may have, we define a binary variable $x_{ij} \in \mathbb{B}^{I \times J}$, $i \in I$, and $j \in J$ such that

$$x_{ij} = \begin{cases} 1 & \text{if product } i \text{ can be assigned to location } j, \\ 0 & \text{otherwise.} \end{cases} \quad (16.1)$$

In a distributed manufacturing system, a product can be processed at more than one manufacturing system; however, different manufacturing systems may not have the same efficiency when processing a particular product. Therefore, the flexibility of the distributed manufacturing system depends on which product is assigned to the various manufacturing systems.

In practice, the assignment of a product to a manufacturing system is prioritised by the real-time efficiency at which a manufacturing system can process the given product. Thus, we consider products individually and assign them to the manufacturing system that has the highest efficiency. This assignment process is continued until every product has been accounted for. To include these entities in our model, we define e_{ij}^{c} to represent the cost-based efficiency, which indicates how well manufacturing system j processes product i with respect to cost, where $0 \leq e_{ij}^{c} \leq 1$, $e_{ij}^{c} \in \mathbb{R}^{I \times J}$, $i \in I$, and $j \in J$.

$$e_{ij}^{c} = \frac{\min_{i \in I} c_{ij} x_{ij}}{c_{ij}}, \qquad \forall i, j. \quad (16.2)$$

For a given manufacturing system j, e_{ij}^{c} is the ratio of the minimum cost to process a product in the possible set of products to the cost of processing product i. Similarly, we define e_{ij}^{t} to represent the time-based efficiency, which indicates how well manufacturing system j processes product i with respect to time, where $0 \leq e_{ij}^{t} \leq 1$, $e_{ij}^{t} \in \mathbb{R}^{I \times J}$, $i \in I$, and $j \in J$.

$$e_{ij}^{t} = \frac{\min_{i \in I} t_{ij} x_{ij}}{t_{ij}}, \qquad \forall i, j. \quad (16.3)$$

For a given manufacturing system j, e_{ij}^{t} is the ratio of the minimum time to process a product in the possible set of products with respect to time to process product i. The priority for assigning a product to a manufacturing system is decided in real time, according to the time-based flexibility, cost-based flexibility, and available technology at each manufacturing system. In order to account for all these aspects of the distributed manufacturing system, we consider the multiplication of cost-based and time-based efficiencies rather than considering an individual efficiency. In order to explain the motivation of such an argument, consider a distributed manufacturing system where a manufacturing system has the most advanced technology and it can process a product within the shortest time compared with other manufacturing systems. However, because of the advanced technology, the cost of processing is much higher compared with other manufacturing systems. In this case, if we consider only the time-based efficiency, the particular manufacturing system is much preferred over the other manufacturing systems. Meanwhile, if we consider the cost-based efficiency, the particular manufacturing system is the least preferred. Therefore, considering one of the efficiencies does not provide an appropriate choice for assigning a product. In order to include the effect of both cost and time, we can consider the multiplication of cost-based and time-based efficiencies as follows:

$$e_{ij} = e_{ij}^{c} e_{ij}^{t}, \qquad \forall i, j \tag{16.4}$$

where $0 \leq e_{ij} \leq 1$, $e_{ij} \in \mathbb{R}^{I \times J}$, $i \in I$, and $j \in J$. In reality, because of external uncertainty such as demand uncertainty for products, it may not be feasible to assign all possible products to a manufacturing system that is capable of processing those products with the highest efficiency. Therefore, we consider the probability of assigning a product to a manufacturing system. The probability of assigning a product to a manufacturing system depends on the probability of that product occurrence (demand of a product) and the efficiency of processing a product in a manufacturing system. The higher the probability of demand of a product, the higher the probability of assigning a product to a manufacturing system is. Similarly, the higher the efficiency of processing a product in a manufacturing system, the higher the probability of assigning a product to that manufacturing system. Therefore, we define the probability of product *i*'s occurrence as p_i and the probability of assigning product *i* to manufacturing system *j* becomes

$$\varphi_{ij} = p_i e_{ij}, \qquad \forall i, j \tag{16.5}$$

where $0 \leq p_i \leq 1$, $p_i \in \mathbb{R}^{I}$, $\sum_i p_i = 1$, and as a consequence $0 \leq \varphi_{ij} \leq 1$. Now, to obtain the properties of an entropy approach, the probability of assigning a product to a manufacturing system given in Equation (16.5) is normalised, *i.e.* for a given product *i*, the efficiency over all possible manufacturing systems should add up to unity. Therefore, we define the following term to normalise the probabilities:

$$\rho_{ij} = \frac{\varphi_{ij}}{\sum_{j} \varphi_{ij}}, \qquad \forall i, j \tag{16.6}$$

where $0 \leq \rho_{ij} \leq 1$, $i \in I$, $j \in J$, and $\sum_j \rho_{ij} = 1$. Then, based on an entropy approach, the routing flexibility (RF_i) of product *i*, which is $0 \leq RF_i \leq 1$, in the distributed manufacturing system is given by

$$RF_i = -\sum_j \rho_{ij} \log \rho_{ij}, \qquad \forall i. \tag{16.7}$$

Total routing flexibility (TRF) of products in the distributed systems can be expressed as

$$TRF = \frac{1}{I} \sum_i RF_i, \tag{16.8}$$

where $0 \leq TRF \leq 1$. The routing flexibility model above considers the cost-based efficiency, time-based efficiency, demand uncertainty of products, and available technology in the distributed manufacturing system.

16.2.1 Numerical Examples

Example 1

In this section we present two numerical examples to explain how the model can be applied and to highlight the performance.

We consider a distributed manufacturing system that consists of four manufacturing systems that process five products. Technical capability of the manufacturing systems is given in Table 16.1, where a value of 1 indicates that a given product can be processed in the manufacturing system (MS); otherwise, it is assigned a value of 0. Cost and time to process the products in each manufacturing system is given in Table 16.2 and Table 16.3, respectively. Demand uncertainty of the products is given in Table 16.4.

Table 16.1. Available technology at manufacturing systems, x_{ij}

Product	MS 1	MS 2	MS 3	MS 4
1	0	1	1	1
2	1	1	0	0
3	1	1	0	1
4	0	0	1	1
5	1	0	1	1

Table 16.2. Cost to manufacture a unit of product at manufacturing systems, c_{ij}

Product	MS 1	MS 2	MS 3	MS 4
1	-	37.50	35.50	33.50
2	20.46	26.86	-	-
3	23.93	16.10	-	20.01
4	-	-	61.64	55.30
5	51.89	-	53.10	39.53

Table 16.3. Time to manufacture a unit of product at manufacturing systems, t_{ij}

Product	MS 1	MS 2	MS 3	MS 4
1	-	5.22	4.68	5.10
2	5.52	5.36	-	-
3	9.90	10.00	-	7.00
4	-	-	4.48	4.04
5	11.57	-	9.88	7.80

Table 16.4. Demand uncertainty of products, p_i

Product	Demand distribution
1	0.30
2	0.20
3	0.15
4	0.10
5	0.25

Table 16.5. Cost-based efficiency, e_{ij}^{c}

Product	MS 1	MS 2	MS 3	MS 4
1	-	0.4293	1.0000	0.5973
2	1.0000	0.5994	-	-
3	0.8550	1.0000	-	1.0000
4	-	-	0.5435	0.3618
5	0.3943	-	0.6309	0.5062

The cost-based efficiency is computed using Equation (16.2). For example, the cost-based efficiency to process product 3 in manufacturing system 1 is equal to 0.8550; as we have min{20.46, 23.93, 51.89}/23.93. Similarly, time-based efficiency for the same product and manufacturing system is computed using Equation (16.3), giving min{5.52, 9.90, 11.57}/9.90 = 0.5576. The quantities of the cost-based efficiency and time-based efficiency are given in Tables 16.5 and 16.6, respectively. Considering the demand distribution, the probability of assigning product 3 to manufacturing system 1 is calculated using Equation (16.5) as 0.15 × 0.8550 × 0.5576 = 0.0715. Then, using Equation (16.6), the normalised probability of processing product 3 at manufacturing system 1 is 0.3025, which is 0.0715/(0.0715 + 0.0783 + 0.0866). The values of the probability of assigning products to machines and their normalised values are given in Tables 16.7 and 16.8, respectively. Subsequently, the routing flexibility of product 1 becomes:

$$\begin{aligned} RF_1 &= -0.2309 \times \log(0.2309) - 0.5147 \times \log(0.5147) - 0.2544 \times \log(0.2544) \\ &= 0.447 \end{aligned}$$

Routing flexibilities of products 2, 3, 4, and 5 are 0.286, 0.476, 0.292, and 0.471, respectively. The average routing flexibility then becomes 0.394, which is computed using Equation (16.8).

Table 16.6. Time-based efficiency, e^t_{ij}

Product	MS 1	MS 2	MS 3	MS 4
1	-	1.0000	0.9573	0.7922
2	1.0000	0.9739	-	-
3	0.5576	0.5220	-	0.5771
4	-	-	1.0000	1.0000
5	0.4771	-	0.4534	0.5179

Table 16.7. Probability of assigning products to manufacturing systems, φ_{ij}

Product	MS 1	MS 2	MS 3	MS 4
1	-	0.1288	0.2872	0.1419
2	0.2000	0.1167	-	-
3	0.0715	0.0783	-	0.0866
4	-	-	0.0543	0.0362
5	0.0470	-	0.0715	0.0655

Table 16.8. Normalised probability, ρ_{ij}

Product	MS 1	MS 2	MS 3	MS 4
1	-	0.2309	0.5147	0.2544
2	0.6314	0.3686	-	-
3	0.3025	0.3312	-	0.3663
4	-	-	0.6003	0.3997
5	0.2555	-	0.3885	0.3560

Example 2

In the second example, we consider a distributed manufacturing system that has very similar configurations to those of example one, except that product 4 can be processed only in manufacturing system 3. The respective technology matrix, cost matrix, and time matrix are given in Tables 16.9, 16.10, and 16.11, respectively. Cost-based efficiency, time-based efficiency, probability of assigning products to manufacturing systems and their normalised values are computed and given in Tables 16.12, 16.13, 16.14, and 16.15, respectively. As a result, the routing flexibility of products 1, 2, 3, 4, and 5 are 0.454, 0.286, 0.469, 0, and 0.466, respectively. By Equation (16.8), the average routing flexibility then becomes 0.335. As one would expect, the first distributed manufacturing system has higher routing

Table 16.9. Available technology at manufacturing systems, x_{ij}

Product	MS 1	MS 2	MS 3	MS 4
1	0	1	1	1
2	1	1	0	0
3	1	1	0	1
4	0	0	1	0
5	1	0	1	1

Table 16.10. Cost to manufacture a unit of product at manufacturing systems, c_{ij}

Product	MS 1	MS 2	MS 3	MS 4
1	-	37.50	35.5	33.50
2	20.46	26.86	-	-
3	23.93	16.10	-	20.01
4	-	-	61.64	-
5	51.89	-	53.1	39.53

Table 16.11. Time to manufacture a unit of product at manufacturing systems, t_{ij}

Product	MS 1	MS 2	MS 3	MS 4
1	-	5.22	4.68	5.10
2	5.52	5.36	-	-
3	9.90	10.00	-	7.00
4	-	-	4.48	-
5	11.57	-	9.88	7.80

Table 16.12. Cost-based efficiency, e_{ij}^{c}

Product	MS 1	MS 2	MS 3	MS 4
1	-	0.4293	1.0000	0.5973
2	1.0000	0.5994	-	-
3	0.8550	1.0000	-	1.0000
4	-	-	0.5435	-
5	0.3943	-	0.6309	0.5062

Table 16.13. Time-based efficiency, e_{ij}^{t}

Product	MS 1	MS 2	MS 3	MS 4
1	-	1.0000	0.9573	1.0000
2	1.0000	0.9739	-	-
3	0.5576	0.5220	-	0.7286
4	-	-	1.0000	-
5	0.4771	-	0.4534	0.6538

Table 16.14. Probability of assigning products to manufacturing systems, φ_{ij}

Product	MS 1	MS 2	MS 3	MS 4
1	-	0.1288	0.2872	0.1792
2	0.2000	0.1167	-	-
3	0.0715	0.0783	-	0.1093
4	-	-	0.0543	-
5	0.0470	-	0.0758	0.0655

Table 16.15. Normalised probability, ρ_{ij}

Product	MS 1	MS 2	MS 3	MS 4
1	-	0.2164	0.4825	0.3011
2	0.6314	0.3686	-	-
3	0.2760	0.3022	-	0.4218
4	-	-	1.0000	-
5	0.2336	-	0.3553	0.4111

flexibility than the second, which our model is able to capture. One can notice that the routing flexibility of product 4 is 0, this is because it can only be assigned to manufacturing system 3 and therefore there is no alternative route available.

16.3 Network Flexibility

A distributed manufacturing system is a network. It consists of a set of nodes and a set of links that connect those nodes with information and product flows. The concept of flexibility in networks has recently gained more attention (*e.g.* [16.19]). Moses [16.19] does not define network flexibility for the manufacturing system, but for an engineering system design in terms of range dimension and cost dimension. Magee and de Weck [16.20] model network flexibility based on the range dimension for an engineering system. In this chapter, we measure the network flexibility of distributed manufacturing systems in terms of three dimensions: range, cost, and time. These are the three dimensions used in the definitions and measures of manufacturing flexibility.

A network is flexible if it is relatively easy to make certain changes to it. Therefore, network flexibility comes into play when we can add a new node and/or connect to an existing node using alternative paths. Hence, in a flexible network, there are multiple paths that connect nodes, and it becomes possible to use alternative paths in the system to reach the other nodes. Therefore, network flexibility increases as the total number of alternative paths in the system increases. Network flexibility measure should take into account the alternative paths, number of nodes, and ease of making changes in the network. One approach to including the ease of making changes is to consider the cost and time related to the alternative links. In other words, the efficiency of each alternative path should be considered.

In a distributed manufacturing system, which is a network, the network flexibility indicates the ability to easily make changes in its manufacturing systems, or the ability to easily modify the technology available in its one or more of the manufacturing systems, so that the distributed manufacturing system can cope with uncertainty in the system. Under uncertainty, a product that has already been assigned to a manufacturing system can be transferred to another manufacturing system that is able to produce the same product. However, it is important to note that when a manufacturing system processes a product transferred from another manufacturing system, it may not process the product with the same efficiency as that of the initially assigned manufacturing system. This can be due to differences in a manufacturing system's technological attributes that can be expressed in terms of cost and time to process a product. Therefore, to consider a manufacturing system's relative efficiency for processing a transferred product, we define μ_{ijk} as the relative efficiency of manufacturing system k when product i is transferred from manufacturing system j (see Figure 16.2).

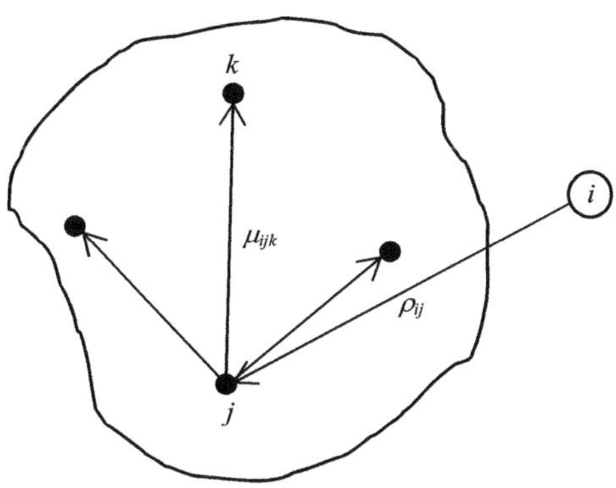

Figure 16.2. A network of a distributed manufacturing system

Here, the relative efficiency indicates the efficiency of processing a product of one manufacturing system with respect to the efficiency of another manufacturing system; which can be measured by a number of methods. One method to measure the relative efficiency is to simply take

$$\mu_{ijk} = \frac{e_{ik}}{e_{ij}}, \tag{16.9}$$

where $i \in I$, $j \in J$, and $k \in J$, representing the ratio of the efficiency of manufacturing system k with respect to the efficiency of manufacturing system j when processing product i. We next define a binary variable, $y_{ijk} \in \mathbb{B}^{I \times J \times J}$, which accounts for the number of possible paths that a product can be transferred to. Therefore, we let

$$y_{ijk} = \begin{cases} 1 & \text{if product } i\text{, which has been assigned to manufacturing} \\ & \text{system } j\text{, can be transferred to manufacturing system } k; \\ 0 & \text{otherwise.} \end{cases} \tag{16.10}$$

The priority for transferring a product from one manufacturing system to another is decided in real time, depending on the relative efficiency. The higher the relative efficiency, the greater the manufacturing system's performance for a given transferred product. Thus, we consider products individually and transfer them to other manufacturing systems based on their relative efficiency. The assignment process is continued until every product of a manufacturing system that has been disturbed is transferred to another manufacturing system. In order to prioritise the transferring process from manufacturing system j to k, we define the weighted relative efficiency of transferring product i from manufacturing system j to manufacturing system k as λ_{ijk}. This equates to the product of the probability of assigning product i to manufacturing system j and the relative efficiency from manufacturing system j to manufacturing system k when processing product i as follows:

$$\lambda_{ijk} = \rho_{ij}\mu_{ijk}y_{ijk}, \qquad \forall j \neq k \tag{16.11}$$

where weight is the probability of assigning product i to manufacturing system j and $0 \leq \lambda_{ijk} \leq 1$. The weighted relative efficiency of transferring product i from manufacturing system j to manufacturing system k depends on the relative efficiency from manufacturing system j to manufacturing system k and the probability of assigning product i to manufacturing system j. A larger value for the demand probability of product i constitutes a greater probability of assigning product i to manufacturing system j. In addition, the higher the value of relative efficiency from system j to system k when processing product i, the higher the weighted relative efficiency of transferring product i from manufacturing system j to manufacturing system k becomes. Then we define the network flexibility (NF) of the distributed manufacturing system as the average weighted relative efficiency of transferring product i from manufacturing system j to manufacturing system k,

$$NF = \frac{1}{I \times J \times (J-1)} \sum_{i} \sum_{j} \sum_{k \neq j} \lambda_{ijk}, \tag{16.12}$$

where $i \in I$, $j \in J$, and $k \in J$. The network flexibility model above considers the relative efficiency of processing a transferred product, assigning a product to a manufacturing system, and demand probability of a product. Weighted efficiency has been used in measuring manufacturing system flexibility in the literature (*e.g.* see [16.9, 16.15, 16.21]).

16.3.1 Numerical Examples

Example 3

We first consider the same distributed manufacturing system as in Example 1 of Section 16.2.1. Hence, the distributed manufacturing system consists of four manufacturing systems and five products, and information about the manufacturing systems and products is given in Tables 16.1–16.4. Based on the information, cost-based and time-based efficiencies are presented in Tables 16.5 and 16.6. The

Table 16.16. Values of e_{ij}

Product	MS 1	MS 2	MS 3	MS 4
1	0	0.4293	0.9573	0.4732
2	1	0.5837	0	0
3	0.4767	0.5220	0	0.5771
4	0	0	0.5435	0.3618
5	0.1881	0	0.2861	0.2622

Table 16.17. Values of relative efficiency (μ_{ijk}) and weighted relative efficiency (λ_{ijk})

	μ_{ijk}				λ_{ijk}			
	MS 1	MS 2	MS 3	MS 4	MS 1	MS 2	MS 3	MS 4
Product 1								
MS 1	-	0	0	0	-	0	0	0
MS 2	0	-	2.230	1.102	0	-	0.515	0.254
MS 3	0	0.449	-	0.494	0	0.231	-	0.254
MS 4	0	0.907	2.023	-	0	0.231	0.515	-
Product 2								
MS 1	-	0.584	0	0	-	0.369	0	0
MS 2	1.713	-	0	0	0.631	-	0	0
MS 3	0	0	-	0	0	0	-	0
MS 4	0	0	0	-	0	0	0	-
Product 3								
MS 1	-	1.095	0	1.211	-	0.331	0	0.366
MS 2	0.913	-	0	1.106	0.303	-	0.000	0.366
MS 3	0	0	-	0	0	0	-	0
MS 4	0.826	0.904	0	-	0.303	0.331	0.000	-
Product 4								
MS 1	-	0	0	0	-	0	0	0
MS 2	0	-	0	0	0	-	0	0
MS 3	0	0	-	0.666	0	0	-	0.400
MS 4	0	0	1.502	-	0	0	0.600	-
Product 5								
MS 1	-	0	1.521	1.394	-	0	0.388	0.356
MS 2	0	-	0	0	0	-	0	0
MS 3	0.658	0	-	0.917	0.255	0	-	0.356
MS 4	0.717	0	1.091	-	0.255	0	0.388	-

multiplications of cost-based and time-based efficiencies are calculated using Equation (16.4) and are presented in Table 16.16, where for example $e_{31} = 0.8550 \times 0.5576 = 0.4767$.

Then, the relative efficiency is computed using Equation (16.9) and given in the left column of Table 16.17. For product 1, relative efficiency from manufacturing system 4 to manufacturing system 2 is computed as 0.4293/0.4732 = 0.907. Considering normalised probability values presented in Table 16.8, weighted relative efficiencies can be determined using Equation (16.11), and are given in the right column of Table 16.17. Once these values are computed, using Equation (16.12), network flexibility (NF) of the distributed manufacturing system is determined as 0.40.

Example 4

To further highlight and compare the characteristics of our model, we consider another example using the same distributed manufacturing system as in Example 2 for routing flexibility in Section 16.2.1. Information about the manufacturing systems and products is given in Tables 16.9–16.11. Based on the information, cost-based and time-based efficiencies are presented in Tables 16.12 and 16.13, respectively. The multiplications of cost-based and time-based efficiencies are calculated using Equation (16.4) and are presented in Table 16.18. The relative efficiency is computed using Equation (16.9) and given in the left column of Table 16.19. Considering normalised probability values presented in Table 16.15, weighted relative efficiencies can be determined using Equation (16.11), and are given in the right column of Table 16.19. Finally, using Equation (16.12), network flexibility of the distributed manufacturing system is determined as 0.35. As one would expect, the first distributed manufacturing system has higher network flexibility than the second one. That characteristic has been captured by our model.

The above two flexibility measures address two important performance indicators for distributed manufacturing systems. The routing flexibility concerns assigning different products to alternative manufacturing systems. However, the network flexibility concerns reassigning different products among alternative manufacturing systems. In distributed manufacturing systems, routing flexibility is very useful to cope with external uncertainties; and network flexibility is important to deal with internal uncertainties. Therefore, it is the investors' choice to give priority to either routing flexibility or network flexibility. For example, one who is more concerned about external uncertainties (*e.g.* customer demand) would give higher priority to the routing flexibility over the network flexibility. One method to express the combined flexibility measure is the weighted aggregation as follows. Let CF be the combined flexibility measure and $CF = \alpha \times TRF + (1-\alpha)NF$, where α is a weight factor between 0 and 1. The weight can be decided based on the preference of the investor.

It is worthwhile reiterating that the value of the above flexibility measures are normalised values that make it possible to compare against other values. Although a single flexibility value that is close to the extreme limits of 0 or 1 can clearly indicate a low or high flexibility level, a non-extreme value may not be so clearly branded unless in comparison with other values. For example, if a flexibility measure is 0.98, then one can strongly conclude that the system is flexible. However, if the value is 0.51, it is hard to label the system as flexible. Nevertheless, one can conclude that the same system is more flexible than another system whose flexibility value is 0.35. Therefore, it is recommended that these flexibility measures be used for comparison purposes.

Table 16.18. Values of e_{ij}

Product	MS 1	MS 2	MS 3	MS 4
1	0	0.4293	0.9573	0.5973
2	1	0.5837	0	0
3	0.4767	0.5220	0	0.7286
4	0	0	0.5435	0
5	0.1881	0	0.2861	0.3310

Table 16.19. Values of relative efficiency (μ_{ijk}) and weighted relative efficiency (λ_{ijk})

	μ_{ijk}				λ_{ijk}			
	MS 1	MS 2	MS 3	MS 4	MS 1	MS 2	MS 3	MS 4
Product 1								
MS 1	-	0	0	0	-	0	0	0
MS 2	0	-	2.230	1.391	0	-	0.483	0.301
MS 3	0	0.449	-	0.624	0	0.216	-	0.301
MS 4	0	0.719	1.603	-	0	0.216	0.483	-
Product 2								
MS 1	-	0.584	0	0	-	0.369	0	0
MS 2	1.713	-	0	0	0.631	-	0	0
MS 3	0	0	-	0	0	0	-	0
MS 4	0	0	0	-	0	0	0	-
Product 3								
MS 1	-	1.095	0	1.528	-	0.302	0	0.422
MS 2	0.913	-	0	1.396	0.276	-	0	0.422
MS 3	0	0	-	0	0	0	-	0
MS 4	0.654	0.716	0	-	0.276	0.302	0	-
Product 4								
MS 1	-	0	0	0	-	0	0	0
MS 2	0	-	0	0	0	-	0	0
MS 3	0	0	-	0	0	0	-	0
MS 4	0	0	0	-	0	0	0	-
Product 5								
MS 1	-	0	1.521	1.759	-	0	0.355	0.411
MS 2	0	-	0	0	0	-	0	0
MS 3	0.658	0	-	1.157	0.234	0	-	0.411
MS 4	0.568	0	0.864	-	0.234	0	0.355	-

16.4 Conclusions

This chapter presented performance measures for distributed manufacturing systems. The proposed model has distinctive features that take into consideration demand uncertainty, routing flexibility and network flexibility. The first set of measures focuses on routing flexibility that is constructed based on cost-based efficiency, time-based efficiency and the probability of assigning a product to a manufacturing system. This routing flexibility reflects the ability of distributed manufacturing systems to continue routing given product mix to alternative manufacturing systems under demand uncertainty.

The second set of performance measures deals with the network flexibility that indicates the ability of a distributed manufacturing system to make changes in the initial assignments of products to manufacturing systems. The proposed network flexibility is based on the relative efficiency when a product is transferred from its initial manufacturing system to a new manufacturing system. The proposed routing flexibility and the network flexibility together help enterprises evaluate the responsiveness of their distributed manufacturing systems to market changes, which can be translated to competitive advantage in a global market.

References

[16.1] Saad, S.M., Perera, T. and Wickramarachchi, R., 2001, "Simulation of distributed manufacturing enterprises: a new approach," In *Proceedings of the 2003 Winter Simulation Conference*, pp. 1167–1173.

[16.2] Shen, W. and Norrie, D.H., 1998, "An agent based approach for manufacturing enterprise integration and supply chain management", In *Globalization of Manufacturing in the Digital Communications Era of the 21st Century: Innovation and Virtual Enterprises*, Jacucci, G. *et al.* (ed.), Kluwer Academic Publishers, pp. 579–590.

[16.3] Leitão, P. and Restivo, F., 2001, "An agile and cooperative architecture for distributed manufacturing systems," In *Proceedings of the IASTED International Conference on Robotics and Manufacturing*, Cancun, Mexico.

[16.4] Maturana, F. and Norrie, D., 1996, "Multi-agent mediator architecture for distributed manufacturing," *Journal of Intelligent Manufacturing*, **7**(4), pp. 257–270.

[16.5] Shen, W., Xue, D. and Norrie, D., 1997, "An agent-based manufacturing enterprise infrastructure for distributed integrated intelligent manufacturing systems," In *Proceedings of the Third International Conference on the Practical Application of Intelligent Agents and Multi-Agents*, London, UK.

[16.6] Wang, L., Feng, H.-Y. and Cai, N., 2003, "Architecture design for distributed process planning," *Journal of Manufacturing Systems*, **22**(2), pp. 99–115.

[16.7] Chandra, P. and Tombak, M.M., 1992, "Model for the evaluation of routing and machine flexibility," *European Journal of Operational Research*, **60**(2), pp. 156–165.

[16.8] Bernado, J.J. and Mohamed, Z., 1992, "The measurement and the use of operational flexibility in loading of flexible manufacturing systems," *European Journal of Operational Research*, **60**(2), pp. 144–155.

[16.9] Das, S.K. and Nagendra, P., 1993, "Investigations into impact of flexibility on manufacturing performance," *International Journal of Production Research*, **31**(10), pp. 2337–2354.

[16.10] Stecke, K.E. and Raman, N., 1995, "FMS planning decisions, operating flexibilities and systems performance," *IEEE Transactions on Engineering Management*, **42**(1), pp.82–89.

[16.11] Sethi, A.K. and Sethi, S.P., 1990, "Flexibility in manufacturing: a survey," *International Journal of Flexible Manufacturing Systems*, **2**(4), pp. 289–328.

[16.12] Sarker, B.R., Krishnamurthy, S. and Kuthethur, S.G., 1994, "A survey and critical review of flexibility measures in manufacturing systems," *Production Planning & Control*, **5**(6), pp. 512–523.

[16.13] Chatterjee, A., Cohen, M.A., Maxwell, W.L. and Miller, L.W., 1984, "Manufacturing flexibility: models and measurements," *First ORSA/TIMS Conference on Flexible Manufacturing Systems*, Ann Arbor, MI, pp. 49–64.

[16.14] Shewchuk, J.P., 1999, "A set of generic flexibility measures for manufacturing applications," *International Journal of Production Research*, **37**(13), pp. 3017–3042.

[16.15] Chen, I.J. and Chung, C.H., 1996, "An examination of flexibility measurements and performance of flexible manufacturing systems," *International Journal of Production Research*, **34**(2), pp. 379–394.

[16.16] Yao, D.D., 1985, "Material and information flows in flexible manufacturing systems," *Material Flow*, **2**, pp. 143–149.

[16.17] Kumar, V., 1987, "Entropic measurement of manufacturing flexibility," *International Journal of Production Research*, **27** (7), pp. 957–966.

[16.18] Yao, D.D. and Pei, F.F., 1990, "Flexible parts routing in manufacturing systems," *IIE Transactions*, **22**(1), pp. 48–55.

[16.19] Moses, J., 2003, "The anatomy of large scale systems," *ESD Internal Symposium*, Cambridge, MA, MIT: 17.

[16.20] Magee, C.L. and de Weck, O.L., 2002, "An attempt at complex system classification", *ESD Internal Symposium*, Cambridge, MA, MIT.

[16.21] Brill, P.H. and Mandelbaum, M., 1989, "On measures of flexibility in manufacturing systems," *International Journal of Production Research*, **27**(5), pp. 747–756.

Index

MIX
Papier aus verantwortungsvollen Quellen
Paper from responsible sources
FSC® C105338

If you have any concerns about our products,
you can contact us on
ProductSafety@springernature.com

In case Publisher is established outside the EU,
the EU authorized representative is:
Springer Nature Customer Service Center GmbH
Europaplatz 3, 69115 Heidelberg, Germany

Printed by Libri Plureos GmbH
in Hamburg, Germany